SICK BUILDINGS

Definition, Diagnosis and Mitigation

Thad Godish, Ph.D., C.I.H.
Department of Natural Resources
and Environmental Management
Ball State University
Muncie, Indiana

LEWIS PUBLISHERS
Boca Raton London New York Washington, D.C.

Library of Congress Cataloging-in-Publication Data

Godish, Thad.
Sick buildings: definition, diagnosis, and mitigation / by
Thaddeus Godish.
p cm.
Includes bibliographical references and index.
ISBN 0-87371-346-X
1. Sick buildings syndrome. 2. Indoor air pollution. I. Title.
RA577.5.G63 1994
613′.5—dc20 94-17544
CIP

Direct all inquiries to CRC Press LLC, 2000 N.W. Corporate Blvd., Boca Raton, Florida 33431.

Lewis Publishers is an imprint of CRC Press LLC

International Standard Book Number 0-87371-346-X
Library of Congress Card Number 94-17544
Printed in the United States of America 4 5 6 7 8 9 0
Printed on acid-free paper

THE AUTHOR

Thad Godish, Ph.D., C.I.H., is Director, Indoor Air Quality Research Laboratory and Professor of Natural Resources and Environmental Management at Ball State University, Muncie, Indiana. He holds a doctorate from Penn State University, where he was affiliated with the Center for Air Environment Studies.

Dr. Godish is responsible for a number of teaching, research, and public service activities at Ball State University. Teaching duties include courses in Air Quality, Industrial Hygiene, and Indoor Air Quality Management. He is the author of numerous research and review articles in his specialty, Indoor Air Pollution. He is widely known for his research on the formaldehyde contamination of buildings and for his two well-regarded books, *Indoor Air Pollution Control* (1989, Lewis Publishers) and *Air Quality*, now in its second edition (1990, Lewis Publishers).

Dr. Godish has had extensive consulting experience conducting both indoor air quality and industrial hygiene investigations. The former has included problem building investigations of offices, schools, commercial establishments and residences. He has served as an expert witness in hundreds of indoor air quality-related product liability civil lawsuits.

Dr. Godish holds a certification from the American Board of Industrial Hygiene; is a Fellow, Indiana Academy of Science; and Member, Sigma Xi, National Honorary Research Society. He is also a member of the Air and Waste Management Association, American Industrial Hygiene Association, American Conference of Governmental Industrial Hygienists, International Society for Indoor Air Quality and Climate, and the American Association for the Advancement of Science.

Dr. Godish has recently served as Visiting Scientist, Environmental Science and Engineering Program, School of Public Health, Harvard University, and Visiting Scientist, Monash University, Gippsland, Australia.

PREFACE

Having had the experience of both conducting research and writing extensively on indoor air quality, the author continues to be awed by a field that was but an infant a short time ago, a field that is in early to mid-adolescence. The author's task has evolved from the limitations of a relatively meager literature base to one that allows expanded treatments of subjects that once were included as only parts of chapters in previous works. This book, which focuses on "sick" or "problem" buildings, could not have been written 5 years ago. It owes its existence to the many who have and continue to "labor in the vineyard"; the scientists and engineers who have conducted increasingly systematic and creative research on the subject. As an American, I would like to acknowledge the leadership and quality research conducted by colleagues in Northern Europe, particularly in the Nordic countries. The publication of their work in "my language" has made it possible for me to glean the fruits of their labor and share them with the readers of *Sick Buildings: Definition, Diagnosis and Mitigation.*

The genesis of this book was longer than planned. In a rapidly evolving field, one must contend with the "moving target", those seminal new references (or sometimes older) which have just come to one's attention. They must be incorporated and previous thoughts on the subject re-examined. The 1993 Helsinki Proceedings (with six volumes) were a mother lode of new studies and information on sick buildings. Their availability late in this effort meant the incorporation of new work, re-examination, and even "gutting" of once "nearly finished" chapters. Such is the way of science and the task of the author in its communication to a broader audience.

A book involves many hundreds of little-noticed decisions. The final product represents the sum of these and the author's unseen struggle between the twin demands of readability and comprehensiveness. Readers may find some chapters (e.g., Chapter 7 on "Diagnosis") to be somewhat laborious. In such cases, the "demon" of comprehensiveness won the struggle. Diagnosis is the key to solving indoor air quality problems. Readers who themselves conduct indoor air quality investigations should be aware of the different protocols that are used and their limitations. Such awareness may expand the scope of what may be included in their investigations, influence how results are interpreted, and increase the probabilty of successful problem building diagnosis and mitigation.

This book is written as a reference work. Its scope is indicated in the subtitle. Though Chapter 1 is designed to define the nature of the problem, the task of definition is also shared by Chapters 2 through 6. Their focus is potential causal phenomena and risk factors for problem building complaints. Chapters 7 and 8 share the diagnosis function; 9 and 10 discuss mitigation.

The book is intended for a variety of readers, including those who conduct indoor air quality investigations, industrial hygienists, consulting engineers, architects, public health and environmental professionals, indoor air quality scientists, and university graduate and undergraduate students. In the last case, the book may prove useful in courses that focus on indoor air quality, particularly within industrial hygiene programs where air quality concerns are of an occupational nature.

CONTENTS

1. Defining the Issues 1
Building-Related Illness 1
Sick Building Syndrome 1
Tight Buildings 4
Sick Buildings/Problem Buildings 5
Field Investigations 5
NIOSH Investigations 6
Canadian Investigations 8
Systematic Building Investigations—Symptom Prevalence 8
U.S. Studies 9
U.K. Studies 10
Danish Town Hall Study 11
Dutch Office Building Study 11
Swedish Sick Building Study 11
Skin Symptoms 14
Work Performance/Productivity 17
Multiple Chemical Sensitivity 18
Recapitulation 22
References 23

2. People-Related Risk Factors 29
Personal Characteristics 29
Gender 30
Atopic History and Other Health Factors 30
Psychosocial Phenomenon 31
Mass Psychogenic Illness 31
Psychosocial Risk Factors 33
Job Stress 33
Job Satisfaction/Dissatisfaction 34
Occupant Density 35
Satisfaction with Physical Work Environment 35
Seasonal Affective Disorder 36
Significance of Psychosocial Factors Related to SBS 36
Tobacco Smoking 37
Environmental Tobacco Smoke 37

ETS Contaminant Concentrations 38
Irritant Effects Determined from Survey Studies 42
Irritant Effects in Controlled Exposure Studies 42
Role of ETS in SBS-Type Complaints 44
Tobacco Smoking and SBS Complaints 45
Recapitulation 46
References 47

3. Environmental Conditions/Environmental Systems 53
Environmental Conditions 53
Thermal Conditions 54
Comfort Relationships 54
Dissatisfaction with Thermal Comfort 55
Building Temperatures and SBS Symptoms 56
Humidity 57
Health and Comfort Relationships 57
Effect on Contaminants 59
Air Flow/Air Movement 59
Lighting 60
Noise/Vibration 61
Air Ions 61
Physiological and Behavioral Studies 62
Clinical Studies 63
Building Studies 63
Electrostatic Charges 64
Electric and Magnetic Fields 65
Environmental Systems 67
Field Investigations 67
Systematic Building Studies 68
Type of Ventilation System 68
Ventilation Conditions — Symptom Prevalence Rates and Occupant Satisfaction/Dissatisfaction with Air Quality 68
Workstation Characteristics 73
Ventilation Conditions and Contaminant Levels 73
The HVAC System as a Source of Contaminants 75
Chemical Contaminants 76
Man-Made Mineral Fibers (MMMF) and Other Particles 78
Biological Contaminants 80
Entrainment/Re-Entry/Cross-Contamination 80
Entrainment/Re-Entry 81
Cross-Contamination 82
Recapitulation 83
References 84

4. Office Materials, Equipment, and Furnishings 93
Office Materials and Equipment 93
Carbonless Copy Paper 93
Product Characterization 93
Complaint Investigations 95
Exposure Investigations 95
Cross-Sectional Epidemiological Studies 98
Potential Causal Factors 98
Other Papers 100
Office Equipment 101
Wet-Process Photocopiers 101
Electrostatic Copying Machines 103
Laser Printers 105
Toners 106
Diazo-Photocopiers 106
Microfilm Copiers 106
Spirit Duplicating Machines 107
Video-Display Terminals/Computers 107
SBS-Type Complaints 107
Skin Symptoms 108
Ergonomic Problems/Job Stress 110
Electromagnetic Radiation 111
Reproductive Hazards 112
Human Exposure Studies 113
Floor Covering 114
Carpeting 114
Complaint Investigations 115
Epidemiological Studies 116
Exposure Studies 116
USEPA Investigations 118
Emission Studies — Carpet Materials 120
Emission Studies — Adhesives 121
Emission Studies — Carpet Systems 123
Carpeting/Substrate Reactions 123
Sink Effects 123
MOD Reservoir and Microbial Amplification 124
Electrostatic Shock Potential 124
Carpet Cleaning Agents 124
Vinyl Floor Covering 125
Complaint Investigations 125
Exposure Studies 125
Indoor Air Contaminants 126
Plasticizers 126

Emission Testing 127
Recapitulation 127
References 129

5. Gas/Vapor- and Particulate-Phase Contaminants 139
Bioeffluents 140
Health and Occupant Comfort Concerns 140
Circumstantial Evidence 140
Systematic Studies 142
Potential Biological Mechanisms 143
Human Pheromones 143
Formaldehyde 145
Complaint Investigations 145
Systematic Studies 146
Potentiation 147
Sources 148
Volatile Organic Compounds 148
TVOC Theory 149
Common Chemical Sense 149
Exposure Studies 149
Irritation and Odor Effects 152
Neurotoxic Effects 154
TVOC Sensitivity 154
Application of TVOC Concept 154
Systematic Building Studies 155
VOC Concentrations 157
"Dust" 158
Expression of "Dust" Concentrations 158
Systematic Studies 159
Airborne Dusts 159
Surface Dusts 160
Dust Concentrations 161
Recapitulation 162
References 163

6. Contaminants of Biological Origin 171
Hypersensitivity Diseases 171
Hypersensitivity Pneumonitis 172
Humidifier Fever 173
Legionnaires' Disease 173
Cooling Waters 174
Potable Water Systems 176

Asthma 176
Chronic Allergic Rhinitis 177
Allergy and Allergens as Risk Factors for SBS Symptoms 177
Dust Mites 178
Systematic Building Studies 178
Dust Mite Populations 180
Mold 180
Presumptive Evidence of a Causal Link 180
Systematic Health Studies 182
Exposure Assessments 183
Macromolecular Organic Dust 185
Other Biological Contaminant-Related Health Concerns 186
Bacteria 186
Systematic Building Studies 187
Assessment of Bacteria in Indoor Spaces 187
Microbial Products 188
Endotoxins and Glucans 188
Mycotoxins 189
Microbial VOCs 192
Recapitulation 194
References 195

7. Diagnosing Problem Buildings 205
Role of Public and Private Groups 205
Role of Industrial Hygiene 206
Problem Building Investigation 207
American Investigative Protocols 207
The NIOSH Protocol 207
USEPA/NIOSH Protocol for In-House Personnel 209
California Protocol 211
AIHA Protocol 215
Investigative Protocols Used by Private Consultants/ Consulting Firms 217
The Building Diagnostics Protocol 217
Environmental Health and Engineering Protocol 221
Canadian Investigative Protocols 224
Public Works Canada Protocol 224
Ontario Interministerial Committee Protocol 226
European Investigative Protocols 227
Danish Building Research Institute Protocol 227
Nordtest Protocol 229
Features of Investigative Protocols 223

Multiple Stages of Investigations 233
Personnel Conducting Investigations 234
Site Visits 234
Assessment of Occupant Symptoms/Complaints 234
Use of Questionnaires/Checklists 236
Assessment of HVAC System Operation and Maintenance 237
Environmental Measurements 237
Source Assessments/Contaminant Considerations 238
Use of IAQ/Comfort Guidelines 239
Procedures for Evaluating Problems with HVAC Systems 243
Types of HVAC Systems 245
All-Air Systems 245
Air-Water Systems 245
All-Water Systems 246
HVAC System Evaluations 246
Assessing Building Ventilation 247
CO_2 Techniques 248
Thermal Balance 250
Using Tracer Gases 251
Recapitulation 252
Appendix A. NIOSH IAQ Survey Questionnaire 254
Appendix B. AIHA Occupant Health and Comfort Questionnaire 256
Appendix C. Building Diagnostics Human Resource Questionnaire 259
Appendix D. Ontario Interministerial Committee IAQ Survey Questionnaire 260
Appendix E. Danish Building Research Institute Indoor Climate Survey Questionnaire 266
References 268

8. Measurement of Indoor Air Contaminants 273
Contaminant Measurement Considerations 273
Sampling Methods 274
Gas Sampling Tubes 275
Electronic Direct-Read Instruments 276
Active Integrated Sampling 277
Passive Integrated Sampling 277
Quality Assurance/Calibration 277
Sampling Decisions 278
Location 278
Time 278
Duration 279
Number 280
Administrative Practices 280

Procedures for Commonly Measured Chemical/Physical Contaminants 280
Carbon Dioxide 280
Carbon Monoxide 282
Formaldehyde 283
Volatile Organic Compounds 285
Airborne and Settled Dusts 288
Sampling Biological Aerosols 288
Sampling Approaches 289
Sampling Devices 290
Viable/Culturable Samplers 290
Comparative Studies 290
Total Spore/Particle Samplers 293
Recommended Bioaerosol Sampling Practices 294
Sampling Objectives 295
Sample Collection—Viable/Culturable Bioaerosol Sampling 296
Area vs. Agressive Sampling 296
Media Selection 296
Sampling Location/Numbers 297
Sampling Volume/Duration 297
Sampler Disinfection 298
Sampler Calibration 298
Sample Handling and Analysis 299
Sample Collection—Total Mold Spore/Particles 300
Quantification 300
Calibration 300
Data Interpretation 300
Recapitulation 302
References 302

9. Source Control 307
Chemical Contaminants 309
Prevention/Avoidance 309
Emission/Source Characterization 309
Bioassays 310
“Bad” Products 310
Prospective USEPA Policies 310
Washington State Initiative 311
Private Initiatives 313
TVOCs and Source Control 315
Product Labeling 316
Design Considerations 317
Implementation of Design Criteria 318
Source Removal/Modification/Treatment 319

Source Removal 319
Source Treatment 320
Building Bake-Out 321
Source Modification 324
Biological Contaminants 325
Legionnaires' Disease 325
Biocidal Treatments 325
Source Modifications 327
Potable Water Systems 327
Effectiveness of Control Measures 328
Hypersensitivity Pneumonitis/Humidifier Fever 330
Remediation Measures 330
Avoidance 331
Asthma/Allergic Rhinitis 332
Microorganisms 332
Dust Mites 334
Surface Dusts 334
Source Removal 334
Reservoir Removal 334
Surface Cleaning 335
Duct Cleaning 335
Recapitulation 336
References 337

10. Contaminant Control 347
General Dilution Ventilation 347
General Dilution Theory 348
Applicability of General Dilution Theory 348
Exceptions to General Dilution Theory 349
Systematic Ventilation Studies 349
Health and Comfort Concerns 349
Contaminant Concentrations 350
Controlling Human Bioeffluents 351
Ventilation Standards/Guidelines 352
Ventilation Rate Procedure 352
Indoor Air Quality Procedure 353
Perceived Air Quality Procedure 354
Ventilation Effectiveness 358
Flush-Out Ventilation 359
Ventilation Innovations 361
Demand-Controlled Ventilation 361
Displacement Ventilation 363
Local Exhaust Ventilation 366
Entrainment/Re-Entry/Cross-Contamination 367

Air Cleaning 368
Dust/Particulate-Phase Contaminants 369
Air Cleaner Performance 370
Evaluation of In-Use Performance 370
Control of Mold Spores/Particles 371
Gaseous Contaminants 372
Combined Filtration Systems 374
Air Cleaners as Contaminant Sources 374
Recapitulation 375
References 378

Index 389

To the children of our union —
may they always know learning's love.

1 DEFINING THE ISSUES

We have become increasingly aware that human health and comfort complaints expressed by occupants of office, institutional, and other public access buildings are in many cases associated with poor indoor air quality (IAQ). When a building is subject to complaints sufficient to convince management to conduct an IAQ investigation, it may be characterized as a "problem" or "sick" building. Health complaints associated with a problem building may have a specific identifiable cause (building-related illness) or, as is true for many problem buildings, no specific causal factor or factors can be identified (sick building syndrome).

BUILDING-RELATED ILLNESS

When causal factors have been identified, the problem is described as "building-related illness" (BRI) or specific building-related illness (SBRI).[1,2] Building-related illnesses are usually characterized by a unique set of symptoms which may be accompanied by clinical signs, laboratory findings, and identifiable pollutants. Included in BRI or SBRI are nosocomial infections, the hypersensitivity diseases (such as hypersensitivity pneumonitis, humidifier fever, asthma, and allergic rhinitis), Legionnaires' disease, fiberglass dermatitis, and direct toxic effects from exposures to contaminants such as carbon monoxide, ammonia, formaldehyde, etc. The term "building-related illness" was initially used to describe health problems associated with the formaldehyde contamination of residences[3] and a variety of general symptoms associated with air quality in nonindustrial buildings.[4]

SICK BUILDING SYNDROME

"Sick building syndrome" (SBS) is used to describe a diffuse spectrum of symptoms in which no specific etiological factor can be identified.[5] A working panel of the World Health Organization (WHO) initially attempted to define this phenomenon in the early 1980s.[6,7] Sick building syndrome was defined on the basis of a group of frequently reported symptoms or complaints including (1) sensory irritation in eyes, nose, and throat; (2) neurotoxic or general health problems; (3) skin irritation; (4) nonspecific hypersensitivity reactions; and (5) odor and taste sensations. Sensory irritation was described as pain, a feeling of dryness, smarting, stinging irritation, hoarseness or voice problems; neurotoxic/general health problems as headache, sluggishness, mental fatigue, reduced memory, reduced capacity to concentrate, dizziness, intoxication, nausea and vomiting, and tiredness; skin irritation as pain, reddening, smarting, or itching sensations or dry skin; nonspecific hypersensitivity reactions as running nose or eyes, asthma-like symptoms among nonasthmatics, or sounds from the respiratory system; odor and taste sensations as changed sensitivity of olfactory or gustatory senses or unpleasant olfactory or gustatory perceptions.

In attempting to define and describe SBS, the WHO panel indicated that (1) mucous membrane irritation of the eyes, nose, and throat should be one of the most frequent symptom expressions; (2) other symptoms involving the lower respiratory airways and internal organs should be infrequent; (3) no evident causality should be identified in relation to occupant sensitivity or to excessive exposures; (4) symptoms should appear especially frequently in one building or part of it; and (5) a majority of occupants should report symptoms.

A sick building is distinguished from a normal one, under the WHO definition, by the prevalence of symptoms; that is, a large percentage of the occupants report symptoms. The reactions of a minority of building occupants who are unusually sensitive to exposures of indoor contaminants are not considered to be due to SBS. More than 30% of new buildings would be sick buildings according to WHO.

A working group of the Commission of European Communities (CEC) has described SBS as a phenomenon being experienced primarily by individuals working in climate-controlled buildings. They suggested that SBS was due to multiple factors and could only be diagnosed by exclusion, that is, by eliminating all other building-related health problems. Characteristic symptoms have been described as including nasal, ocular, oropharyngeal, cutaneous, and general (headaches, lethargy, general fatigue) manifestations, with a characteristic periodicity increasing with severity during the workday and resolving rapidly on leaving the building environment.

A working group of the American Thoracic Society (ATS) characterized SBS as having a subtle onset and progression, with cases being clustered within a building or areas of a building.[2] Symptom onset would occur as a consequence of being in a problem building with asymptomatic periods when absent. Characteristic symptoms would include eye irritation; headaches, clearing overnight and on weekends; recurrent fatigue, drowsiness, or dizziness; throat irritation without microbial etiology; chest burning, cough, sputum production in the absence of smoking or exposure to environmental tobacco smoke (ETS); wheezing or chest tightness with paroxysmal cough tending to continue but slowly improving on leaving the offending environment; malaise which may be associated with poor attention span and short-term memory problems; and rhinitis and nasal congestion without atopic history.

Though no specific etiological factors can be identified, SBS may be attributed to such factors as insufficient ventilation or thermal control, inadequate maintenance of building systems, changes in thermal or contaminant loads, changes in building operation to meet new objectives such as energy conservation, and inadequate design. Other contributing factors were described as being physical, chemical, biological, and psychosocial.

The constellation of symptoms characterizing SBS varies somewhat among the three working groups. Common to all are sensory irritation of eyes, nose, and throat and symptoms affecting the central nervous system (headaches, lethargy, fatigue). Both European panels identify skin symptoms as a part of the SBS complex; the American panel does not. The WHO and ATS working groups include pulmonary effects; the CEC working group does not. The WHO panel lists odor and taste sensations; the other two do not. Differences in symptomatologies described for SBS are likely due to differences in panel composition and differences in SBS symptom reporting in Europe and in North America. Symptom complexes described by WHO, CEC, and ATS working groups represent a range of human health responses. This apparent all-inclusive nature suggests that SBS is being defined on the basis of all symptoms of unknown causality which have been reported in sick building investigations.

The WHO characterization of SBS appears to be based on the theory that sick-building complaints of a sensory nature are a consequence of the nonspecific irritation or overstimulation of trigeminal nerves in mucous membranes, our so-called chemical sense. Trigeminal nerves respond to chemical odors that cause irritation, tickling, or burning. The major function of the trigeminal nerve system is to stimulate reflex actions such as sneezing or interruption of breathing when the human body is exposed to potentially harmful odor-producing substances. Similar physiological responses may be elicited by many different chemical substances.[9]

A subgroup of the WHO IAQ panel has suggested that indoor air contains a complex of sensory stimuli which produce irritant responses not specific to the individual contaminant exposures. As a consequence, no single contaminant can be said to be responsible. Additionally, reactions of the "referred pain type" may take place, so that facial skin sensations, headaches, etc. may be due to the irritation of trigeminal nerves in nasal mucosa. Upper airway symptoms, according to this theory, would be the net result of a summation of numerous subthreshold stimuli involving sensory systems after absorption of contaminants on nasal mucosa.[10]

The broad inclusion of symptoms/symptom types in definitions of SBS has been criticized by Hodgson[11] who suggests that each of the WHO symptom categories may represent various pathophysiologic entities, each recognized on its own. He suggests, for example, that eye and nose irritation may be due to allergenic contaminants, central nervous system symptoms to solvent neurotoxicity, skin complaints to photodermatitis from monovalent light or irritation from volatile organic compounds (VOCs) and low relative humidity, and odor complaints from contaminants or misperception of odors due to solvent neurotoxicity.

One of the problems in defining phenomena such as BRI and SBS is that such terms have been used interchangeably. The WHO subgroup[10] indicated that SBS is synonymous with such terms as "building-related illness", "building illness syndrome", "ill buildings", "stuffy offices", and "tight office building syndrome". Other synonyms have also included "tight building syndrome" and "stuffy building syndrome".[11,12]

A major difficulty in defining SBS is that one attempts to define that which is at best poorly understood. Any definition will, therefore, reflect the limits of our understanding. The WHO definition suggests a single phenomenon resulting from exposure to multiple substances which produce a similar effect, while multifactoral causal phenomena are suggested by the CEC panel.

TIGHT BUILDINGS

The term "tight building syndrome" began to be used in the late 1970s and early 1980s when it was widely believed that BRI and SBS phenomena were due to the implementation of energy conservation measures in the design, construction, and management of buildings. It is not clear what users of this term actually mean by it. Strictly speaking, a tight building is one that has been designed and constructed to attain and maintain low air infiltration conditions regardless of the percentage of outside air used for ventilation. Of course, when "tight buildings" were initially being designed and constructed, there was a concomitant move by building managers to reduce energy consumption and associated costs by reducing the amount of outdoor air used for ventilation in mechanically ventilated buildings. In many cases, buildings were operated on nearly 100% recirculated air. Such energy conservation practices were implemented in a broad range of mechanically ventilated buildings regardless of their "tightness" relative to air infiltration. Low outdoor exchange rates were being promoted as an energy conservation measure by a number of federal governments in northern Europe and North America. Ventilation standards[13-15] reflected the appeal of low outdoor air requirements to minimize energy costs. For many users of the term, it appears that a tight building is one with low regulated inflows of outdoor air regardless of whether it is, in the strict sense of the word, a "tight" building.

Several investigators[16,17] have suggested that sick-building complaints are for the most part associated with modern sealed buildings described as air-tight shells[16] in which individual occupants cannot directly control their personal comfort (temperature, humidity, ventilation). It is for such individuals "a tight building". Those who have advocated the use of the term suggest that the preponderance of sick-building complaints have arisen in new, energy-efficient buildings,[17] though there appears to be little documented evidence of this.

Both the initial and continued use of the "tight building" concept is unfortunate. It is simplistic at best to attempt to characterize problem-building complaints as being caused by the design, construction, and operation of new, energy-efficient buildings. Such buildings represent only a fraction of the total building population, and there is considerable evidence to suggest that sick building-type symptoms are common to many buildings, including many which are not "tight buildings". Intuitively, the use of low outdoor ventilation rates as an energy conservation measure would be expected to be a significant contributing factor to SBS. However, there is little scientific evidence to support a direct relationship between SBS symptoms and ventilation rates (see Chapters 3 and 8).

SICK BUILDINGS/PROBLEM BUILDINGS

It is common to describe buildings subject to complaints as being sick or problem buildings. Molhave[18] suggests that the term "problem building" be applied to those buildings in which occupants are dissatisfied with air quality for any reason, with the term "sick building" reserved for a subpopulation of a larger set of "problem buildings". Molhave[18] describes sick buildings as having no exposure factor exceeding any generally accepted dose-response threshold, with causality being multifactoral. Causality could be due to an unknown exposure factor or factors and/or to an unknown reaction to the already known exposure factor.

Molhave's[18] characterization of sick buildings as being a subset of a larger population of problem buildings characterized by unknown etiological factors makes good scientific sense. Unfortunately, the terms "problem" and "sick building" are often used interchangeably. The term "sick building" also implies that a building is sick. Being nonhuman, of course, a building can only be sick in a metaphorical sense.

Field Investigations

An apparent relationship between building environments and complaints of illness by occupants became evident from what has been described as "sick building investigations". Such investigations have been conducted by a variety of governmental agencies and private consultants who provide industrial hygiene (IH) and/or IAQ services.

Field investigations are typically conducted as a consequence of occupant/building management requests. They have varied considerably in methodologies employed, the training and capabilities of those conducting the investigation, and success in identifying and mitigating building-related health problems.

NIOSH Investigations

A significant number of problem building investigations have been conducted in the United States by health hazard evaluation teams of the National Institute of Occupational Safety and Health (NIOSH). These teams, commonly comprised of industrial hygienists, epidemiologists, and HVAC system experts, have conducted over 700 health hazard evaluations of problem buildings since 1978. Summary reports of findings have been published periodically.[19-21] Buildings investigated have included governmental and private offices, schools, colleges, and health care facilities.

In many NIOSH investigations reported symptoms were diverse and not specific enough to identify causal agents easily. These have included headache; eye, nose, throat, and skin irritation; fatigue; respiratory problems such as sinus congestion, sneezing, cough, and shortness of breath; and less frequently, nausea and dizziness. Symptom prevalence as a percentage of buildings investigated[19] is indicated in Table 1.1. In a large percentage of investigated buildings (>50%), occupants reported symptoms of eye irritation, dry throat, headache, fatigue, and sinus congestion. These symptoms describe, for the most part, both mucous membrane irritation and neurotoxic effects.

Reported causal and contributing factors/sources of health complaints in NIOSH investigations for the period 1971–1988 are indicated in Table 1.2.[21] NIOSH investigators identified inadequate ventilation as the primary contributor to or cause of the problem in at least 53% of the 529 investigations conducted. Adequacy of ventilation was determined by reference to a guideline value of 1000 ppm CO_2. Other ventilation problems included poor air distribution and mixing (causing stratification), draftiness and pressure differences between office spaces, extremes of temperature and humidity, and filtration problems caused by inadequate maintenance.

Air contamination from sources within the building itself were reported in 15% of cases investigated. Office equipment, particularly copying machines, was reported to be a major cause of such contamination. Contaminants included methanol from spirit duplicators, butyl methacrylate from signature machines, and ammonia and acetic acid from blueprint machines. Other contamination problems included misapplied pesticides such as chlordane, diethyl ethanolamine from boiler additives, improperly diluted cleaning agents, tobacco smoke, combustion gases from cafeterias and laboratories, and cross-contamination from one building zone to another.

Indoor contamination from outdoor sources was identified as the major problem in 10% of NIOSH investigations. This included intake air entrainment of motor vehicle exhaust, boiler gases, previously exhausted air, contaminants from construction or road work such as asphalt, solvents and dusts, and gasoline fumes infiltrating basements or sewerage systems.

Contamination due to materials and products was reported as the causal factor in less than 5% of investigations conducted. Causal contaminants included formaldehyde emitted from urea-formaldehyde bonded wood products

Table 1.1 Frequency of reported symptoms in NIOSH building investigations.

Symptom	% of Buildings
Eye irritation	81
Dry throat	71
Headache	67
Fatigue	53
Sinus congestion	51
Skin irritation	38
Shortness of breath	33
Cough	24
Dizziness	22
Nausea	15

From Wallingford, K. M. and J. Carpenter. 1986. *Proceedings IAQ '86: Managing Indoor Air for Health and Energy Conservation.* American Society of Heating, Refrigerating and Air-Conditioning Engineers, Atlanta.

Table 1.2 Problem types identified in NIOSH building investigations.

Problem type	Buildings investigated	%
Contamination from indoor sources	80	15
Contamination from outdoor sources	53	10
Building fabric as contaminant source	21	4
Microbial contamination	27	5
Inadequate ventilation	280	53
Unknown	68	13
Total	529	100

From Seitz, T.A. 1989. *Proceedings Indoor Air Quality International Symposium: The Practitioner's Approach to Indoor Air Quality Investigations.* American Industrial Hygiene Association. Akron, OH. With permission.

such as particleboard, hardwood plywood, and medium-density fiberboard, fiberglass particles from the erosion of duct liner materials, various organic solvents from glues and adhesives, acetic acid from silicone caulking, liquid and vapor contamination from asphalt, and PCBs from fluorescent lighting ballast failure.

Microbial contamination was reported to be the cause of health complaints in approximately 5% of all NIOSH investigations. Hypersensitivity pneumonitis, a severe health condition caused by exposure to very high levels of fungal spores or thermophilic actinomycetes, was the major health problem reported. It was identified as the probable cause of reported health problems based on excessive microbial contamination rather than on medical or epidemiological evaluations.

NIOSH health hazard evaluations of problem buildings represent a significant resource of documented building investigations. Though investigators attempted to identify causal factors and make recommendations for their mitigation, the accuracy of their assessments is not known. Although NIOSH now uses a standard investigative protocol and has for the latter part of the 1980s, much of the data presented in Table 1.2 includes investigations which did not employ a standardized approach. Many of these investigations were reviewed retrospectively, and a potential exists for misclassification of causal factors because of vague reports.

Canadian Investigations

Investigations of building-related health complaints in Canada are conducted by federal environmental health officers. A total of 82 of these individuals were surveyed by Health and Welfare Canada in 1990. Using the NIOSH model for classifying causes of IAQ problems, Kirkbride et al.[22] reported the prevalence of IAQ problem types in 1362 buildings investigated during the period from 1984–1989. These included inadequate ventilation, 52%; indoor contaminants, 5%; outdoor contaminants, 9%; building fabric, 2%; biological contaminants, 0.4%; and no problem identified, 24%. Questionnaire responses revealed a wide range in investigative methodologies, contaminants measured, and guidelines used to determine whether a problem was associated with specific contaminants.

Reprise

In comparing Health and Welfare Canada's data to those reported by NIOSH, it is interesting to note the remarkable similarity in the percentage of investigations attributed to inadequate ventilation—roughly 50% in both cases. On the other hand, significant differences were observed in the categories "biological contaminants" and "no problem identified". In the former case, NIOSH investigations report biological contaminants causing hypersensitivity pneumonitis-type ailments ten times more frequently than Canadian investigators. This may be due to differences in expertise in identifying biological contamination as a cause of BRI or differences which are climate-related (e.g., less need for air-conditioning in Canada). Canadian investigators twice as frequently reported that they were unable to identify the cause of building health complaints than their NIOSH counterparts.

Systematic Building Investigations—Symptom Prevalence

Field investigations have served to identify and initially define the nature of problem building phenomena. Such investigations have limited scientific usefulness due to the inherent bias involved in conducting a study of a building which is subject to occupant health and comfort complaints and, in general, the relatively unsystematic approaches used in the conduct of such investigations. Reports of field investigations are, in many cases, anecdotal in nature.

Systematic studies are required to take the phenomenon of problem/sick buildings from an initial identification stage based on field investigations to the level of "good science". Such studies typically include scientific assessments of symptom prevalences and statistically valid analyses of data. They may in many cases also use a control or reference population.

U.S. Studies

The studies of Turiel et al.[23] in San Francisco office buildings represent one of the earliest reported attempts to systematically evaluate the nature of sick building complaints. The study was primarily designed to evaluate the effects of reduced ventilation on the air quality of an office building subject to complaints. However, a medical survey of symptoms/complaints of occupants was carried out by one of the authors. A similar survey was conducted on a "noncomplaint" building nearby, the latter serving as a control. Significant differences were observed in the prevalence of four symptom categories in the two buildings. Higher prevalences were reported in the complaint building for eye irritation/itching (54.9 vs. 36.1%), irritation of nose or throat (52.5 vs. 23.5%), shortness of breath (18.9 vs. 3.0%), and chest tightness (20.6 vs. 3.4%). Notably, relatively high frequencies of eye and nose/throat irritation were reported for both buildings.

Systematic studies of symptom prevalence rates and a variety of environmental and work-related factors have been conducted in the United States Environmental Protection Agency (USEPA) headquarters buildings and in the Madison Building of the Library of Congress.[24-28] The former had been the focus of occupant complaints of poor air quality and has received considerable notoriety because of the irony of the situation and the failure of USEPA, NIOSH, and a host of consultants to identify and mitigate the cause or causes. In 1989, USEPA and NIOSH staff scientists conducted a study of the health and comfort concerns of building occupants and physical and environmental characteristics of the USEPA and Library of Congress buildings.

There are three buildings at USEPA headquarters. These include Waterside Mall (WM), a large building divided into sectors, and two other smaller buildings, Crystal City (CC) and Fairchild (FC). Symptom prevalences for IAQ-type symptoms and respiratory or flu-like symptoms have been reported[26] for different sectors of the Waterside Mall, the Waterside Mall building as a whole, and the Crystal City and Fairchild buildings (Table 1.3). Symptoms of USEPA employees reported to occur often or always and resolve when leaving the building varied from 7–21%.

Similar health symptom prevalence rates were reported for employees surveyed in the John Madison building of the Library of Congress.[25] Indoor air quality-type symptoms varied from a low of 10% for runny nose to 25% for sleepiness and 1–7% for respiratory or flu-like symptoms.

Mendell et al.[29] conducted an epidemiological survey of approximately 900 individuals in 12 non-complaint California office buildings.

Table 1.3 Symptom prevalence rates (%) in USEPA headquarters buildings.

	WM Sectors						Building		
	1	2	3	4	5	6	WM	CC	FC
Symptoms	N = 772	600	400	500	435	223	N = 3070	445	407
IAQ-type									
Headache	14	13	18	19	16	18	16	11	16
Runny nose	7	9	9	10	8	8	8	9	7
Stuffy nose	15	13	16	21	16	16	16	17	15
Dry eyes	14	15	21	18	13	20	17	12	15
Burning eyes	9	10	13	11	9	10	10	8	11
Dry throat	8	9	15	12	8	14	10	7	9
Fatigue	12	15	17	17	12	15	15	14	11
Sleepiness	13	14	18	17	14	20	15	19	13
Respiratory/flu-like									
Cough	4	5	6	6	4	2	4	5	4
Wheezing	1	1	1	2	1	2	1	1	2
Short of breath	1	2	3	3	3	2	2	1	2
Chest tightness	1	1	3	2	2	2	1	1	0
Fever	4	0	0	1	1	5	1	1	0
Aching muscles/joints	3	4	5	5	4	6	4	4	2

Note: N denotes number of persons in a sector or building. WM = Waterside Mall; CC = Crystal City; FC = Fairchild.

From Fidler, A.T. et al., 1990. *Proceedings Fifth International Conference on Indoor Air Quality and Climate.* Vol. 4. Toronto.

Building-related symptom prevalence rates were runny nose, 16.6%; stuffy nose, 25.2%; dry, irritated throat, 17.7%; dry, irritated, or itching eyes, 22%; all eye, nose, throat symptoms, 40.3%; chest tightness, 3.7%; difficulty breathing, 6.5%; chills or fever, 4.5%; fatigue/tiredness, 25.4%; sleepiness, 24.9%; either fatigue or sleepiness, 33.2%; headache, 19.8%; and dry/itchy skin, 10.8%.

U.K. Studies

Several systematic sick building surveys have been conducted in the United Kingdom. In an early study,[30] a random sample of the work population of nine office buildings was surveyed by the use of a physician-administered questionnaire. Seven of the nine buildings had no previous history of occupant complaints. Symptom prevalence rates varied widely both among symptom types and among buildings. Symptoms with relatively high prevalences in one or more of the buildings investigated included nasal (5.1–21.8%), eye (4.1–35.6%), mucous membrane (7.1–55.1%), tight chest (0–17.9%), headache (15.2–57.7%), dry skin (5.4–23.1%), and lethargy (13.4–76%). Symptom prevalence rates were highest in the two complaint buildings.

In a much larger study, the U.K. group[31] surveyed 4373 workers in 47 office buildings, in most of which occupants had not expressed any previous dissatisfaction with air quality or their work environment. Ten symptom categories were assessed by means of a self-administered questionnaire. The most commonly reported building/work-related symptoms were lethargy (57%), blocked nose (47%), dry throat (46%), and headache (46%). The least frequent were chest tightness and difficulty breathing (9%).

Danish Town Hall Study

The Danish Town Hall Study[32] represented a major effort to assess occupant symptom prevalence rates and indoor climate conditions in buildings in which dissatisfaction with air quality was not previously expressed. The study design included the use of a self-administered questionnaire and measurements of various parameters of indoor climate in 14 town halls and affiliated buildings in Greater Copenhagen, Denmark.

Prevalence rates for 12 work-related symptoms and 4 symptom groups are summarized for males and females in Table 1.4. Symptom prevalence rates were relatively similar to those reported for the noncomplaint buildings of the nine-building U.K. study of Finnegan et al.[30] but were considerably less than those reported in the 47-office building study of Burge et al.[31] In the latter case, prevalence rates for such symptoms as lethargy, blocked nose, throat irritation, and headache were in excess of 40%.

Nasal irritation was the most frequently reported mucous membrane irritation symptom (12% males/20% females), with fatigue the most frequently reported general symptom (21% males/31% females). The prevalence of skin and irritability symptoms were low (<10%). Significant correlations were observed for mucous membrane irritation (eye, nose, throat irritation) and general symptoms in females but not in males, mucous membrane irritation and skin symptoms (dry skin/rash) in females but not in males. Both males and females showed highly significant correlations between general symptoms and skin symptoms.

Dutch Office Building Study

Zweers et al.[33] conducted studies of work-related health and indoor climate complaints in 60 office buildings in the Netherlands. Mean prevalence and ranges for the five symptom categories were reported as skin (6.8%, range 0.0–17.0%), eye (19.5%, range 3.2–39.5%), oronasal (23.5%, range 0.0–45.5%), nervous system (20.3%, range 3.8–51.3%) and fever (8.8%, range 0.0–33.3%). Symptom prevalence rates among the 61 office buildings surveyed varied widely with relatively high mean prevalence rates for eye, oronasal, and nervous system symptoms.

Swedish Sick Building Study

Norback et al.[34] conducted systematic investigations of 11 office buildings presumed to be "sick buildings" due to occupant complaints. A number of individual symptoms were assessed by means of a self-administered

Table 1.4 Symptom prevalences among male and female employees in Danish town hall buildings.

	Prevalence rates (%)	
	Males (N = 1093–1115)	**Females (N = 2280–2345)**
Symptoms		
Eye irritation	8.0	15.1
Nasal irritation	12.0	20.0
Blocked, runny nose	4.7	8.3
Throat irritation	10.9	17.9
Sore throat	1.9	2.5
Dry skin	3.6	7.5
Rash	1.2	1.6
Headache	13.0	22.9
Fatigue	20.9	30.8
Malaise	4.9	9.2
Irritability	5.4	6.3
Lack of concentration	3.7	4.7
Symptom groups		
Mucous membrane irritation	20.3	
Skin reactions	4.2	
General symptoms	26.1	
Irritability	7.9	

From Skov, P. et al. 1987. *Environ. Int.,* 13:399–349. With permission.

Table 1.5 Symptom prevalence rates in 11 Swedish "sick" office buildings.

Symptom	**Total mean prevalence (%)**	**Range**
Eye irritation	36	13–67
Swollen eyelids	13	0–32
Nasal catarrh	21	7–46
Nasal congestion	33	12–54
Throat dryness	38	13–64
Sore throat	18	8–36
Irritative cough	15	6–27
Headache	36	19–60
Abnormal tiredness	49	19–92
Sensation of getting a cold	42	23–77
Nausea	8	0–23
Facial itch	12	0–31
Facial rash	14	0–38
Itching on hands	12	5–31
Rashes on hands	8	0–23
Eczema	15	5–26

From Norback, D. et al., 1990. *Scand. J. Work Environ. Health,* 16:121–128. With permission.

questionnaire. Considerable variation in symptom prevalence rates were observed over the range of buildings surveyed (Table 1.5). Notably high (>30%) prevalence rates were reported for eye irritation, nasal congestion, throat dryness, sensation of getting a cold, headache, and abnormal tiredness.

Table 1.6 Symptom prevalence indices for ten selected IAQ symptoms in seven university buildings.

Building	Symptom prevalence index	
	Often-always	Sometimes-often-always
A	0.44	1.31
B	0.50	1.02
C	0.50	1.23
D	0.74	1.68
E	0.46	1.23
F	0.67	1.38
G	0.93	2.16

From Godish, T. and J. Dittmer, 1993. *Proceedings Sixth International Conference on Indoor Air Qualtiy and Climate.* Vol. 2. Helsinki.

Reprise

The systematic office building studies described above have reported a broad range of symptom prevalence rates in buildings surveyed. This variation in reported SBS symptoms is likely to have been due to a number of factors including differences in symptom expression and reporting among both individuals and buildings, inclusion of "complaint" buildings in building populations surveyed, and differences in assessment methodology.

Symptom prevalence rates can be significantly affected by assessment methodology employed. In the U.K. studies of Burge et al.,[31] prevalence rates were based on work-related symptoms (resolved after leaving work) that occurred as few as twice in the past 12 months. In the Swedish studies of Norback et al.,[34] apparently all symptoms which occurred on the job in a previous 6-month period were assessed. In contrast, only work- or building-related symptoms which occurred often or always were assessed in the Danish Town Hall Study,[32] the Dutch Office Building Study,[33] and the USEPA headquarters buildings[28] and John Madison/ Library of Congress building studies.[27] The effect of assessment methodology is most evident when symptom prevalence rates reported for U.K. office building studies[31] are compared to other studies. The former tend to be much higher. The U.K. studies included symptoms which would be reported as occurring "sometimes" in the USEPA assessment methodology. However, only those symptoms which were reported "often" or "always" were used to determine symptom prevalence rates in USEPA studies.[28] This would also be, for the most part, true for the Danish Town Hall Study[32] and the Dutch Office Building Study.[33]

Differences in symptom prevalence based on classification criteria can be seen in Table 1.6.[35] In this case, personal symptom indices (PSI) based on the number of reported building-related symptoms from ten IAQ symptoms are reported for categories described as often-always and sometimes-often-always. Not surprising, the PSI which reflects symptom prevalence was significantly higher in the latter as compared to the former.

Despite differences in assessment methodologies, the results of all reported systematic studies indicate that building/work-related symptoms are common in buildings including those in which occupants expressed no previous dissatisfaction: buildings which would ordinarily be considered as "healthy". This suggests that most office-type buildings have, to some extent, occupants who experience building/work-related health symptoms, that most office buildings and likely other public access buildings have air quality problems to some degree.

The observation that building/work-related symptoms are common among occupants in office buildings in general is significant in light of attempts to distinguish between "sick" buildings and those which are "healthy". The WHO panel[6] described SBS as a condition where a majority of the occupants are affected. Under what has been described as the building diagnostics investigative protocol (see Chapter 7), air quality dissatisfaction which is reported by less than 20% of building occupants is acceptable,[36,37] and such buildings would be described as being "healthy". This would also be true for investigative protocols used by the Danish Building Research Institute[38] and the Nordic Ventilation Group.[39]

It is difficult to distinguish between "sick" and "healthy" buildings both conceptually and on a practical basis. More buildings are likely to fall into the "sick" category if assessed using the methodologies of Burge et al.[31] and Norback et al.[34] as compared to those used in Danish Town Hall,[32] Dutch Office Building,[33] the California Healthy Building Study,[29] and USEPA studies.[27,28] There is, of course, a problem in using any guideline to separate "healthy" and "sick" buildings. In using symptom prevalence rates in defining sick buildings or SBS, those buildings which fall below the "magic" number or numbers are by implication in some way "normal" or otherwise "healthy". This would suggest that no IAQ problems exist in such buildings even though a relatively significant portion (up to 20%) of the building population may have building/work-related symptoms.

Skin Symptoms

A variety of skin symptoms have been reported in problem building investigations. Because of this apparent association, the WHO[6] panel has included dry and eczematous skin among the constellation of symptoms used to define SBS. The characterization of skin symptoms as a part of SBS is not universal. Neither CEC nor ATS panels included skin symptoms in their definition or concept of SBS.

In NIOSH investigations, skin irritation was reported in 38% of buildings investigated.[21] In one of these investigations, skin symptoms characterized by localized swelling, eczema, and pinhead papules and itching were associated with exposure to glass fibers released from duct liner materials in an air-handling system.[40] In another NIOSH investigation, skin symptoms were reported among office workers which developed after a change in chemicals used in water treatment for a steam humidification system.[41] Complaints consisted of burning and itching of the exposed skin with a red rash. Skin examinations revealed an irritant-type rash on the exposed areas of the face, neck, and hands. The pattern suggested a phototoxic skin reaction.

Godish[42] reported high prevalence rates (5–65%) of transient facial erythema among staff and students in a problem school building. Prevalence rates varied from day to day and from classroom to classroom. Facial erythema was observed to be related to the operation of HVAC systems.

Skin symptoms characterized as dry skin or rash were reported in 4.2% of the men and 8.3% of the women surveyed in the Danish Town Hall Study.[32] Finnegan et al.[30] reported symptoms including dry skin, rash, and itchy skin with prevalence rates which varied from 5.4–23.1%, 0.9–6.4%, and 1.8–10.4%, respectively, in seven buildings assessed. Highest prevalence rates were reported for dry and itchy skin in humidified buildings (circa 15.5 and 7.3%, respectively). Norback et al.[34] reported average prevalence rates of 12% for facial itch, 14% for facial rash, 12% for itching on hands, 8% for rashes on hands, and 15% for eczema in 11 sick buildings. Skin symptoms were reported in 6.8% of the surveyed population in the Dutch Office Building Study.[33]

Stenberg[43] conducted clinical investigations of 77 patients from seven workplaces in the period from 1982–1986. These included four office buildings, one museum, the composing room of a newspaper office, and one hospital. Skin symptoms based on patient histories were clearly associated with the work environment. Symptoms diagnosed included seborrheic dermatitis, a scaling dermatitis of the scalp, ears, and face; facial erythema, a reddening or flushing on cheeks, chin, and side of neck; periorbital eczema; rosacea, characterized by facial erythema, ectatic small blood vessels, papules, and pustules; urticaria/pruritis; and itching folliculitis, a relapsing, itching acne-like rash appearing mostly on the upper chest and back.

Stenberg et al.[44] conducted case-referent studies of populations of office workers affected by general, mucosal, and skin symptoms, and office VDT users having sensory and visual skin symptoms. Many of the SBS and VDT subjects associated skin symptoms with their work environment. These associations can be seen in Table 1.7. A relatively large percentage (>30%) of SBS cases associated symptoms of dry skin (48.7%), erythema (71.4%), rosacea/perioral dermatitis (36.4%), pruritus/urticaria (36%), and atopic dermatitis (44.4%) with their work environment.

Skin symptoms among workers, particularly industrial workers, represents one of the single largest categories (74%) of occupational disease in the United States.[45] The major cause of occupational skin disease is the direct contact of body parts with either primary irritants or irritants which cause an allergic reaction. Primary irritants account for approximately 80% of occupational skin disease, with allergic reactions the remainder. Typical skin responses associated with exposure to primary irritants or allergenic substances range from erythema (reddening) with burning and/or itching to various-sized blisters (vesicles) or small bumps (papules). The dermatitis resolves once exposure

Table 1.7 Relative rates of office worker association between skin symptoms and their work environment.

Clinical findings	SBS cases	VDT cases
Dry skin	48.7	62.1
Seborrheic dermatitis/ external otitis	27.0	25.8
Erythema	71.4	73.8
Rosacea/perioral dermatitis	36.4	61.5
Acne/facial folliculitis	9.8	9.1
Folliculitis on trunk/scalp	12.8	8.7
Pruritis/urticaria	36.0	14.3
Atopic dermatitis	44.4	14.3

From Stenberg, B. et al., 1990. *Proceedings Fifth International Conference on Indoor Air Quality and Climate.* Vol. 4. Toronto.

ceases. Eczema (dry, scaly, rough, thickened skin) may occur as a consequence of prolonged or chronic exposure.

Skin symptoms are commonly reported in investigations of problem buildings and in systematic building studies. The occurrence of skin symptoms together with other typical SBS symptoms would suggest that mucous membrane, general, and skin symptoms may have a common cause. Stenberg,[43,44] however, suggests that an association between skin symptoms and SBS is unlikely.

It is probable in many cases that the occurrence of skin symptoms is coincidental to IAQ problems. Several factors suggest that skin symptoms reported in problem building investigations and SBS surveys may be due to unique exposures. These include studies in which specific causal agents have been identified (e.g., fiberglass as a cause of dermatitis) or where skin symptoms have been implicated with working with office equipment and materials such as VDTs,[46-52] carbonless copy paper,[53-55] and wet-process photocopiers.[56] Such exposures and resultant effects may be independent of general building air quality.

It is widely believed in the IAQ research community that the heretofore unidentified cause or causes of SBS-type complaints is/are gas phase materials. Historically, skin symptoms have been associated with direct contact with liquid or solid phase materials or exposure to ultraviolet light or other skin-damaging forms of electromagnetic radiation. The effect of gas phase substances on the skin, particularly at the low concentrations found in building air, is for the most part unknown. At least a few substances found in indoor air are known skin irritants, the most prominent of which is formaldehyde. Although there is no direct evidence that exposures to formaldehyde in the gas phase can cause skin symptoms, field investigations have implicated formaldehyde exposures with skin symptoms in residences at levels above 0.10 ppm.[57] If formaldehyde can cause skin symptoms from gas phase exposures, it is probable that other contaminants can as well.

Work Performance/Productivity

Symptoms associated with problem building or SBS-type complaints are for the most part relatively minor and as such may not constitute any significant health concern. Their primary effect may be to reduce an individual's quality of life. These quality of life affects may not, however, be without consequence. Of particular concern is their potential effect on work performance, both in terms of productivity while at work and overall productivity which would be affected by increased absenteeism. If poor IAQ does contribute to decreased productivity as a consequence of either (or both) increased absenteeism or reduced work performance, it would impose potentially significant economic costs on employers of affected building occupants.

A qualitative relationship between building air quality, work performance, or absenteeism has been reported in a number of studies. In the USEPA headquarters building, one third of respondents reported that poor IAQ reduced their ability to work at least some of the time, and approximately one fourth indicated that it caused them to stay home from work or leave early.[25] Similar responses have also been reported for the Library of Congress building.[26] In an analysis of Library of Congress data, Hall et al.[58] observed that 18% of the variability in productivity measures could be explained by building/work-related mucous membrane symptoms and 30% by all building/work-related symptoms. In the telephone survey of office workers by Woods et al.,[59] approximately 40% of respondents claimed that building air quality hampered their work.

Raw et al.[60] conducted analyses of worker productivity based on the 47-office building survey of Burge et al.[31] Productivity was assessed by subjective ratings of office workers. A decline in perceived productivity was associated with workers reporting more than two symptoms. Productivity appeared to improve when environmental conditions were perceived to be better (e.g., more comfortable or satisfactory, fresher air, neither hot nor cold, not stuffy nor draughty). Lower productivity ratings were reported for individuals in offices with tobacco smoking.

Preller et al.[61] conducted a survey of sick leave and work-related health complaints among 7000 workers in 61 office buildings in the Netherlands. Significant increases in sick leave were associated with working in a building with steam or spray humidification or working more than 4 hours per day at a video display terminal.

Robertson et al.[62] studied the effect of moving from a naturally ventilated to an air-conditioned building and from an air-conditioned building (ACB) to a naturally ventilated one (NVB) on symptom prevalence rates and absenteeism. In addition to a significant rise in symptom prevalence rates, there was a small increase in overall absenteeism associated with moving to an ACB from an NVB and a slight decline in moving from an ACB to an NVB. In the latter case, no significant decline in symptom prevalence rates was observed. Trends for all categories of illness absence were not consistent with the prevalence of

SBS symptoms. Observed effects were attributed to physical and psychological factors related to the building and its internal environment and not due to irritant or toxic responses.

Zyla-Wisensale and Stolwijk[63] evaluated the potential effect of environmental factors on the quantitative productivity of 228 office workers in a relatively new office building. The daily contribution of each worker to increases in data entry file size associated with the processing of insurance claim forms was used to measure productivity. Productivity per person-day fluctuated little over the 6-month study period. Spontaneous output for individual workers was observed to vary considerably from day to day. Day-to-day variation in productivity was suggested to be due to factors internal to employees rather than external factors related to the building or work environment. Building-related factors that appeared to be associated with productivity included worker proximity to supply and return air vents (negative effect for closeness) and fluorescent lights (positive effect for closeness).

The potential effect of SBS symptoms on worker productivity is a significant concern because of associated economic costs and the policy implications relative to the design, construction, and operation of new buildings and the management of existing ones. Woods[64] has attempted to assess costs associated with productivity losses, including those to individual occupants (lost wages as a result of absenteeism, medical costs associated with seeking medical assistance, costs associated with medical insurance coverage) and to employers (lost productivity). Woods[64] postulated that significant individual costs could be expected in 7–10% of office and institutional building stock. Costs of medical care associated with major health-effecting cases of indoor air pollution were projected to exceed $1 billion annually, with estimated annual costs of medical visits in excess of $500 million.

Productivity losses associated with 3 days' lost work among half of the 20% of office workers who reported that air quality hampered their work performance[59] would be expected to cost employers in the U.S. approximately $10 billion annually. Along the same line, if office workers were to lose an average of 6 minutes productive concentration per day, the national impact on productivity would be projected to be on the order of $10 billion a year.

Poor IAQ has the potential for imposing significant economic costs on society. Though the use of the above projections of Woods[64] to influence public policy on IAQ is attractive, it is important to note that productivity studies conducted to date are for the most part qualitative in nature and, as such, are far from being definitive.

MULTIPLE CHEMICAL SENSITIVITY

A side issue of sick building phenomena are claims by some affected individuals that they have become chemically sensitive as a result of exposure to building contaminants. Such individuals may be diagnosed as having multiple chemical sensitivities. Multiple chemical sensitivity (MCS) has been described as an acquired disorder characterized by recurrent symptoms affecting multiple organs, symptoms which occur as a result of exposure to many unrelated compounds at doses far below those reported to cause harmful effects.[65] MCS symptoms are suggested to be elicited by exposures to diverse chemical structural classes and toxicological modes of action at exposure levels lower than 1% of established threshold limit values.

The concept of MCS has been developed by a small but growing number of occupational physicians who in their clinical practice have observed patients who appear to have acute symptomatic reactions to low concentrations of workplace contaminants after what is believed to be an initial high-level exposure. The basic characterization of MCS as being a polysymptomatic phenomenon provoked by multiple offending agents at very low exposure concentrations is somewhat similar to disease syndromes previously described by clinical ecologists. Clinical ecologists are medical practitioners who specialize in treating patients who they characterize as suffering from heightened reactivities to chemical substances in their environment. Such putative hyperreactive responses to environmental exposures have been described as multiple chemical sensitivity syndrome, chemical hypersensitivity syndrome, universal allergy, 20th century disease, environmental illness, ecologic illness, cerebral allergy, environmental maladaptation syndrome, and food and chemical sensitivity. As defined by the American Academy of Environmental Medicine (a professional society founded by clinical ecologists), ecologic illness (EI) is a chronic multi-system, usually polysymptomatic disorder caused by adverse reactions to environmental incitants, modified by individual susceptibility and specific adaptation.[66]

The concept of MCS as defined by occupational physicians and EI as defined by clinical ecologists has not generally been accepted by the larger medical community. This has been, for the most part, due to the fact that traditional medical practitioners are skeptical of the various theories and practices of clinical ecology. The skepticism and even strong opposition to clinical ecology is evidenced in panel reports of the California Medical Association,[67] the Ontario Ministry of Health,[68] and the American Academy of Allergy and Immunology,[69] which concluded that the practice of clinical ecology is based on unproven and experimental methodology and is lacking scientific validation.

The controversies associated with the practice of clinical ecology, or environmental medicine as it is sometimes called, have, unfortunately, confused the issues associated with the putatively hypersensitive. Physicians skeptical of both clinical ecology and/or the theory of MCS contend that illness symptoms reported by so-called MCS or ecologically ill patients are psychosomatic in nature and are not due to physiological responses to low-level chemical exposures.[70-74]

Despite widespread skepticism in the medical community and general belief that the root cause of health complaints reported by MCS or EI patients is psychological, a number of recent studies have contributed a variety of evidences that support the MCS theory. Several biologically plausible mechanisms have also been proposed to explain symptom expression in MCS subjects exposed to low concentrations of chemical contaminants.

Morrow et al.[75] conducted a study of 22 male workers who had a history of exposure to mixtures of organic solvents. Evaluations using the Minnesota Multiphasic Personality Inventory (MMPI) revealed a high rate of somatic disturbance, anxiety, depression, social isolation, and fear of losing control. Those with the longest exposure history had the largest increases in MMPI scores measuring thinking disturbances, social alienation, poor concentration, and anxiety. This study is significant in that chemical exposures appeared to cause neurobehavioral changes, some of which, because of their psychological nature, could be easily diagnosed as being psychosomatic.

Simon et al.[76] conducted studies of plastics workers in an aerospace facility who reportedly experienced chemically induced illness. Of 37 symptomatic individuals, 13 had been diagnosed with EI, that is, being unusually sensitive to chemicals with varied symptom responses. Although self-reported measures of somatization and hypochondria were strongly correlated with the development of EI, no subjects met the diagnostic criteria for somatization disorder. Additionally, there was no apparent association between EI and current anxiety and depression. However, significant differences were observed between exposed subjects diagnosed with EI and those who were not so diagnosed. Environmental illness subjects did, however, have a significant prior history of anxiety and depressive disorder (54 versus 4%) and medically unexplained physical symptoms before exposure (6.2 versus 2.9%). These findings indicated that EI was related more to an underlying trait of symptom amplification and psychological distress than to current psychiatric symptoms or diagnoses. Subjects with EI may, as a consequence, have experienced heightened adverse reactions to chemical substances because of an underlying vulnerability to noxious events.

Meggs and Cleveland[77] conducted a nasal pathology study of ten patients who met the case definition for MCS. All 10 patients were observed to have objective nasal abnormalities determined by rhinolarynogoscopy. These included edema, excessive mucus, a cobblestone appearance of the posterior pharynx (throat) and base of the tongue, mucosal injection and blanched mucosa that surrounded a prominent blood vessel. Edema of the adenoids and tonsillar hyperplasia were also frequently observed.

Bell et al.[78] conducted a survey study of young adult college students that evaluated the prevalence of self-reported illness from the smell of pesticides, automobile exhaust, paint, new carpet, and perfume. Sixty-six percent of 643 students surveyed reported feeling ill to a moderate or marked degree from the smell of one or more of the five chemical exposures; 15% indicated that the smell of four of the five chemical exposures made them ill. Both daytime tiredness and grogginess were significantly correlated with illness ratings for four of the five self-reported exposures. Ratings of illness associated with pesticide odor were weakly associated with ratings for the largest number of symptoms. Psychological variables could not account for any of the variance in self-reported illness associated with chemical odor exposures. The authors concluded that their data were similar to patterns described clinically for MCS patients.

A heightened aversion responsiveness to chemical odors appears to be common to individuals diagnosed with MCS or EI. Many such individuals report that symptoms occur in time after exposure to odiferous substances such as perfumes, automobile exhausts, etc. Such aversion responses would be consistent with trigeminal nerve stimulation associated with the common chemical sense. Trigeminal nerves, which are found in the nasal cavity, throat, mouth, tongue, and eyes, are responsible for the perception of irritation or pungency in both taste and smell.[79] As indicated previously, stimulation of the trigeminal nerves by a chemical exposure may result in a number of nonspecific symptoms. Similar symptoms can be caused by a range of different chemical substances which stimulate the trigeminal nerves, not unlike responses reported by MCS or environmental illness patients.

A putative mechanism for chemical sensitivity has been proposed by Ashford and Miller[66] and Bell et al.[80] They suggest that MCS is caused by a sensitization or kindling of olfactory-limbic pathways in the brain by acute or chronic exposures to solvents and pesticides. The limbic system, which includes the hypothalmus, governs the interaction of an individual with his or her environment. The limbic system has been described as the "primitive smell brain". It is involved in a variety of human emotions associated with self-preservation such as food finding and feeding, fighting, and personal protection. The olfactory bulbs are in close proximity to the limbic area and supply much of its neural input. Olfactory nerves link the limbic system to the external chemical environment. Since there is no blood-brain barrier in the limbic region, various substances can enter the olfactory bulbs and be transported in the olfactory neurons.

Kindling is described as a special type of time-dependent sensitization in which repeated, intermittent, subthreshold stimuli induce an amplification of nerve cell responses to a conclusive endpoint. Once kindling has occurred, the same low-level stimulus that originally caused little electrophysiological response is suggested to trigger a full "seizure" manifested by symptom expression.

Though not common to problem buildings, the issue of MCS does occasionally arise when one or more office workers/building employees are diagnosed as being chemically sensitive or environmentally ill. Such employees may, as a consequence, seek special accommodations for purposes of coping with illness symptoms associated with the building or work environment. Individuals so diagnosed claim to be sensitive to all synthetic chemicals. Building managers typically view accommodation requests from such individuals as not only extreme but also difficult if not impossible to satisfy. Such individuals may on medical advice quit their employment and, as a consequence, file claims for workmen's compensation, disability discrimination, personal injury, etc. In such cases, MCS, or whatever it is called, is no longer a peripheral issue to building management. Neither is it a minor concern to those providing building diagnostic services. As a practical matter, it is difficult for a building diagnostician to recommend mitigation measures which will improve IAQ sufficiently to satisfy individuals who claim that they are "sensitive to everything".

MCS or EI on occasion does become an issue in buildings in which dissatisfaction with air quality has not been previously expressed. This commonly occurs in schools where a child diagnosed with EI is enrolled or when the child's parent has been diagnosed with EI. In such cases, parents insist that special accommodations be made for the child or that home learning be available. Cases of MCS/EI pose unique challenges to school administrators and other building managers.

RECAPITULATION

The phenomenon of illness and dissatisfaction with air quality in office and institutional buildings is relatively complex. Building-related illness describes those cases where causal factors of building/work-related health complaints have been identified. If no causal factor or agent can be identified, the phenomenon is described as SBS and the building, a sick building. A variety of organizations have attempted to define SBS and sick buildings. Though definitions are for the most part similar, there are individual differences. In general, SBS describes a case of building-related health complaints where the etiology is unknown. It is typically associated with a high prevalence of mucous membrane and general symptoms. Sick buildings have been suggested to be a subpopulation of a larger group of problem buildings. The sick building phenomenon is often referred to as "tight building syndrome," though there is little evidence to support the thesis that either tight building construction or reduced outdoor volumetric ventilation rates are responsible for the reported occupant health complaints.

The nature of the sick or problem building phenomenon has been elucidated from field investigations of complaint buildings and from systematic building studies. In the latter case, significant research has been conducted in the United States, the United Kingdom, Denmark, the Netherlands, and Sweden. Systematic research studies have shown significant prevalence rates of mucous membrane, general, and skin symptoms in both complaint and noncomplaint office buildings. They have shown that building/work-related health complaints occur in all buildings surveyed with prevalence rates varying significantly from building to building.

Skin symptoms have been reported in a number of problem building investigations and systematic building studies. Since skin symptoms are most commonly associated with physical contact with solid or liquid phase materials or with exposure to electromagnetic radiation such as ultraviolet light, it is not clear whether they are associated with building air contaminants, that is, whether they are, in fact, related to SBS phenomena. There is evidence that skin symptoms may be associated with handling carbonless copy paper and working with VDTs. Evidence to link skin symptoms with exposures to air contaminants is quite limited.

Symptoms associated with the building or work environment have the potential for decreasing worker productivity as manifested in lost time from work (absenteeism) and quantitative performance. Such losses of worker productivity in office environments are projected to cause significant economic losses to employers. Studies of worker productivity and sick building phenomenon have, for the most part, been qualitative and suggestive of a potentially significant effect. Additional definition of effects on productivity based on systematic studies are, however, required.

The issue of MCS has occasionally been raised in relationship to exposures which occur in industrial and office workplace environments. Multiple chemical sensitivity has been defined as an unusual sensitivity to low-level exposures of a variety of chemicals which resulted from a previous initial exposure to high concentrations of one or more generally gas phase chemical substances. Symptom manifestation is typically subjective and broad in terms of organs/organ-systems which may be affected. The concept of MCS has been defined by a small group of occupational physicians and is similar to unusual chemical sensitivities described as ecologic or environmental illness (EI), etc. by medical practitioners described as clinical ecologists.

The issue of MCS or EI is a subject of considerable controversy. A number of medical writers have suggested that symptom responses reported by MCS and EI subjects are psychosomatic in origin and, by implication, not due to a physiological response to chemical exposures. However, a number of recent investigations have linked chemical exposures with neurobehavioral symptoms, nasal pathologies determined objectively, and self-reported illness responses associated with apparent exposures to chemical odors.

Several biologically plausible mechanisms may explain the MCS phenomenon in whole or in part. These include the non-specific stimulation of trigeminal nerves by irritant chemicals and an olfactory limbic model which proposes that a sensitization of the limbic region of the brain occurs as a consequence of exposures to odorant chemicals which are transported to the brain through the olfactory system.

REFERENCES

1. Kreiss, K. 1989. "The Epidemiology of Building-Related Complaints and Illness." 575–592. In: Cone, J.E. and M.J. Hodgson (Eds.). *Problem Buildings: Building-Associated Illness and the Sick Building Syndrome. Occupational Medicine: State of the Art Reviews.* Hanley and Belfus, Inc., Philadelphia.

2. American Thoracic Society. 1990. "Environmental Controls and Lung Disease." Report of the ATS Workshop on Environmental Controls and Lung Disease. Santa Fe, NM, March 24–26, 1988. *Am. Rev. Respir. Dis.* 142:915–939.
3. Godish, T. 1981. "Formaldehyde and Building-Related Illness." *J. Environ. Health.* 44:116–121.
4. Rafferty, P.J. and P.J. Quinlan. 1990. "The Practitioner's Guide to Indoor Air Quality." 73–86. In: Weekes, D.M. and R.B. Gammage (Eds.). *Proceedings of the Indoor Air Quality International Symposium: The Practitioner's Approach to Indoor Air Quality.* American Industrial Hygiene Association. Akron, OH.
5. Spengler, J.D. and K. Sexton. 1983. "Indoor Air Pollution: A Public Health Perspective." *Science.* 221:9–17.
6. World Health Organization. 1983. "Indoor Air Pollutants, Exposure and Health Effects Assessment." Euro-Reports and Studies No. 78. World Health Organization Regional Office for Europe. Copenhagen.
7. World Health Organization. 1984. "Indoor Air Quality Research." Euro-Reports and Studies No. 103. World Health Organization Regional Office for Europe. Copenhagen.
8. Molina, C. et al. 1989. "Sick Building Syndrome—A Practical Guide." Report No. 4. Commission of the European Communities, Brussels-Luxembourg.
9. Anderson, I. 1986. "Sick Buildings: Physical and Psychosocial Features, Effects on Humans and Preventative Measures." 77–81. In: *Proceedings of the Third International Conference on Indoor Air Quality and Climate.* Stockholm. Vol. 6.
10. Akimenko, V.V. et al. 1986. "The Sick Building Syndrome." 87–97. In: *Proceedings of the Third International Conference on Indoor Air Quality and Climate.* Vol. 6. Stockholm.
11. Hodgson, M.J. 1989. "Clinical Diagnosis and Management of Building-Related Illness and the Sick Building Syndrome." 593–606. In: Cone, J.E. and M.J. Hodgson (Eds.). *Problem Buildings: Building-Associated Illness and the Sick Building Syndrome. Occupational Medicine: State of the Art Reviews.* Hanley and Belfus, Inc., Philadelphia.
12. Stolwijk, J. 1984. "The Sick Building Syndrome." 23–30. In: *Proceedings of the Third International Conference on Indoor Air Quality and Climate.* Vol. 1. Stockholm.
13. ASHRAE Standard 62–73. 1977. "Standards for Natural and Mechanical Ventilation." American Society of Heating, Refrigerating and Air-Conditioning Engineers. Atlanta.
14. ASHRAE Standard 62–1981. 1981. "Ventilation for Acceptable Indoor Air Quality." American Society of Heating, Refrigerating and Air-Conditioning Engineers. Atlanta.
15. Colthorpe, K. 1990. "A Review of Building Airtightness and Ventilation Standards." Technical Note 30. Air Infiltration and Ventilation Centre. Great Britain.
16. Sterling, T.D. et al. 1983. "Building Illness in the White Collar Workplace." *Int. J. Health Services.* 13:277–287.
17. Bernard, J.M. 1986. "Building-Associated Illness in an Office Environment." 44–52. In: *Proceedings of IAQ '86: Managing Indoor Air for Health and Energy Conservation.* American Society of Heating, Refrigerating and Air-Conditioning Engineers. Atlanta.

18. Molhave, L. 1987. "The Sick Buildings—A Subpopulation Among the Problem Buildings." 469–473. In: *Proceedings of the Fourth International Conference on Indoor Air Quality and Climate*. Vol. 2. West Berlin.
19. Melius, J. et al. 1984. "Indoor Air Quality—The NIOSH Experience." *Ann. ACGIH: Evaluating Office Environmental Problems*. 10:3–8.
20. Wallingford, K.M. and J. Carpenter. 1986. "Field Experience Overview: Investigating Sources of Indoor Air Quality Problems in Office Buildings." 448–453. In: *Proceedings of IAQ '86: Managing Indoor Air for Health and Energy Conservation*. American Society of Heating, Refrigerating and Air-Conditioning Engineers. Atlanta.
21. Seitz, T.A. 1989. "NIOSH Indoor Air Quality Investigations 1971–1988." 163–171. In: Weekes, D.M. and R.B. Gammage (Eds.). *Proceedings of the Indoor Air Quality International Symposium: The Practitioner's Approach to Indoor Air Quality Investigations*. American Industrial Hygiene Association. Akron, OH.
22. Kirkbride, J. 1990. "Health and Welfare Canada's Experience in Indoor Air Quality Investigation." 99–106. In: *Proceedings of the Fifth International Conference on Indoor Air Quality and Climate*. Vol. 5. Toronto.
23. Turiel, I. et al. 1983. "The Effects of Reduced Ventilation on Indoor Air Quality in an Office Building." *Atmos. Environ.* 17:51–64.
24. U.S. Environmental Protection Agency. 1989. "Indoor Air Quality and Work Environment Study, EPA Headquarters Building, Vol. 1: Employee Survey." Office of Administration and Resource Management. Washington D.C.
25. U.S. Environmental Protection Agency. 1990. "Indoor Air Quality and Work Environment Study, EPA Headquarters Building. Vol. 2 and Supplement: Results of Indoor Air Environmental Monitoring Study." Office of Administration and Resource Management. Washington D.C.
26. Fidler, A.T. et al. 1990. "Library of Congress Indoor Air Quality and Work Environment Study: Health Symptoms and Comfort Concerns." 603–608. In: *Proceedings of the Fifth International Conference on Indoor Air Quality and Climate*. Vol. 4. Toronto.
27. Nelson, C.J. et al. 1990. "Environmental Protection Agency Indoor Air Quality and Work Environment Study: Health Symptoms and Comfort Concerns." 615–620. In: *Proceedings of the Fifth International Conference on Indoor Air Quality and Climate*. Vol. 4. Toronto.
28. Crandall, M.S. et al. 1990. "Library of Congress and USEPA Indoor Air Quality and Work Environment Study: Environmental Survey Results." 597–602. In: *Proceedings of the Fifth International Conference on Indoor Air Quality and Climate*. Vol. 4. Toronto.
29. Mendell, M.J. 1991. "Risk Factors for Work-Related Symptoms in Northern California Office Workers." Ph.D. thesis, Lawrence Berkeley Laboratory, LBL-32636 UC-350.
30. Finnegan, M.J. et al. 1984. "The Sick Building Syndrome: Prevalence Studies." *Br. Med. J.* 289:1573–1575.
31. Burge, S. et al. 1987. "Sick Building Syndrome: A Study of 4373 Office Workers." *Ann. Occup. Hyg.* 31:493–504.
32. Skov, P. et al. 1987. "The Sick Building Syndrome in the Office Environment: The Danish Town Hall Study." *Environ. Int.* 13:339–349.

33. Zweers, T. et al. 1992. "Health and Indoor Climate Complaints of 7043 Office Workers in 61 Buildings in the Netherlands." *Indoor Air*. 2:127–136.
34. Norback, D. et al. 1990. "Indoor Air Quality and Personal Factors Related to the Sick Building Syndrome." *Scand. J. Work Environ. Health*. 16:121–128.
35. Godish, T. and J. Dittmer. 1993. "Sick Building Survey of Seven Non-complaint University Buildings." 339–344. In: *Proceedings of the Sixth International Conference on Indoor Air Quality and Climate*. Vol. 2. Helsinki.
36. Woods, J.E. et al. 1989. "Indoor Air Quality Diagnostics: Qualitative and Quantitative Procedures to Improve Environmental Conditions." 80–98. In: Nagda, N.L. and J.P. Harper (Eds.). *Design and Protocol for Monitoring Indoor Air Quality*. ASTM STP 1002. American Society for Testing and Materials. Philadelphia.
37. Lane, C.A. et al. 1989. "Indoor Air Quality Diagnostic Procedures for Sick and Healthy Buildings." 237–240. In: *Proceedings of IAQ '89: The Human Equation: Health and Comfort*. American Society of Heating, Refrigerating and Air-Conditioning Engineers. Atlanta.
38. Valbjorn, O. et al. 1990. "Indoor Climate and Air Quality Problems. Investigation and Remedy." SBI Report 212. Danish Building Research Institute.
39. Kukkonen, E. et al. 1993. "Indoor Climate Problems. Investigation and Remedial Measures." Nordtest Report NT Tech. Report 204. Nordic Ventilation Group. Espoo, Finland.
40. National Institute of Occupational Safety and Health. 1980. Technical Assistance Report HETA 80–80. Ellis Hospital, Schenectady, NY.
41. National Institute of Occupational Safety and Health. 1981. HETA 81–247–958. Boehringer Ingeleim, Ltd., Ridgefield, CT.
42. Godish, T. 1992. "HVAC System Materials as a Potential Source of Contaminants and Cause of Symptoms in a School Building." 208–217. In: *Proceedings of the Eleventh International Conference of the Clean Air Society of Australia and New Zealand*. Brisbane, Australia.
43. Stenberg, B. 1989. "Skin Complaints in Buildings with Indoor Climate Problems." *Environ. Int*. 15:81–84.
44. Stenberg, B. et al. 1990. "The Office Illness Project in Northern Sweden II. Case Referent Study of Sick Building Syndrome (SBS) and VDT-Related Skin Symptoms—Clinical Characteristics of Cases and Predisposing Factors." 683–688. In: *Proceedings of the Fifth International Conference on Indoor Air Quality and Climate*. Vol. 4. Toronto.
45. Adams, R.M. 1983. *Occupational Skin Diseases*. Grune and Stratton. New York.
46. Linden, V. and S. Rolfsen. 1981. "Video Computer Terminals and Occupational Dermatitis." *Scand. J. Work Environ. Health*. 7:62–67.
47. Nilsen, A. 1982. "Facial Rash in Visual Display Unit Operators." *Contact Derm*. 8:25–28.
48. Liden, C. and J.E. Walberg. 1985. "Does Visual Display Work Provoke Rosacea?" *Contact Derm*. 13:235–241.
49. Berg, M. and I. Langlet. 1987. "Defective Video Displays, Shields and Skin Problems." *Lancet*. 1:800.
50. Berg, M. 1988. "Skin Problems in Workers Using Visual Display Terminals. A Study of 201 Patients." *Contact Derm*. 19:335–341.

51. Feldman, L. et al. 1985. "Letter to the Editor. Terminal Illness." *J. Am. Acad. Dermatol.* 12:366.
52. Fisher, A.M. 1986. "Terminal Dermatitis Due to Computers (Visual Display Units)." *Cutis.* 37:153–154.
53. Calnan, C.D. 1979. "Carbon and Carbonless Copy Paper." *Acta Dermatovener 59.* Suppl. 85:27–32.
54. Menne, T. et al. 1981. "Skin and Mucous Membrane Problems from 'No Carbon Required' Paper." *Contact Derm.* 7:72–76.
55. Kleinman, G. and S.W. Horstman. 1982. "Health Complaints Attributed to the Use of Carbonless Copy Paper (A Preliminary Report)." *Am. Ind. Hyg. Assoc. J.* 43:432–435.
56. Jensen, M. and J. Rold-Petersen. 1979. "Itching Erythema Among Post Office Workers Caused by a Photocopying Machine with a Wet Toner." *Contact Derm.* 5:389–391.
57. Ritchie, I.M. and R.H. Lehnen. 1987. "Formaldehyde-Related Health Complaints of Residents Living in Mobile and Conventional Houses." *Am. J. Public Health.* 77:323–328.
58. Hall, H.I. et al. 1991. "Influence of Building-Related Symptoms on Self-Reported Productivity." 33–35. In: *Proceedings of IAQ '91: Healthy Buildings.* American Society of Heating, Refrigerating and Air-Conditioning Engineers. Atlanta.
59. Woods, J.E. et al. 1987. "Office Worker Perceptions of Indoor Air Quality Effects on Discomfort and Performance." 464–468. In: *Proceedings of the Fourth International Conference on Indoor Air Quality and Climate.* Vol. 2. West Berlin.
60. Raw, G. J. et al. 1990. "Further Findings from the Office Environment Survey: Productivity." 231–236. In: *Proceedings of the Fifth International Conference on Indoor Air Quality and Climate.* Vol. 1. Toronto.
61. Preller, L. et al. 1990. "Sick Leave Due to Work-Related Health Complaints among Office Workers in the Netherlands." 227–230. In: *Proceedings of the Fifth International Conference on Indoor Air Quality and Climate.* Vol. 1. Toronto.
62. Robertson, A.S. et al. 1990. "The Effect of Change in Building Ventilation Category on Sickness Absence Rates and the Prevalence of Sick Building Syndrome." 237–242. In: *Proceedings of the Fifth International Conference on Indoor Air Quality and Climate.* Vol. 1. Toronto.
63. Zyla-Wisensale, N.H. and J.A.J. Stolwijk. 1990. "Indoor Air Quality as a Determinant of Office Worker Productivity." 249–254. In: *Proceedings of the Fifth International Conference on Indoor Air Quality and Climate.* Vol. 1. Toronto.
64. Woods, J.E. 1989. "Cost Avoidance and Productivity in Owning and Operating Buildings." 753–770. In: Cone, J.E. and M.J. Hodgson (Eds.). *Problem Buildings: Building-Associated Illness and the Sick Building Syndrome. Occupational Medicine: State of the Art Reviews.* Hanley and Belfus, Inc. Philadelphia.
65. Cullen, M.R. 1987. "The Worker with Multiple Chemical Sensitivities: An Overview." 655–661. In: Cullen, M. R. (Ed.). *Workers with Multiple Chemical Sensitivity. State of Art Reviews.*
66. Ashford, N.A. and C.S. Miller. 1991. *Chemical Exposures. Low Levels and High Stakes.* Van Nostrand Reinhold. New York.

67. California Medical Association Scientific Board. Task Force on Clinical Ecology. 1986. "Clinical Ecology — A Critical Appraisal." *West. J. Med.* 144:239–245.
68. Ontario Ministry of Health. 1985. *Report of the Ad Hoc Advisory Panel on Environmental Hypersensitivity Disorders.* Toronto.
69. Executive Committee of the American Academy of Allergy and Immunology. 1986. "Clinical Ecology." *J. Allergy Clin. Immunol.* 78:269–271.
70. Brodsky, C.M. 1983. "'Allergic to Everything': A Medical Subculture." *Psychosomatics.* 24:731–742.
71. Stewart, D.E. and J. Raskin. 1985. "Psychiatric Assessment of Patients with '20th Century Disease' ('Total Allergy Syndrome')." *Can. Med. Assoc. J.* 133:1001–1006.
72. Terr, A.I. 1986. "Environmental Illness. A Clinical Review of 50 Cases." *Arch. Int. Med.* 146:145–149.
73. Rosenberg, S.J. et al. 1990. "Personality Styles of Patients Asserting Environmental Illness." *J. Occup. Med.* 32:679–681.
74. Black, D.W. et al. 1990. "Environmental Illness: A Controlled Study of 26 Subjects with '20th Century Disease'." *J.A.M.A.* 264:3166–3170.
75. Morrow, L.A. et al. 1989. "A Distinct Pattern of Personality Disturbance Following Exposure to Mixtures of Organic Solvents." *J. Occup. Med.* 31:743–746.
76. Simon, G.F. et al. 1990. "Allergic to Life: Psychological Factors in Environmental Illness." *Amer. J. Psych.* 147:901–906.
77. Meggs, W.J. and C.H. Cleveland. 1993. "Rhinolaryngoscopic Examination of Patients with the Multiple Chemical Sensitivity Syndrome." *Arch. Environ. Health.* 48:14–18.
78. Bell, I.R. et al. 1993. "Self-Reported Illness from Chemical Odors in Young Adults without Clinical Syndromes or Occupational Exposures." *Arch. Environ. Health.* 48:6–13.
79. Shusterman, D. 1992. "Critical Review: The Health Significance of Odor Pollution." *Arch. Environ. Health.* 47:76–87.
80. Bell, I.R. et al. 1992. "An Olfactory-Limbic Model of Multiple Chemical Sensitivity Syndrome: Possible Relationships to Kindling and Affective Spectrum Disorders." *Biol. Psych.* 32:218–242.

2 PEOPLE-RELATED RISK FACTORS

There is increasing evidence from field investigations, case reports, and systematic studies of building populations and potential contaminant sources to indicate that health and comfort complaints reported to be associated with problem or sick building environments are due to a variety of factors which may have direct or indirect effects. This multifactoral nature of building-related health/comfort complaints makes it difficult to diagnose a problem building accurately and recommend appropriate mitigation or remedial measures. It also makes it difficult to both construct and maintain "healthy" buildings.

In this and in the following four chapters, risk factors for building-related health and comfort complaints will be discussed in detail. We begin here with the occupants themselves, the individuals who because of something intrinsic to themselves or elements of their own behavior and/or that of others may be at increased risk for experiencing/reporting SBS-type complaints. These people-based risk factors include personal characteristics, psychosocial phenomenon, and tobacco smoking.

PERSONAL CHARACTERISTICS

A variety of personal characteristics have been evaluated as potential contributing factors to SBS symptom prevalence rates in systematic building studies. These include gender, age, marital status, atopic status, and a variety of lifestyle factors such as smoking, alcohol consumption, coffee consumption, regular exercise, use of contact lenses, etc.

Personal characteristics which have been shown to be major SBS risk factors are gender and atopic history. Mixed results have been reported for the lifestyle factor of tobacco smoking which will be discussed in a separate section in this chapter.

Gender

One of the notable observations made in most problem building investigations and in systematic epidemiological studies of complaint and noncomplaint buildings is that females consistently report higher rates of SBS symptoms than do males.[1-12] Differences in reporting rates may be as high as three to one.

The reasons for differences in symptom reporting rates are not known. Hedge et al.[13] suggest that females may be more sensitive to environmental influences or may be more aware of physical symptoms. Such differential symptom reporting appears to be more universal. Pennebaker and Skelton[14] have observed that females across all age groups report more physical symptoms in general, take more prescribed medications, and visit a physician more frequently. Changes in social roles involving stresses associated with combining family and work responsibilities have been suggested as a contributing factor to increased illness in adult females.[15,16]

Raw and Grey[11] have proposed that gender differences in symptom prevalence are due to the fact that males tend to under-report symptoms. This is based on observations that gender differences in symptom-reporting rates are not observed in office staff who (1) report extreme dissatisfaction with the office environment, (2) have a high level of control over their work, (3) work less than 6 hours/day, and (4) have worked in the same office for more than 8 years. They suggest that gender differences are not due to an increased sensitivity of females to environmental exposures. Raw and Grey[11] cite unpublished Danish studies in which no gender differences in objectively measured eye symptoms were observed. Males, nevertheless, reported a lower prevalence of eye symptoms than females. No differences between male and female responses to eye-irritating exposures of CO_2 were found in the studies of Molhave et al.[17] Raw and Grey[11] suggest that males under-report because they perceive symptom reporting as an admission of weakness which they have been culturally conditioned not to accept.

In addition to these factors, females may be subject to different psychosocial and environmental conditions. Females are more often exposed to tobacco smoke, hold a lower position in the office, work more often in open-plan offices, and handle more paper than males.[12] Females are also more likely to use video-display terminals for extended periods of time, handle carbonless copy paper, and use office equipment identified as risk factors for SBS symptoms.

Atopic History and Other Health Factors

A large number of investigations conducted in many different countries have shown significant relationships between a history of atopy (genetic predisposition to allergic manifestations of exposure to common allergens such as dust mites, mold, pollen, and animal dander) and prevalence of SBS symptoms.[7,8,18-24] These associations are based on respondents' self-reports of allergy, hay fever, asthma, or eczema and not on clinical diagnosis. Only one study which has evaluated this relationship has reported negative results.[25] The relationship between atopy and SBS symptoms is particularly significant for mucous membrane irritation.

In addition to atopic history, Skov et al.[8] have reported a relationship between the prevalence rate of general symptoms and respondent reports of migraine headache.

PSYCHOSOCIAL PHENOMENON

In conducting a building evaluation, the investigator will quickly come to recognize the dual nature of "problem" buildings. There are, of course, aspects of the investigation which are of the expected technical nature and the more problematic "people" aspect. The latter involves a variety of interactions among employees, supervisors, and building management.

The psychosocial dynamics of problem buildings may suggest to some investigators that alleged problems are not due to environmental exposures but to some type of contagious human hysteria. In the less extreme sense, psychosocial phenomena may be risk factors for SBS symptoms or at least symptom-reporting rates.

Mass Psychogenic Illness

Outbreaks of illness with high prevalence rates among individuals in a number of indoor environments reported in the medical literature have been variously described as hysterical contagion,[26] mass hysteria,[27-29] epidemic hysteria,[30] psychosomatic illness,[31] epidemic psychogenic illness,[32] and mass psychogenic illness.[33-36] These are presently considered to be synonymous, with mass psychogenic illness (MPI) being the most commonly accepted and widely used term.

The concept of MPI developed from field investigations of illness episodes among occupants of a variety of buildings. The following observations are common to all reported cases of MPI: (1) illness onset was sudden, (2) investigators believed that illness problems became worse as a consequence of verbal and visual contact among those affected, (3) females exhibited very high prevalence rates, and (4) investigators were unable to identify an etiological agent of an infectious or toxic nature. Interestingly, a relatively high percentage (circa 75%) of MPI episodes have been reported in schools.[31]

No clear or consistent symptom patterns are associated with MPI reports. Symptoms have included fainting, headaches, dizziness, abdominal pain, shortness of breath, hyperventilation, eye itchiness, cough, weakness, sore throat, numbness, nausea, vomiting, diarrhea, skin itching, rash, and unpleasant odor. With few exceptions, reported symptoms were nonspecific with few physical signs and no laboratory findings of disease. Exceptions were several outbreaks of skin symptoms which, in the absence of any positive findings of a causal agent, were diagnosed as psychogenic in origin. By psychogenic, it is meant that the initiation and continued development of the outbreak was due to the emotions and anxiety of those involved.

One of the major tenets of MPI is that medical and industrial hygiene investigations have been unable to identify any potential causal agents in the reported outbreaks. Diagnoses of MPI have been challenged by Faust and Brilliant[37] as excuses for incomplete investigations of low-level environmental contamination.

Though the term MPI and its synonyms, such as mass hysteria, are widely used and often carry the aura of medical authority, these putative phenomenon are based on field investigations which are anecdotal in nature. No credible scientific evidence has been reported to establish that MPI is a real medical phenomenon.

Though MPI episodes differ from SBS (as described in the previous chapter) in their sudden onset, they are remarkably similar to cases of building-related illness (BRI) identified as a consequence of field investigations or where building-related symptoms have been observed but the causal agent or agents were not identified. The inability of investigators to find specific causal factors for health complaints is the norm rather than the exception in problem building investigations. It reflects the complexities involved in identifying and measuring causal agents and the inability of investigators to identify and solve the problem.

Problem buildings in all instances have some form of dynamic related to the behavior of occupants. There is some level of individual and group emotion. The extremity of emotion varies depending on the circumstances involved. These include, for example, the suddenness of problem onset, the inability of building occupants to convince management a problem exists, and individual liability. The inability of investigators to identify the problem may also have an emotional aspect to it as well. It is not unusual for building investigators to conclude that a real problem does not exist, that the problem is in the heads of the occupants. It is often easier to blame building occupants than oneself for having failed in what is often a very difficult task. Management typically responds in a similar way.

In problem buildings where there is no dramatic onset of symptoms, occupants conduct their daily activities unaware that either mucous membrane, general, or skin symptoms being experienced are related to their work or building environment. This is apparent in epidemiological investigations of buildings in which no dissatisfaction with indoor air quality (IAQ) had been previously reported.[1-3,7,9] In such buildings and those described as problem buildings, an awareness of the apparent relationship between symptoms and the building environment develops among some occupants which is rapidly communicated to others who "compare notes", so to speak, and in so doing conclude that previously undisclosed symptoms show a pattern of resolving on leaving the building and on weekends. This "contagion" of awareness can easily be construed as psychogenic since it develops as a consequence of the awareness of others. Emotions may at this point run high as occupants begin to engage (and sometimes enrage!) an often disbelieving management in an effort to identify and mitigate the problem.

Psychosocial Risk Factors

A number of psychosocial factors have been reported to be significantly associated with building/work-related symptoms. These have included job category,[2,8] dissatisfaction with superiors or colleagues,[3,8] quantity of work,[8,12,33-46] job stress,[2,38,40-42] role conflict,[38,40] general job dissatisfaction,[7,38,43,44] perceived degree of environmental control,[13,43] the type of organization an individual works for,[13] occupant density,[7,43] and perceived satisfaction with one's physical environment.[13,45]

In the study of Burge et al.,[2] managers and professional/technical staff reported fewer symptoms than clerical staff even when male/female differences for symptom reporting were taken into account. Reasons for these differences were suggested to be due to (1) the enhanced accommodation of the managers/professionals in the building, (2) the greater degree of control over their job, and (3) their ability to have changes made in the operation of building HVAC systems.

Job Stress

Hedge et al.[41] concluded that once an employee's level of self-reported job stress and job satisfaction are taken into account, symptoms do not appear to correlate with a worker's job category. In an extensive statistical analysis of the office building survey data previously reported by Burge et al.,[2] symptom prevalences were observed to be strongly correlated with job stress, with older workers, those in more responsible positions, and those working with VDTs for long periods of time reporting higher levels of job stress, which in turn were associated with increased symptom reports.[41]

In a study of five smoke-prohibited air-conditioned office buildings, Hedge et al.[42] initially observed that work stressors such as role conflict and workload were significantly related to overall symptom prevalence rates. However, when other aspects of job stress (work pace and general stress rating) were taken into consideration, role conflict and workload were no longer significantly related to symptom prevalence rates.

Hodgson[46] has criticized the work of investigators[2,3,44] who have observed relationships between psychosocial factors and SBS symptom prevalence rates because they failed to use validated questionnaires. In his studies,[47] work stress explained twice as much of the variability in symptoms as measured environmental factors. Work-related tension was the component of work stress more clearly associated with symptoms. Bauer et al.[48] suggest that persisting external causes of SBS complaints such as inadequate environmental control may contribute to increased work stress because of building management's failure to remediate the problem.

In the initial statistical analysis of USEPA Headquarters Building data, Nelson et al.[38] observed that males who reported higher external stress levels and role conflicts also reported higher throat and central nervous system symptoms. Females with high perceived work loads reported more central nervous system symptoms. Wallace et al.,[40] using pooled data with a more extensive statistical analysis, observed that the prevalence of depression, tension, and difficulty concentrating/remembering were significantly associated with job pressure/conflict and other psychosocial variables.

Job Satisfaction/Dissatisfaction

In USEPA Headquarters Building studies,[38,40] some symptoms were observed to be associated with job satisfaction/dissatisfaction parameters such as underutilization and career frustration. Other studies have also observed significant associations between SBS symptoms and psychosocial factors related to job satisfaction or dissatisfaction. Negative correlations were reported between job satisfaction and the prevalence of SBS symptoms in the Danish Town Hall Study,[8] the United Kingdom office building study of Hawkins and Wang,[43] the Dutch Office Building Study,[7] the Office Illness Project of Northern Sweden,[24,39] and an 18-office building study of Hedge.[49] In a previous study by Hedge et al.[41] and studies by Norback et al.[50] and Lenvick,[51] no significant relationships between job satisfaction and SBS symptom prevalence rates were observed.

In the 18-office building study, Hedge[49] reported that managerial workers were more satisfied with their jobs than professional/technical workers who were, in turn, more satisfied with their jobs than were clerical workers. Managers reported the highest job satisfaction and job stress. The effect of these two factors on symptom reporting rates can be seen in Figure 2.1. As job stress increases, SBS symptom reporting rates increase, and as job satisfaction increases, symptom reporting rates decrease.

Differences in symptom prevalence rates between office workers in the public and private sectors have been suggested to be due to differences in job satisfaction, with public employees reporting higher symptom prevalences.[13] This suggestion is based on what appears to be higher job satisfaction ratings for private sector managers as compared to those in the public sector.[52] The effect of job satisfaction may be to influence an individual's propensity to report symptoms. Differences in symptom prevalence rates between public and private employees were not observed in the Office Illness Project of Northern Sweden.[24]

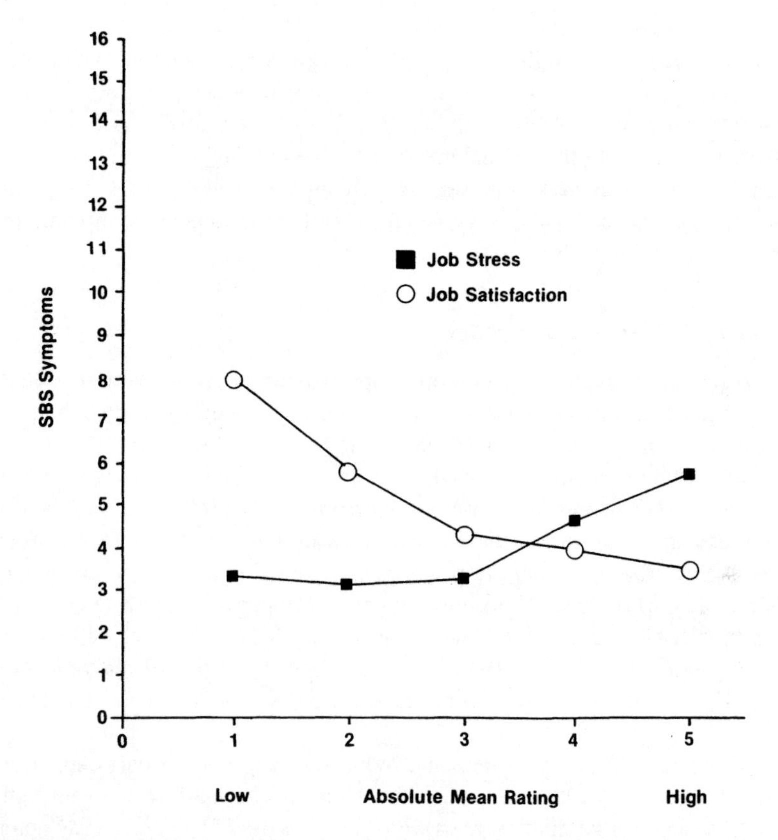

Figure 2.1 Relationship between ratings of job stress, job satisfaction, and SBS symptom prevalence rates. (Courtesy of Hedge, A.,1988.)

Occupant Density

Occupant density or perceived occupant density may be another potential psychosocial factor which may be related to symptom prevalence or complaint rates in buildings. In the Dutch Office Building Study,[7] the number of individuals in a work space was significantly associated with higher prevalence rates of oronasal, nervous system, temperature, air quality, and dry air complaints. In the U.K. studies of Hawkins and Wang,[43] individuals who perceived their office spaces to be crowded had significantly higher building symptom scores.

Satisfaction with Physical Work Environment

Satisfaction or perceived satisfaction with one's physical work environment appears to be related to symptom prevalence. Hedge et al.[13] observed that workers report more SBS symptoms when dissatisfied with environmental conditions in their workplace. Environmental satisfaction appears to be affected by several factors including perceived control over temperature, ventilation, and lighting and perceived ambient conditions. Broder et al.[45] reported significant relationships between symptom prevalence and perceptions of environmental factors rather than actual measured values of environmental conditions such as CO_2. Perceived odor was significantly related to nasal symptoms, comfort to tiring easily, and perceived fresh air to the number of symptoms, eye irritation, and throat discomfort.

Seasonal Affective Disorder

A potential psychogenic or more properly neurophysiologic effect on SBS symptom prevalence rates has been reported. In conducting a problem building investigation in Toledo, Ohio, NIOSH staff[53] evaluated the potential effect of seasonal affective disorder (SAD), a recurring mood-changing phenomenon characterized by depressive symptoms with onset in late fall continuing through the winter and remitting in the spring, on symptom reporting rates. Seasonal affective disorder is thought to result from seasonal light intensity and photoperiod changes (reduced light intensity and photoperiod). Building occupants with high SAD scores were observed to have 3.8 times the risk of reporting building-related symptoms. Hedge et al.[42] in their five-building study observed a significant but weak association between SAD scores and the number of symptoms.

The potential effect of seasonal mood changes may in part explain the anomalous results of Menzies et al.[19,54] in Canada whose efforts to evaluate the effects of varying ventilation rates were compromised by decreased symptom prevalence rates observed over a 6-week period extending from mid-February when SAD effects would be expected to begin remitting.

Significance of Psychosocial Factors Related to SBS

As described above, a number of studies report significant associations between symptom prevalence rates and psychosocial factors. These associations can be variously interpreted. For some, they would reinforce previously held beliefs that sick-building complaints are more psychological than real. Skov et al.,[8] however, have concluded that despite the pronounced effect of psychosocial factors on symptom prevalence, they could not explain the overall variation in symptom prevalence observed between buildings. This is consistent with the observations of Norback et al.[50] that there were no differences in psychosocial factors between sick and reference buildings. In addition, Hedge et al.[42] failed to find any significant effect of personality variables such as neuroticism or stability and introversion or extroversion or depression on SBS symptoms reported, suggesting that SBS symptoms were not psychogenic in origin.

The most scientifically plausible explanation of the observed relationships (other than potential effects of SAD) between psychosocial factors and symptoms would be effects on reporting rates. Individuals experiencing work-related stress may be more likely to report symptoms when they are being experienced than others subject to little or no stress; similarly, those experiencing high job satisfaction may be less likely to report symptoms even when present. The effect of psychosocial factors may, therefore, be indirect (reflecting reporting rates) rather than direct as is often supposed.

TOBACCO SMOKING

The smoking of tobacco products in indoor environments represents a potentially significant environmental exposure and SBS symptom risk. Effects on IAQ are a direct result of human behavior and social acceptance/nonacceptance of tobacco smoking in indoor spaces. In North America, slightly less than 30% of the adult population smokes, with higher rates in European countries[55] and in countries in Southeast Asia. Smokers expose both themselves and nonsmokers to a large variety of both gas- and particulate-phase materials. Several thousand different chemical species have been identified in tobacco smoke with approximately 400 quantified. Some of the most notable gas-phase species include carbon monoxide (CO), carbon dioxide (CO_2), nitrogen oxides (NO_x), acrolein, formaldehyde, benzene, hydrogen cyanide, and hydrazine; particulate-phase materials include tar, nicotine, phenol, 2-naphthylamine, benzo-α-pyrene, benzo-α-anthracene, *n*-nitrosodiethanolamine, cadmium, and nickel.[56]

Environmental Tobacco Smoke

The contamination of indoor air from tobacco smoking results from exhaled mainstream smoke (MS) and sidestream smoke (SS) which is produced by burning cigarettes between puffs. Together, exhaled MS and SS comprise what is commonly described as environmental tobacco smoke (ETS). Exposure to ETS is referred to as passive or involuntary smoking.[56]

MS and SS, for the most part, contain the same chemical constituents. Because of differences in cigarette cone temperatures and oxidation potential during active smoking and the periods in between, significant quantitative differences in emissions result. Emissions of selected gas- or particulate-phase smoke constituents in SS may vary by a factor of two to an order of magnitude or more higher than in MS. However, because of the diluting effect of room air, nonsmoker exposure to tobacco combustion byproducts may only be a fraction of that experienced by smokers. This dilution effect depends, of course, on the volume of room air available for dilution and outdoor air exchange rates.

ETS Contaminant Concentrations

A number of studies have reported concentrations of ETS in a variety of public-access buildings. Data from some of these studies are summarized in Table 2.1.[55,57-62] Data in Table 2.1 are reported for environments where smoking density was relatively high. Potential ETS exposures in office environments have been studied by several investigators. Sterling et al.[63] compared contaminant levels in office buildings where smoking was permitted to those in which smoking was restricted (Table 2.2). Based on these comparisons they concluded that under normal conditions of ventilation and occupancy, ETS-related contaminant concentrations varied little between office areas where smoking was permitted or restricted. The only significant observations made from these data were the absence of measurable nicotine levels in the smoking-restricted environments and differences found in CO_2 levels. Carbon dioxide levels were significantly higher in the nonsmoking (759 ppm) compared to the smoking-permitted office areas (507 ppm), suggesting that the latter had lower occupant densities or higher ventilation rates. If the latter is true, it may in part explain why no apparent differences in contaminant levels were observed between smoking and smoking-restricted office environments.

As a part of the same study, Sterling and Mueller[64] measured CO, CO_2, respirable suspended particles (RSP), and nicotine in smoking and nonsmoking sections of cafeterias, a smoking lounge, and several offices. Significantly higher levels of CO, CO_2, RSP, and nicotine were observed in smoking areas. They observed no differences in CO, CO_2, and RSP levels in nonsmoking office areas as compared to those which received recirculated air. Nicotine levels in the former were reported to be less than 1.0 $\mu g/m^3$.

Because of the large variations in environmental factors in the multiple-building study reported by Sterling et al.[63] and the lack of specificity of CO, CO_2, and RSP for tobacco smoke, the role of tobacco smoking in causing air contamination in the office buildings studied may have been obscured. The "before" and "after" study of Vaughn and Hammond[65] may, therefore, be more illuminating. They conducted a series of measurements of nicotine and particle concentrations in a modern office building before and after a smoking-restricted policy was implemented. Observed reductions in nicotine levels ranged from 80–98% with the lower reductions associated with spillage of tobacco smoke into nonsmoking areas. Significant reductions in particulate-phase matter were observed in areas where smoking was restricted and significant increases in smoking-permitted areas such as a snack bar and cafeteria.

Turner et al.[66] conducted measurements of contaminants commonly associated with ETS in 585 offices, comparing areas where smoking was observed to those where it was not. Strong correlations between RSP, nicotine, and smoking density, and RSP and nicotine, were observed. Weaker but significant correlations were observed between CO, CO_2, and smoking density, and CO_2 and nicotine. There were significant differences in smoking and non-smoking

Table 2.1 Mean tobacco-related contaminant levels in public access buildings.

Contaminant	Type of environment	Levels	Nonsmoking controls	Ref.
		ppm		
CO	15 restaurants	4	2.5	57
	Arena (11,806 people)	9	3.0	58
	44 work rooms	2.8	2.0	59
		$\mu g/m^3$		
RSP/TSP	Bar and grill[a]	589		60
	Bingo hall[a]	1140		60
	Fast food restaurant[a]	109		58
	44 work rooms	117		59
	Restaurant	200		62
	Cocktail lounge	400		61
	Restaurant	240		55
	Restaurant	144		55
		ppb		
NO_2	Restaurant	63	50	62
NO	Bar	21	48	62
	44 work rooms	84	62	59
		$\mu g/m^3$		
Nicotine	Restaurant	5.2		62
	Cocktail lounge	10.3		62
	44 work rooms	1.1		59
		ng/m^3		
Benzo-α-pyrene	Arena	9.9	0.69	58

[a] Short-term sampling (<0.5 hr).

rooms in levels of RSP, nicotine, and CO. The overall concentrations of ETS components were postulated to have been significantly lower than those estimated a decade previously.

Hedge et al.[67] conducted studies of contaminant concentrations in 27 air-conditioned office buildings with different smoking policies. These included buildings in which smoking was prohibited (SP) and those where smoking was spatially restricted. The latter included smoking areas with air filtration (RF), no air treatment (RNT), separate exhaust ventilation (RSV), and enclosed offices and open-plan cubicle workstations (RWS). Though the last case is described as a restricted smoking policy, it does not appear to be so.

Results of contaminant measurements in these buildings as related to smoking policy are summarized in Table 2.3. Differences in contaminant levels between smoking and nonsmoking areas in buildings with spatially restricted policies are summarized in Table 2.4.

Carbon monoxide levels were on the average relatively low and appeared to be somewhat higher in smoking areas in buildings with spatially restricted policies. Smoking policies did not affect CO_2 levels. The relatively small differences in CO_2 levels indicated that ventilation conditions in the buildings being studied were relatively similar and in comparison to ASHRAE guidelines,[68] well-ventilated.

Table 2.2 Pollutant levels in smoking permitted and smoking restricted buildings.

	Smoking permitted		
Pollutant	**Measurement (N)**	**Median**	**Range**
Aldehydes	10	ND	ND-0.03 mg/m^3
Amines	6	ND	ND-404 ppb
Aromatic HCs	113	Trace	ND-12.5 mg/m^3
Carbon dioxide	94	506.5 ppm	ND-2300 ppm
Carbon monoxide	194	3.1 ppm	ND-242 ppm
Formaldehyde	200	0.016 ppm	ND-0.6 ppm
Hydrocarbons	124	Trace	ND-28 ppm
Nicotine	10	8.5 ng/m^3	ND-53 ng/m^3
Nitrogen oxides	92	ND	ND-160 ppb
Nitrogen dioxide	43	2 ppb	ND-100 ppb
Ozone	76	ND	ND-90 ppb
Particulate matter	81	0.038 mg/m^3	ND-0.7 mg/m^3
Sulfur dioxide	39	ND	ND-0.17 ppm
	Smoking restricted		
Pollutant	**Measurement**	**Median**	**Range**
Aldehydes	—	—	—
Amines	3	ND	ND
Aromatic HCs	6	0.012 mg/m^3	ND-104 mg/m^3
Carbon dioxide	6	759.4 ppm	ND-2000 ppm
Carbon monoxide	15	3.4 ppm	ND-75 ppm
Formaldehyde	7	ND	ND-0.22 ppm
Hydrocarbons	5	ND	ND-Trace
Nicotine	—	—	—
Nitrogen oxides	5	26.5 ppb	ND-70 ppb
Nitrogen dioxide	2	ND	ND
Ozone	6	18 ppb	ND-95 ppb
Particulate matter	20	0.038 mg $/m^3$	0.014-0.32 mg/m^3
Sulfur dioxide	1	ND	ND

From Sterling, T.D. et al. 1987. *J. Occup. Med.* 292:57–62. With permission.

Hedge et al.[67] reported that smoking policy did not affect formaldehyde levels. There were no differences among sites for SP and RWS policies. However, formaldehyde levels were slightly higher in smoking areas compared to nonsmoking areas in buildings with smoking-restricted policies. These results were, however, mixed: there were no differences in formaldehyde between smoking and nonsmoking areas for the RSV policy; levels in smoking areas were double those in the nonsmoking areas for the RF policy; and the RNT policy had the highest formaldehyde levels for both the nonsmoking and smoking areas.

Smoking policies were observed to affect the concentration of particulate-phase materials such as $RSP_{3.5}$ (respirable particles with a cutoff diameter of 3.5 μm), ultraviolet light measured particulate matter (UVPM), and nicotine. For $RSP_{3.5}$ significant differences were only observed between the RSV and RF smoking policies. Significant differences in $RSP_{3.5}$ levels were observed between smoking and nonsmoking areas in buildings with spatially restricted smoking policies. Using UV measurements of particulate-phase materials, Hedge et al.[67] observed much lower particle levels in smoking-prohibited

Table 2.3 Contaminant levels associated with different building smoking policies.

Contaminant	Smoking policy				
	Smoking prohibited (SP)	Restricted separate ventilation (RSV)	Restricted filtration (RF)	Restricted no treatment (RNT)	Restricted work-station (RWS)
CO (ppm)	0.1 (0.01–0.5)	0.7 (0.1–5.2)	0.4 (0.1–2.1)	0.6 (0.1–3.3)	0.1 (0.0–0.5)
CO_2 (ppm)	561 (478–657)	600 (494–729)	674 (575–791)	634 (533–755)	573 (488–671)
Formaldehyde (ppm)	0.01 (0.008–0.028)	0.02 (0.009–0.46)	0.02 (0.013–0.047)	0.05 (0.025–1.103)	0.01 (0.008–0.028)
$RSP_{3.5}$ ($\mu g/m^3$)	14.8 ± 14.0	28.9 ± 17.2	2.7 ± 14.0	5.3 ± 15.3	2.6 ± 14.0
UVPM ($\mu g/m^3$)	0.2 ± 22.9	44.1 ± 28.1	114.9 ± 22.9	19.8 ± 25.1	10.2 ± 22.9
Nicotine ($\mu g/m^3$)	ND	6.9 ± 3.0[a]	20.8 ± 2.4	3.8 ± 2.6	2.4 ± 2.4

[a] Only measured in separately ventilated smoking areas.

From Hedge, A. et al. 1993. *Proceedings Sixth International Conference on Indoor Air Quality and Climate.* Vol. 1. Helsinki.

Table 2.4 Differences in contaminant levels in nonsmoking and smoking areas of spatially restricted smoking policies.

Contaminant	Smoking policy					
	RSV nonsmoking	RSV smoking	RF nonsmoking	RF smoking	RNT nonsmoking	RNT smoking
CO (ppm)	0.3 (0.2–0.7)	1.5 (.07–3.4)	0.2 (0.1–0.3)	1.2 (0.7–2.1)	0.4 (0.2–0.8)	1.0 (0.4–2.2)
CO_2 (ppm)	573 (523–627)	635 (571–707)	673 (626–722)	676 (550–644)	597 (555–647)	698 (624–781)
Formaldehyde (ppm)	0.02 (0.013–0.024)	0.02 (0.017–0.034)	0.02 (0.014–0.022)	0.04 (0.03–0.049)	0.05 (0.038–0.063)	0.06 (0.039–0.079)
$RSP_{3.5}$ ($\mu g/m^3$)	11.9 ± 10.1	53.7 ± 12.0	30.2 ± 8.0	126.1 ± 8.8	19.3 ± 9.0	35.6 ± 12.5
UVPM ($\mu g/m^3$)	3.2 ± 23.1	109.1 ± 27.3	14.5 ± 18.3	236.6 ± 20.0	9.3 ± 20.5	35.6 ± 28.6
Nicotine[a] ($\mu g/m^3$)	ND	8.2 ± 5.1[a]	0.9 ± 3.3	44.2 ± 3.6	0.3 ± 3.8	10.3 ± 5.2

[a] Nicotine only measured in separately ventilated smoking areas.

From Hedge, A. et al. 1993. *Proceedings Sixth International Conference on Indoor Air Quality and Climate.* Vol. 1. Helsinki.

buildings as compared to buildings with other smoking policies. Higher UVPM levels were observed in smoking vs. nonsmoking areas in smoking-restricted buildings. Differences varied among smoking policies and were greater for the RF policy. Local filtration did not appear to be effective in reducing UVPM levels. Comparisons of all non-smoking office areas showed that UVPM levels were significantly lower in SP buildings than the average of buildings with RSV, RF, and RNT policies. Though $RSP_{3.5}$ and UVPM are both measures of particulate-phase materials, the relatively modest correlation (r = 0.50) indicates signficant differences between them.

As expected, significant differences in nicotine levels were observed in buildings with different smoking policies. In SP buildings, nicotine could not be detected. Buildings with RF had the highest nicotine levels consistent with higher particulate matter concentrations. Differences in nicotine levels were observed between smoking and nonsmoking areas in smoking-restricted buildings. The relatively low nicotine levels in nonsmoking areas indicated some spillage of tobacco-related contaminants from smoking to nonsmoking areas.

Irritant Effects Determined from Survey Studies

Environmental tobacco smoke has been a pervasive contaminant of indoor air. In a study of 663 never- and ex-smokers, Cummings et al.[69] reported that 91% had detectable levels of cotinine, a metabolite of nicotine, in their urine. Of currently employed study participants, 77% reported exposure to ETS at work.

Though a variety of ETS-related health effects have been reported (including lung cancer and heart disease), it is the irritant and annoyance potential of ETS which is most relevant to SBS. As such, the foregoing discussion will be limited to these concerns.

A number of irritants have been identified in MS and SS tobacco smoke. Among the more significant of these are gas-phase substances such as acrolein and formaldehyde.[56]

In one of the first investigations of the subjective effects of exposure to ETS, Speer[70] observed that groups of allergic and nonallergic subjects had relatively high prevalences rates of symptoms in response to commonly encountered ETS exposures (Table 2.5). Eye irritation was the most frequent complaint. Not surprisingly, allergic nonsmokers reported significantly higher frequencies of respiratory complaints such as nasal symptoms, cough, wheezing, sore throat, and hoarseness. Cummings et al.[71] conducted a survey of acute reactions to ETS by adult nonsmokers. Seventy-five percent of study subjects were "moderately" or "very much" bothered by ETS. Higher frequencies of "being bothered" by ETS were reported by (1) those who have never smoked and (2) those who had a tendency to atopy or history of respiratory illness. The most common symptoms associated with ETS exposures were watery eyes (45%), nose irritation (36%), coughing episodes (26%), sore throat (16%), and sneezing (14%).

Irritant Effects in Controlled Exposure Studies

Weber et al.[72] conducted controlled exposure studies of 60 subjects to evaluate the potential irritation effects of ETS. Responses were assessed by questionnaire and related to concentrations of CO, formaldehyde, and acrolein. Significant dose-response relationships were observed between symptoms of eye and nasal irritation and concentrations of tobacco smoke components. Expressions of annoyance such as air quality, desire to open windows, visible

Table 2.5 Subjective symptoms reported by allergic nonsmokers (N = 191) and nonallergic nonsmokers (N = 250) in response to common exposures to ETS.

	Prevalence (%)	
Symptom	**Allergic subjects**	**Nonallergic subjects**
Eye irritation	73	69
Nasal symptoms	67	29
Headache	46	32
Cough	46	25
Wheezing	23	4
Nausea	15	9
Hoarseness	16	4
Dizziness	5	6

From Speer, F. 1968. *Arch. Environ. Health.* 16:443–446. With permission.

smoke, and urge to leave the room appeared to be more sensitive to relatively low concentrations of tobacco smoke than eye or nasal irritation. Effects on the throat and respiratory system, as well as general complaints were small. Acrolein was suggested to be the most significant irritant component of tobacco smoke because its concentration was the highest relative to its toxicity.

Weber[73] in another study conducted both field and laboratory investigations of the acute irritating effects of smoke components. She concluded that 30–70% of the indoor concentrations of CO, NO, and particulate matter in Swiss workrooms were due to tobacco smoke and that 25–40% of the employees were disturbed and/or annoyed by such exposures. Additionally, nonsmokers reported significantly more reactions to ETS than smokers, and those who reported more eye irritation had problems with hay fever.

Weber[73] measured the degree of acute irritation to SS tobacco smoke by subjective assessment of eye, nose, and throat irritation and by objectively measuring eye blink rates. Blink rate as well as subjective eye irritation was observed to increase with increasing smoke concentration and duration. A similar but less pronounced observation was reported for nose and throat irritation. The pattern of annoyance differed. It increased rapidly after exposure began and became constant after 10 15 minutes, indicating a measure of accommodation.

Weber[73] suggested that gas-phase substances such as acrolein and formaldehyde were, for the most part, responsible for annoyance but that the particulate phase was largely responsible for both the objective and subjective eye irritation observed. She reported a strong increase in eye irritation corresponding to CO levels of 1.3 and 2.5 ppm, suggesting this to be a critical range of overall tobacco smoke concentration.

The observations of Weber[73] which associated irritation with particulate-phase materials and annoyance with gas-phase substances were not consistent with the studies of Hugood[74] and Cain et al.[75] Hugood[74] observed that filtration of tobacco smoke diminished neither irritation nor annoyance in exposed subjects. Cain et al.[75] reported that filtration did reduce irritation, but that the decrease was small. Despite these inconsistencies, the results of the studies reported by Cain et al.[75] were similar in most respects to those of Weber.[73] Cain et al.[75] also investigated the role of tobacco odor in relation to irritation and annoyance effects. Pinching off the nasal passages of test subjects neither diminished nor magnified the degree of eye irritation. They suggested that at smoke levels equal to 2 ppm CO, odor rather than irritation was the major cause of subject dissatisfaction.

The observations of Cain et al.[75] were confirmed by Clausen[55] who reported that about 10 times more ventilation is needed to control odor than to control irritation. Odor perceived in rooms where smoking occurs is apparently caused by both present tobacco smoking and substances adsorbed on room surfaces (from previous smoking) and desorbed to indoor air.

Role of ETS in SBS-Type Complaints

The results of survey studies of nonsmokers indicate that ETS is bothersome, and those exposed report experiencing a variety of symptoms of the eyes, nose, and throat.[69-71] Controlled chamber studies of relatively high ETS exposures typical of bars and some restaurants have confirmed that nonsmokers experience significant irritation symptoms and annoyance.[72-75] Levels of ETS components found in typical work environments subject to sick building complaints are, for the most part, only a fraction of those associated with bars and restaurants.

Despite the lower ETS exposures which may be experienced in office buildings, a number of investigators have, nevertheless, observed significant relationships between reported exposures to ETS and SBS symptoms. Robertson et al.[77,78] in a U.K. study of 4373 office workers observed that SBS symptoms were more prevalent in individuals who had reported significant exposure to ETS. In a larger study of 7043 office workers, Dutch investigators[7] reported significantly higher prevalence rates of eye, oronasal, and central nervous system symptoms as well as IAQ and organoleptic environmental complaints associated with exposure to ETS. Significant associations between SBS symptoms and ETS exposure were observed in the 15-building U.K. study of Hawkins and Wang[43] and the Finnish Office Building Study of Jaakkola et al.[20] Raw et al.[79] in the U.K. reported an association between self-assessed worker productivity among nonsmokers sharing an office with smokers. Productivity was perceived to decrease.

Each of the above studies was based on perceived ETS exposures rather than objectively measured ETS levels. In a cross-sectional study of 875 office workers, Raynal et al.[80] observed a weak positive correlation between symptoms and environmental nicotine which was significantly associated with saliva cotinine levels. Symptoms were more strongly correlated, however, with perceived ETS exposure.

Hedge et al.[67] in their study on smoking policies, observed no significant differences among buildings with different smoking policies and SBS symptoms. They also observed no significant associations between ETS exposure at work, at home, at other locations, or total passive smoking exposure and the number of SBS symptoms per worker. They did, however, observe that stuffy, congested nose and hoarseness were associated with increased concentrations of nicotine.

Reprise

Studies which have attempted to evaluate passive smoking as a risk factor for SBS symptoms have consistently shown a relationship between office workers' perception of exposure to ETS and increased symptom prevalence. Several studies have shown weaker relationships between objective measurements and symptom prevalence. These studies indicate that passive smoke may contribute to SBS symptom prevalence rates. In addition, the trained panel studies of Fanger et al.[81] indicate that about 25% of the dissatisfaction with perceived air quality in Danish office buildings is due to ETS. Their studies, as well as survey and controlled chamber studies, indicate a significant irritant potential associated with ETS that may contribute directly to increases in symptom prevalence.

Exposures to ETS may also have an indirect effect. Because of the subjective annoyance expressed by nonsmokers relative to ETS exposures, it is quite possible that the presence of ETS in the office environment may affect reporting rates. It is also plausible that both direct and indirect effects are associated with increased prevalence rates associated with ETS exposure.

Most of the studies described above have been conducted in European countries where smoking rates are higher than in the United States and where there are fewer restrictions on smoking in building environments. The exposure risk of ETS to U.S. office workers would be expected to be less and be dependent on smoking policies which exist in a given building.

Tobacco Smoking and SBS Complaints

Smoking by building occupants may itself have an effect on SBS symptom prevalence. The results of reported studies are, however, mixed. Several studies indicate that smoking increases SBS prevalence rates: Norback et al.[44] and Hawkins and Wang[43] have reported that overall SBS symptom scores were higher among smokers; Skov et al.[3] have reported higher prevalence rates of general symptoms among female smokers, and Hodgson and Collopy,[82] higher symptom rates among male smokers; and Blum et al.[22] reported that smoking was significantly associated with one or more symptoms in any of the symptom groups in their study. Several studies[9,24] have failed to observe any relationship between smoking and symptom reporting rates. On the other hand, Raynal et al.[80] observed that smokers consistently report fewer symptoms than nonsmokers.

RECAPITULATION

A variety of "people-related" risk factors have been shown to be associated with increased prevalence of SBS symptoms. These include gender, atopy or allergic predisposition, a variety of psychosocial factors, and tobacco smoking.

Numerous studies have reported that symptom prevalence or symptom reporting rates are several times higher among females. The reasons for these gender differences are not known. There is evidence to suggest that female-male differences may be due to the under-reporting of symptoms by males because of cultural factors. Males and females may also be subject to different exposures. Most studies report atopy to be a significant risk factor for SBS symptoms, suggesting that allergy-prone individuals as a group may be more sensitive to environmental exposures in general.

Most problem buildings are subject to significant psychosocial dynamics. In early investigations of "problem buildings", medical and industrial hygiene evaluations failed to identify causal factors. Investigators concluded that reported illness symptoms were due to a type of hysteria among those apparently affected. Such illness episodes were described as mass hysteria or, more recently, mass psychogenic illness. With increased understanding of problem building phenomena, diagnoses of mass psychogenic illness have largely been discredited. Nevertheless, a variety of psychosocial factors have been reported to be associated with SBS symptoms. The most notable of these have been factors that define work stress and job satisfaction. Symptom prevalence rates increase with increasing workplace stress and decrease with increasing job satisfaction.

Tobacco smoke exposures to both smokers and nonsmokers occur as a consequence of human behavioral choices. The irritant properties of ETS have been assessed in human survey and controlled-chamber exposure studies. Most systematic building studies have observed significant relationships between perceived exposure to ETS and SBS symptoms. Several studies have reported smaller but significant relationships between objective measures and symptoms. It appears that ETS is a risk factor for SBS symptoms in building occupants where smoking is not prohibited or severely restricted. The effect of ETS may be both direct and indirect, in the latter case perceived ETS may result in an increase in symptom reporting rates.

The effects of smoking (on the smoker) on symptom prevalence rates have been evaluated in a number of studies. Results have varied from increased prevalence rates among smokers, to no differences, to decreased prevalence rates among smokers.

REFERENCES

1. Finnegan, M.J. and C.A.C. Pickering. 1987. "Prevalence of Symptoms of the Sick Building Syndrome in Buildings Without Expressed Dissatisfaction." 542–546. In: *Proceedings of the Fourth International Conference on Indoor Air Quality and Climate*. Vol. 2. West Berlin.
2. Burge, P.S. et al. 1987. "Sick Building Syndrome: A Study of 4373 Office Workers." *Ann. Occup. Hyg.* 31:493–504.
3. Skov, P. et al. 1987. "The Sick Building Syndrome in the Office Environment: The Danish Town Hall Study." *Environ. Int.* 13:339–349.
4. Robertson, A.S. et al. 1985. "Comparison of Health Problems Related to Work and Environmental Measurements in Two Office Buildings with Different Ventilation Systems." *Br. Med. J.* 291:373–376.
5. Hedge, A. 1984. "Suggestive Evidence of a Relationship Between Office Design and Self Reports of Ill-Health among Office Workers in the United Kingdom." *J. Arch. Plan. Res.* 1:163–174.
6. Hedge, A. et al. 1987. "Indoor Climate and Employee Health in Offices." 492–496. In: *Proceedings of the Fourth International Conference on Indoor Air Quality and Climate*. Vol. 2. West Berlin.
7. Zweers, T. et al. 1992. "Health and Indoor Climate Complaints of 7043 Office Workers in 61 Buildings in the Netherlands." *Indoor Air*. 2:127–136.
8. Skov, P. et al. 1989. "Influence of Personal Characteristics, Job-Related Factors, and Psychological Factors on the Sick Building Syndrome." *Scand. J. Work Environ. Health*. 15:286–295.
9. Trauter, R.M. et al. 1993. "Results of a Sick Building Syndrome Prevalence Study in Johannesburg." 411–416. In: *Proceedings of the Sixth International Conference on Indoor Air Quality and Climate*. Vol. 1. Helsinki.
10. Levy, F. et al. 1993. "Gender and Hypersensitivity as Indicators of Indoor-Related Health Complaints in a National Reference Population." 357–362. In: *Proceedings of the Sixth International Conference on Indoor Air Quality and Climate*. Vol. 1. Helsinki.
11. Raw, G.J. and A. Grey. 1993. "Sex Differences in Sick Building Syndrome." 381–386. In: *Proceedings of the Sixth International Conference on Indoor Air Quality and Climate*. Vol. 1. Helsinki.
12. Stenberg, B. and S. Wall. 1993. "Why Do Females Report 'Sick Building' Symptoms More Often Than Males?" 399–404. In: *Proceedings of the Sixth International Conference on Indoor Air Quality and Climate*. Vol. 1. Helsinki.
13. Hedge, A. et al. 1989. "Work-Related Illness in Offices: A Proposed Model of the 'Sick Building Syndrome'." *Environ. Int.* 15:143–158.
14. Pennebaker, C.A.C. and J.A. Skelton. 1981. "Selective Monitoring of Bodily Sensations." *J. Person. Soc. Psychol.* 41:213–223.
15. Verbrugge, L.M. 1976. "Females and Illness: Recent Trends in Sex Differences in the United States." *J. Health Soc. Behav.* 17:387–403.
16. Verbrugge, L.M. 1989. "Empirical Explanations of Sex Differences in Health and Mortality." *J. Health Soc. Behav.* 30:282–304.
17. Molhave, L. et al. 1993. "Integration and Adaptation in Eye Irritation." 35–40. In: *Proceedings of the Sixth International Conference on Indoor Air Quality and Climate*. Vol. 1. Helsinki.

18. Mendell, M. 1992. "Health Effects in Office Workers Associated with Ventilation Systems: Evidence from the First U.S. Study." Dissertation Abstracts P.N. 92–28,770. University Microfilms, Inc. Ann Arbor, MI.
19. Menzies, R. et al. 1991. "The Effect of Varying Levels of Outdoor Ventilation on Symptoms of Sick Building Syndrome." 90–96. In: *Proceedings of IAQ '91: Healthy Buildings*. American Society of Heating, Refrigerating and Air-Conditioning Engineers. Atlanta.
20. Jaakkola, J.J.K. et al. 1991. "Mechanical Ventilation in Office Buildings and the Sick Building Syndrome: An Epidemiological Study." *Indoor Air*. 2:111–121.
21. Norback, D. et al. 1993. "Sick Building Syndrome in the General Swedish Population — The Significance of Outdoor and Indoor Air Quality and Seasonal Variation." 273–278. In: *Proceedings of the Sixth International Conference on Indoor Air Quality and Climate*. Vol. 1. Helsinki.
22. Blum, P. et al. 1993. "The Frequency of Building-Related Health Complaints in Norway." 321–326. In: *Proceedings of the Sixth International Conference on Indoor Air Quality and Climate*. Vol. 1. Helsinki.
23. Molina, C. et al. 1993. "Sick Building Syndrome and Atopy." 369–373. In: *Proceedings of the Sixth International Conference on Indoor Air Quality and Climate*. Vol. 1. Helsinki.
24. Stenberg, B. et al. 1993. "The Office Illness Project in Northern Sweden — An Interdisciplinary Study of the Sick Building Syndrome (SBS)." 393–398. In: *Proceedings of the Sixth International Conference on Indoor Air Quality and Climate*. Vol. 1. Helsinki.
25. Muzi, G. et al. 1993. "Prevalence of Irritative Symptoms in Non-problem Air-Conditioned Buildings." 375–380. In: *Proceedings of the Sixth International Conference on Indoor Air Quality and Climate*. Vol. 1. Helsinki.
26. Kerchoff, A.C. and K.W. Back. 1968. "The June Bug: A Study of Hysterical Contagion." Appleton-Century Crofts, New York.
27. Stahl, S.M. and M. Lebedien. 1974. "Mystery Gas: An Analysis of Mass Hysteria." *J. Health Soc. Behav*. 15:44–50.
28. Small, G.W. and A.M. Nicoli. 1982. "Mass Hysteria among School Children." *Arch. Gen. Psychiatry*. 39:721–724.
29. Small, G.W. and J.F. Boris. 1983. "Outbreaks of Illness in a School Chorus: Toxic Poisoning or Mass Hysteria." *New Eng. J. Med*. 308:632–635.
30. Robinson, P. et al. 1984. "Outbreaks of Itching and Rash: Epidemic Hysteria in an Elementary School." *Arch. Intern. Med.* 144:1959–1962.
31. Levine, R.J. and D.J. Sexton. 1974. "Outbreak of Psychosomatic Illness in a Rural Elementary School." *Lancet*. 2:1500–1503.
32. Alexander, R.A. and M.J. Fedoruk. 1986. "Epidemic Psychogenic Illness in a Telephone Operators Building." *J. Occup. Med.* 28:42–45.
33. Colligan, M.J. et al. 1979. "An Investigation of Apparent Mass Psychogenic Illness in an Electronics Plant." *J. Behav. Med.* 2:297–304.
34. Colligan, M.J. and L.R. Murphy. 1979. "Mass Psychogenic Illness in Organizations: An Overview." *J. Occup. Psych.* 52:77–90.
35. Elesh, E. et al. 1979. "Mass Psychogenic Illness in Industry—NIOSH's Role." Presented at the American Industrial Hygiene Association Conference—Chicago, IL, May 20–June 1.

36. Boxer, P.A. 1985. "Occupational Mass Psychogenic Illness." *J. Occup. Med.* 27:867–872.
37. Faust, H.S. and L.B. Brilliant. 1981. "Is the Diagnosis of 'Mass Hysteria' an Excuse for Incomplete Investigation of Low Level Environmental Contamination?" *J. Occup. Med.* 23:22–26.
38. Nelson, C.J. et al. 1991. "EPA's Indoor Air Quality and Work Environment Survey: Relationships of Employees' Self- Reported Health Symptoms with Direct Indoor Air Quality Measurements." 22–32. In: *Proceedings of IAQ '91: Healthy Buildings.* American Society of Heating, Refrigerating and Air-Conditioning Engineers. Atlanta.
39. Eriksson, N. and J. Hoog. 1993. "The Office Illness Project in Northern Sweden. The Significance of Psychosocial Factors for the Prevalence of 'Sick Building Syndrome.' A Case Referent Study." 333–338. In: *Proceedings of the Sixth International Conference on Indoor Air Quality and Climate.* Vol. 1. Helsinki.
40. Wallace, L. et al. 1993. "Association of Personal and Workplace Characteristics with Reported Health Symptoms of 6771 Government Employees in Washington, D.C." 427–432. In: *Proceedings of the Sixth International Conference on Indoor Air Quality and Climate.* Vol. 1. Helsinki.
41. Hedge, A. et al. 1990. "The Roles of Job Stress and Job Satisfaction in the Etiology of the Sick Building Syndrome in Offices." Presented at the American Psychological Society/NIOSH Conference—Washington, D.C., November 15–17.
42. Hedge, A. et al. 1993. "Psychosocial Correlates of Sick Building Syndrome." 345–350. In: *Proceedings of the Sixth International Conference on Indoor Air Quality and Climate.* Vol. 1. Helsinki.
43. Hawkins, L.H. and T. Wang. 1991. "The Office Environment and the Sick Building Syndrome." 365–370. In: *Proceedings of IAQ '91: Healthy Buildings.* American Society of Heating, Refrigerating and Air-Conditioning Engineers. Atlanta.
44. Norback, D. et al. 1990. "Indoor Air Quality and Personal Factors Related to the Sick Building Syndrome." *Scand. J. Work Environ. Health.* 16:121–128.
45. Broder, I. et al. 1990. "Building-Related Discomfort Is Associated with Perceived Rather than Measured Levels of Indoor Environmental Variables." 221–226. In: *Proceedings of the Fifth International Conference on Indoor Air Quality and Climate.* Vol. 1. Toronto.
46. Hodgson, M.J. 1993. "Work Stress, Illness Behavior, and Mass Psychogenic Illness and the Sick Building Syndrome." 773–777. In: *Proceedings of the Sixth International Conference on Indoor Air Quality and Climate.* Vol. 1. Helsinki.
47. Hodgson, M.J. et al. 1993. "Work Stress, Microenvironmental Measures and the Sick Building Syndrome." 47–58. In: *Proceedings of IAQ '93: Environments for People.* American Society of Heating, Refrigerating and Air-Conditioning Engineers. Atlanta.
48. Bauer, R. et al. 1992. "The Role of Psychological Factors in the Report of Building-Related Symptoms in Sick Building Syndrome." *J. Clin. Consult. Neuropsych.* 60:213–219.
49. Hedge, A. 1988. "Job Stress, Job Satisfaction and Work-Related Illness in Offices." 777–779. In: *Proceedings of the 32nd Annual Meeting of the Human Factors Society.* Vol. 2. Anaheim, CA.

50. Norback, D. et al. 1989. "The Prevalence of Symptoms Associated with Sick Buildings and Polluted Industrial Environments as Compared to Unexposed Reference Groups Without Expressed Dissatisfaction." *Environ. Int.* 15:85–94.
51. Lenvik, K. 1990. "Comparisons of Working Conditions and 'Sick Building Syndrome' Symptoms among Employees with Different Job Function." 507–512. In: *Proceedings of the Fifth International Conference on Indoor Air Quality and Climate*. Vol. 1. Toronto.
52. Solomon, E. E. 1986. "Private and Public Sector Managers: An Empirical Investigation of Job Characteristics and Organizational Climate." *J. Appl. Psychol.* 71:244–259.
53. Alderfer, R. et al. 1993. "Seasonal Mood Changes and Building-Related Symptoms." 309–314. In: *Proceedings of the Sixth International Conference on Indoor Air Quality and Climate*. Vol. 1. Helsinki.
54. Menzies, R.I. et al. 1990. "Sick Building Syndrome: The Effect of Changes in Ventilation Rates on Symptom Prevalence: The Evaluation of a Double Blind Experimental Approach." 519–524. In: *Proceedings of the Fifth International Conference on Indoor Air Quality and Climate*. Vol. 1. Toronto.
55. Clausen, G.H. et al. 1986. "Background Odor Caused by Previous Tobacco Smoking." 119–125. In: *Proceedings of IAQ '86: Managing Indoor Air for Health and Energy Conservation*. American Society of Heating, Refrigerating and Air-Conditioning Engineers. Atlanta.
56. U.S. Surgeon General. 1986. "Health Consequences of Involuntary Smoking." DHHS (CDC) 87–8398.
57. Chappel, S.B. and R.J. Parker. 1978. "Smoking and Carbon Monoxide in Enclosed Public Places." *Can. J. Pub. Health.* 68:159–161.
58. Elliot, L.P. and D.R. Rowe. 1975. "Air Quality During Public Gatherings." *JAPCA.* 25:635–636.
59. Weber, A. et al. 1979. "Passive Smoking in Experimental and Field Conditions." *Environ. Res.* 20:205–216.
60. Repace, J.L. and A.H. Lowrey. 1982. "Tobacco Smoke Ventilation and Indoor Air Quality." *ASHRAE Trans.* 88. Pt. 1:895–914.
61. Repace, J.L. and A.H. Lowrey. 1980. "Indoor Air Pollution, Tobacco Smoke, and Public Health." *Science.* 208:464–472.
62. Hinds, W.C. and M.W. First. 1975. "Concentrations of Nicotine and Tobacco Smoke in Public Places." *New Eng. J. Med.* 292:844–845.
63. Sterling, T.D. et al. 1987. "Environmental Tobacco Smoke and Indoor Air Quality in Modern Office Work Environments." *J. Occup. Med.* 29:57–62.
64. Sterling, T.D. and B. Mueller. 1988. "Concentrations of Nicotine, RSP, CO, and CO_2 in Non-smoking Areas of Offices Ventilated by Air Recirculated from Smoking-Designated Areas." *Am. Ind. Hyg. Assoc. J.* 49:423–426.
65. Vaughn, W.M. and S.K. Hammond. 1990. "Impact of 'Designated Smoking Area' Policy on Nicotine Vapor and Particle Concentrations in a Modern Office Building." *J. Air Waste Manag. Assoc.* 40:1012–1017.
66. Turner, S. et al. 1992. "The Measurement of Environmental Tobacco Smoke in 585 Office Environments." *Environ. Int.* 18:19–28.
67. Hedge, A. et al. 1993. "Effects of Restrictive Smoking Policies on Indoor Air Quality and Sick Building Syndrome: A Study of 27 Air-Conditioned Offices." 517–522. In: *Proceedings of the Sixth International Conference on Indoor Air Quality and Climate*. Vol. 1. Helsinki.

68. ASHRAE Standard 62–1989. 1989. "Ventilation for Acceptable Indoor Air Quality." American Society of Heating, Refrigerating and Air-Conditioning Engineers. Atlanta.
69. Cummings, K.M. et al. 1986. "Measurement of Current Exposure to Environmental Tobacco Smoke." *Arch. Environ. Health.* 45:74–79.
70. Speer, F. 1968. "Tobacco and the Nonsmoker: A Study of Subjective Symptoms." *Arch. Environ. Health.* 16:443–446.
71. Cummings, K. M. et al. 1991. "Variation in Sensitivity to Environmental Tobacco Smoke among Adult Nonsmokers." *Int. J. Epidemiol.* 20:121–125.
72. Weber, A. et al. 1976. "Irritating Effects on Man of Air Pollution Due to Cigarette Smoke." *Am. J. Pub. Health.* 66:672–676.
73. Weber, A. 1984. "Annoyance and Irritation by Passive Smoking." *Prev. Med.* 13:618–625.
74. Hugood, C. 1984. "Indoor Air Pollution with Smoke Constituents—An Experimental Investigation." *Prev. Med.* 13:582–588.
75. Cain, W.S. et al. 1987. "Environmental Tobacco Smoke: Sensory Reactions of Occupants." *Atmos. Environ.* 21:347–353.
76. Clausen, G.H. 1988. "Comfort and Environmental Tobacco Smoke." 267–274. In: *Proceedings of IAQ '88: Engineering Solutions to Indoor Air Problems.* American Society of Heating, Refrigerating and Air-Conditioning Engineers. Atlanta.
77. Robertson, A.S. et al. 1988. "Relation Between Passive Cigarette Smoke Exposure and 'Building Sickness'." *Thorax.* 43:263.
78. Robertson, A.S. et al. 1988. "The Relationship Between Passive Smoke Exposure in Office Workers and Symptoms of 'Building Sickness'." 320–326. In: Perry, R. and P.W. Kirk (Eds.). *Indoor Air and Ambient Air Quality.* Seliper Ltd. London.
79. Raw, G.J. et al. 1990. "Further Findings from the Office Environment Survey: Productivity." 231–236. In: *Proceedings of the Fifth International Conference on Indoor Air Quality and Climate.* Vol. 1. Toronto.
80. Raynal, A. et al. 1993. "How Much Does Environmental Tobacco Smoke Contribute to the Building Symptom Index?" 529–534. In: *Proceedings of the Sixth International Conference on Indoor Air Quality and Climate.* Vol. 1. Helsinki.
81. Fanger, P.O. et al. 1988. "Air Pollution Sources in Offices and Assembly Halls Quantified by the Olf Unit." *Energy and Buildings.* 12:7–19.
82. Hodgson, M.J. and P. Collopy. 1991. "Symptoms and the Micro-Environment in the Sick Building Syndrome." 8–16. In: *Proceedings of IAQ '91: Healthy Buildings.* American Society of Heating, Refrigerating and Air-Conditioning Engineers. Atlanta.

3 ENVIRONMENTAL CONDITIONS/ ENVIRONMENTAL SYSTEMS

ENVIRONMENTAL CONDITIONS

Environmental conditions extant in building spaces at any given moment are a product of a number of physical factors. These include temperature, relative humidity, air movement, ventilation, lighting, noise, vibration, and a variety of electrical and magnetic phenomena. Many of these have been evaluated for their potential effects or contributions to health and/or comfort complaints in problem buildings and buildings in which no previous dissatisfaction with air quality has been reported.

In most modern buildings environmental conditions which have significant effects on human comfort, such as temperature, ventilation, and, in some cases, relative humidity, are mechanically controlled (or relatively so) by a variety of environmental/climate control systems. Increasingly, new buildings are being designed to provide year-round climate control, and older buildings are being retrofitted to do so. These environmental systems provide heating, ventilation, and air-conditioning (cooling) and are described by the acronym HVAC. Many factors associated with design, construction, maintenance, and operation of HVAC and other environmental control systems have the potential for causing or contributing to building-related health and comfort complaints. Problems associated with environmental systems include ventilation adequacy, system type, contaminant generation, and entrainment/re-entry/cross-contamination.

Much of the focus of problem building investigations centers on identifying causal factors responsible for health and/or comfort concerns. Though many investigators attempt to identify causal contaminants putatively responsible for building-related health complaints, many in the engineering community view indoor air quality (IAQ) problems as comfort concerns, with comfort determined by a variety of environmental factors as well as the irritative effects of gas- and potentially particulate-phase exposures.[1,2]

On a day-to-day basis, human comfort is determined by physical factors which dominate our sense of personal well-being, the most important of which are thermal conditions resulting from surrounding air temperatures, relative humidity, air movement, and the radiant effects of indoor surfaces. Relative humidity and air velocity may also have independent effects on human comfort or health. Other factors which may affect human comfort and possibly human health include lighting conditions, sound, vibration, air ions, and electrostatic and electric and magnetic phenomena.

Occupants of buildings reporting symptoms and expressing dissatisfaction with air quality typically work in environments where a number of coexisting factors influence their sense of well-being and personal comfort. Health or air quality problems do not occur in isolation from other environmental concerns. Though a very high percentage (93%) of respondents in the telephone survey of office workers conducted by Woods et al.[3] rated air quality as very important or somewhat important, higher percentages (98%) were reported for temperature and lighting conditions. Temperature was identified as the most serious problem by 47% of respondents, followed by noise (41%), air quality (39%), lighting (32%), and housekeeping (32%). Dissatisfaction with temperature (too hot in summer, too cold in winter) and relative humidity (too humid in summer) was reported with high frequency (>50%) among those indicating that the building environment affected their ability to do their jobs.

Thermal Conditions

Comfort Relationships

Fanger[4] has attempted to develop a thermal comfort standard. He defines thermal comfort as that condition of mind which expresses satisfaction with the thermal environment. Dissatisfaction may be due to a general bodily discomfort to warm or cool conditions or to unwanted heating or cooling of one particular part of the body (local discomfort). Parameters which affect the thermal balance of the body as a whole include air temperature, mean radiant temperature, air velocity, and relative humidity. Relative humidity affects thermal comfort by its effect on evaporative heat losses from the body. The higher the relative humidity, the less steep the water vapor pressure gradient between the skin and the surrounding air. This reduces perspiration and heat loss as well. Air velocity affects heat loss by convection. Increasing air velocities result in increased convective heat losses. Under warm conditions velocity-induced convective heat losses are desirable to maintain thermal comfort.

Local thermal discomfort occurs as a result of heating or cooling of a limited portion of the body. Such phenomena occur when a significant vertical air temperature difference exists between head and ankles, for example, when the floor is too warm or too cool. Local thermal discomfort also occurs as a consequence of excessive air velocity (drafts) or radiant temperature asymmetries. Fanger[4] indicates that drafts are the most common reason for thermal comfort complaints and that fluctuating air velocities are more uncomfortable than constant ones.

Table 3.1 Thermal comfort requirements during summertime conditions under light, mainly sedentary conditions.

Air temperature between 23 and 26°C.
Vertical air temperature difference between 1.1 m and 0.1 m above floor less than 3°C.
Mean air velocity less than 0.25 m/s.

From Fanger, P.O., 1984. *Proceedings Third International Conference on Indoor Air Quality and Climate.* Vol. 1. Stockholm.

Table 3.2 Thermal comfort requirements during wintertime conditions under mainly light sedentary conditions.

Air temperature between 20 and 24°C.
Vertical temperature difference between 1.1 m and 0.1 m (head and ankle level) less than 3°C.
Surface temperature of floor between 19 and 26°C.
Mean air velocity less than 0.15 m/s.
Radiant air temperature assymetries from windows or other cold vertical surfaces: less than 10°C in relation to a small vertical plane 0.6 m above floor.
Radiant temperature assymetry from a warm or heated ceiling: less than 5°C in relation to a small horizontal plane 0.6 m above the floor.

From Fanger, P.O., 1984. *Proceedings Third International Conference Indoor Air Quality and Climate.* Vol. 1. Stockholm.

Radiant field asymmetries are a common cause of thermal comfort complaints. These may occur despite the fact that air temperatures are in the comfort range. When a wall is warm, it radiates heat toward an individual who absorbs it. When a wall is cold, heat will be readily lost from that portion of the body facing it, and the person may experience a localized chilling effect. Radiant asymmetries are commonly noticeable on north and south faces of buildings. Thermal comfort requirements for winter and summertime conditions recommended by Fanger[4] are summarized in Tables 3.1 and 3.2. These are meant to satisfy a minimum of 80% of occupants.

Dissatisfaction with Thermal Comfort

Significant dissatisfaction with temperature and other factors which may affect thermal comfort have been reported in studies conducted in USEPA headquarters[5] and Library of Congress buildings.[6] The percentage of dissatisfied occupants relative to temperature, air movement, and relative humidity in Library of Congress and USEPA headquarters buildings is summarized in Table 3.3. As can be seen, dissatisfaction with environmental conditions was relatively high, with temperature and lack of air movement being particularly significant sources of dissatisfaction in the Waterside Mall building of the USEPA headquarters complex. Waterside Mall occupants also reported higher prevalences of SBS-type symptoms.[5] Dissatisfaction percentages summarized in Table 3.3 are similar to those reported by occupants who "wanted to adjust" temperature, air movement, and relative humidity.

Table 3.3 Percentage of occupants dissatisfied with temperature, relative humidity, and air movement in USEPA headquarters buildings and the Library of Congress.

	Environmental parameter (% dissatisfied)		
Location	Temperature	Relative humidity	Air movement
Library of Congress	39	26	43
USEPA buildings			
Waterside Mall	57	36	53
Crystal City	40	37	48
Fairchild	40	33	41
Total	53	36	49

From Selfridge, O. J., et al., 1990. *Proceedings Fifth International Conference Indoor Air Quality and Climate.* Vol. 4. Toronto.

Building Temperatures and SBS Symptoms

A number of epidemiological studies have attempted to evaluate potential associations between workspace temperatures and SBS symptom prevalence rates. A significant initial association between temperature and prevalence of general and mucous membrane symptoms was observed in the Danish Town Hall Study.[7] When multivariate statistical models were applied to the data, temperature was observed to vary in a co-linear fashion with other factors which investigators concluded were more directly related to SBS complaints.[8] However, in a follow-up study of a portion of the buildings studied, Skov et al.[9] observed a strong relationship between office temperatures and prevalence of general symptoms. The relative risk of general symptoms increased by a factor of three for every 3°C rise in temperature.

In an epidemiological study of a modern eight-story office building in Helsinki, Finland, Jaakkola et al.[10,11] observed that air temperature appeared to be the most important indoor parameter affecting SBS symptoms and the sensation of "dryness". A linear relationship was observed between SBS symptom prevalence and temperatures above 22°C. An excess of SBS symptoms was also observed when occupants considered the temperature to be too warm or too cold. Jaakkola et al.[10,11] suggested that SBS symptoms may reflect general satisfaction with building temperatures. A significant linear relationship over the temperature range of 21–25°C was observed for SBS symptoms, allergic symptoms, and dryness symptoms (Table 3.4) in a follow-up longitudinal study conducted by Reinikainen and Jaakkola[12] on the same building. Positive associations were also observed between temperature and the sensation of dryness and warmth.

Tamblyn et al.[13] conducted a study of 10 Canadian office buildings to determine whether environmental conditions were different for individuals who complained or did not complain of poor air quality. A variety of relationships between perceptions of air quality and environmental parameters were observed. Mean workspace temperatures were less significant than deviations from optimal temperature conditions relative to increased perceptions of poor air quality.

Table 3.4 Means of symptom scores adjusted for sex, atopy, ETS exposure, and psychosocial factors.

Symptoms/symptom score	Room temperature °C		
	21–<22	22–<23	23–≤25
Dryness	0.961	1.251	1.335
Allergic	0.755	0.793	1.082
Asthma	0.258	0.283	0.343
General	0.231	0.295	0.334
SBS-type	1.032	1.382	1.387

From Reinikainen, L.M., and J.J. Jaakkola, 1993. *Proceedings Sixth International Conference Indoor Air Quality and Climate.* Vol. 1. Helsinki.

In a study of 15 office buildings in the U.K., Hawkins and Wang[14] observed significant associations between SBS symptoms and dissatisfaction with workspace temperatures and other thermal comfort parameters. In the Dutch Office Building Study,[15] the absence of personal control of temperature in the workspace was related to nearly all health and indoor climate complaints. However, in the Swedish school study of Norback et al.,[16] there were no associations between SBS symptoms and temperature.

Humidity

Health and Comfort Relationships

Humidity may have a variety of relationships relative to health, comfort, and perceived satisfaction/dissatisfaction with air quality. As previously indicated, humidity significantly affects thermal comfort. Low humidity has also been reported to have direct effects on mucous membranes and human skin. Humidity may also affect human perceptions of air quality and levels of some contaminants.

Associations between humidity and SBS symptoms have been observed in several epidemiological studies. Hawkins and Wang[14] reported a significant association between SBS symptoms and dissatisfaction with building humidity. In the Canadian winter-time study of Tamblyn et al.,[13] low relative humidity was observed to be the most important risk factor for air quality complaints. This is consistent with the experimental studies of Reinikainen et al.[17] in Finland who reported lower prevalence rates of dry skin, nose, and throat and obstruction of the nose in humidified (30–40%) as compared to nonhumidified (20–30%) rooms. Humidification appeared to have a positive effect in reducing SBS-type complaints. In additional studies[18] humidities in the range of 30–40% were observed to be associated with significantly lower prevalence rates of skin and eye symptoms (but not symptoms of the nose or throat) than humidities in the range of 20–30%. Allergy-type (nasal congestion and sneezing) symptoms were less prevalent under humidified conditions. Increased airway irritation (nose and throat) has also been reported to be associated with elevated humidity levels in a sick library building.[19]

Reinikainen et al.[18] reported that the sensation of "stuffiness" was more commonly reported in humidified rooms. In another study of the same building, Reinikainen[20] observed that both the perception of "stuffiness" and unpleasant building odor increased when room air was humidified. From this perspective, humidified air (30–40% RH) was less acceptable than nonhumidified air (20–30% RH). Increased "stuffiness" associated with increasing humidity has also been reported by Berglund and Cain.[21]

In contrast to the studies described above which show a relationship between "dryness" symptoms and air quality complaints with low relative humidities, other studies have failed to show any association between SBS or "dryness" symptoms and relative humidity. In the large building study of Burge et al.,[22] sick buildings had higher prevalences of "dry-type" symptoms even though "healthy" buildings had higher relative humidities. In the Office Building Project of Northern Sweden (which included 4943 office workers in 160 buildings), sensation of dryness was related to SBS symptoms, but neither were related to measured humidity levels.[23] Both groups of investigators concluded that dryness symptoms was unrelated to the water vapor content of building air. The studies of Norback et al.[16] and Hodgson et al.[24] also failed to show any association between SBS symptoms and building humidity levels.

Building humidity levels have been reported to be associated with increases in acute respiratory infections.[25] Schools with higher relative humidities have reportedly had lower rates of absenteeism. Decreases in the prevalence of colds with increasing relative humidity have been reported. This apparent effect of relative humidity was suggested to be due to several factors. One of the most significant of these is the apparent reduced survival of infective organisms in the midrange of relative humidities. Higher humidities appear to produce larger particles with reduced infectivities. Additionally, low humidities have been reported to cause drying and cracking of mucous membranes in nasal passages which, in conjunction with infective organism survival, may increase the probability of infection.[25]

An association between low relative humidity and dryness and possibly cracking of mucous membranes is widely held. This is the primary basis for the use of humidification in buildings and physician recommendations of humidification for relief of upper respiratory symptoms. The scientific evidence is quite mixed. Low relative humidities (<20%) have been reported to cause eye irritation,[26] and several clinical reports have suggested that relative humidities above 30% are needed to prevent drying of mucous membranes and the maintenance of adequate mucous transport and ciliary activity.[27,28] In the experimental studies of Kay et al.,[29] eye blink frequency and perceptions of annoyance and nasal irritation were observed to be significantly affected by relative humidity. However, in the controlled exposure studies of Anderson et al.,[30] no relationship between low relative humidities and mucous membrane dryness was observed.

Effect on Contaminants

Humidity levels can significantly affect concentrations of formaldehyde in buildings. Typically, a 1% rise in relative humidity will result in a 1% increase in formaldehyde levels.[31] In most cases, however, formaldehyde levels are considered to be too low to cause symptoms in problem buildings. In the 1-year follow-up of the Danish Town Hall Study, Skov et al.[9] observed that a 10% increase in relative humidity levels appeared to double the risk for general symptoms (headache, fatigue) but not for mucous membrane symptoms. Reported formaldehyde levels in the four buildings investigated were in the range of 0.03–0.11 ppm (0.04–0.13 mg/m^3) with levels in the upper end of this range above the European threshold guideline[32] for formaldehyde of 0.082 ppm (0.10 mg/m^3).

Air Flow/Air Movement

The movement of air in building spaces may affect human comfort, perceived air quality, and maybe even symptoms directly. Air movement is a major factor in human thermal comfort as moving air helps to remove heat from body surfaces. When air movement is perceived as excessively cool, it is usually described as being a draft and thermally uncomfortable. Uncomfortable drafts may result from air leakage through the building envelope during cold outdoor weather or during the cooling season when chilled air flows at high velocity through ceiling diffusers directly onto individuals situated in its trajectory.

In addition to thermal effects, warm dry air flows have the potential for drying the mucous membranes of the eyes, a not too uncommon effect experienced by motor vehicle operators during cold weather driving and heater fan operation. Similar complaints have been observed by the author in problem building investigations.

It is common for building occupants to complain of inadequate air movement.[5,6] This suggests that occupants perceive that air circulation is inadequate and/or that ventilation is not adequate in some way.

Only a relatively few studies have attempted to address the relationship between SBS symptoms and/or dissatisfaction with air movement. In the initial statistical evaluations of Danish Town Hall Study data, a significant relationship was observed between SBS symptoms and air velocity.[8] However, after adjusting for covarying factors, the relationship was no longer significant.

Office workers reporting higher air quality complaint rates were exposed to slightly lower air flow velocities in the study of Tamblyn et al.[13] However, in the sick library building study of Lundin,[19] high total air flows in the building were associated with increased symptom prevalence rates. Lundin[19] suggested that high air flows may deteriorate air quality by producing a surplus of VOCs and humidity which was suggested to be more directly related to symptom complaints.

Lighting

Lighting conditions may affect personal comfort in several ways. If lighting is poor, it may cause eyestrain and fatigue. Factors associated with poor illumination may include inadequate lighting, excessive lighting, both direct and reflected glare, and dark shadows.[33] Some investigators have suggested a role for lighting in sick building complaints in addition to eye fatigue which can be expected for poor lighting conditions. Rask et al.,[34] for example, indicate that lighting is a major environmental stressor in buildings. They describe common symptoms alleged to be associated with poor luminosity to be headache, dizziness, drowsiness, fatigue, nausea, and eye irritation. Lighting described as harsh, dim, producing glare, etc. was reported to be a problem in 20% of 51 problem building investigations conducted. Excessive luminosity was also reported to be a major cause of complaints in the sick building investigations of Abbritti et al.[35]

Lighting complaints in the Dutch Office Building Study[15] were observed to be significantly related to personal, workplace, and job characteristics including exposure to ETS, allergy/respiratory symptoms, low job satisfaction, work with VDTs, handling of carbonless copy paper, and lack of control over workplace temperature. Lighting complaints averaged 30% with a range of 8–53%.

Landrus and Larkin[36] conducted a study of student responses to lighting conditions based on the hypothesis that fluorescent lighting types would have significant effects on a variety of parameters related to learning, health, and personal attitudes. Only a few differences were observed among the many-fold parameters evaluated. These included observations that fewer students felt closed-in or claustrophobic under daylight-simulating light conditions as compared to warm or cool-white fluorescent lighting. Additionally, fewer students reported dissatisfaction with warm-white as compared to the cool-white lighting.

Several investigators have attempted to determine whether lighting conditions contributed to SBS symptoms in cross-sectional epidemiological studies. Hodgson et al.[24] reported an association between increased lighting intensity and complaints of chest tightness in a sick building investigation. In a later report, no association between lighting intensity and SBS symptoms was observed.[37] No associations between lighting and SBS symptoms were observed in the U.K. Sick Building[22] and the Danish Town Hall Studies.[8,9]

Robertson et al.[38] conducted a questionnaire study of randomly selected occupants of two large air-conditioned and naturally ventilated office buildings. Office workers in the former had significantly higher prevalence rates of work or building-related headache and lethargy. Though individuals in both buildings reported there was too little light, those in the air-conditioned building indicated that lighting was less comfortable and expressed significant negative ratings of fluorescent lighting. Overall visual quality was assessed as significantly worse in the air-conditioned building as was discomfort with glare. Desk-top readings of luminosity were below recommended levels of 500 lux and were one third of those found in the naturally ventilated building.

Despite these marked differences in illumination, the adequacy of lighting was rated similarly by individuals in both buildings.

A relatively interesting role for lighting in contributing to SBS symptoms has been proposed by Sterling and Sterling[39] who observed that eye irritation in office workers declined substantially when ventilation was increased and sunlight-simulating fluorescent lamps were replaced by lamps emitting less ultraviolet light. They hypothesized that the effect of these mitigation measures was to reduce the level of photochemical smog components formed by the interaction of VOCs in the building and the catalytic effect of ultraviolet-radiating fluorescent lighting. Though the hypothesis is an interesting one, no monitoring studies which would confirm the presence of precursors or products of photochemical smog were conducted.

An interesting role for light in contributing to SBS prevalence rates has been proposed by NIOSH investigators who evaluated the potential effect of seasonal mood changes clinically described as seasonal affective disorder (SAD) on SBS prevalence in a problem building.[40] SAD is a recurring mood-change phenomenon characterized by depressive symptoms in fall and winter that remit in the springtime. SAD is apparently caused by reduced light intensity and shortened photoperiod (day length). Office employees with high SAD scores had 3.8 times the risk of reporting building-related symptoms compared to those with low scores.

Noise/Vibration

Noise has been described as an environmental stressor in problem buildings with excessive exposure suggested to cause headaches, dizziness, drowsiness, fatigue, and nausea. Rask et al.[34] reported excessive noise or acoustical problems to be a concern in 16% of building investigations conducted. These included air noise from diffusers, fans, and nearby processes. Sverdrup et al.[41] suggested that both noise and infrasound are potential contributors to higher prevalence of SBS symptoms in mechanically ventilated buildings. They observed a significant association between fatigue and noise levels as well as to "stuffy" or "bad air". No associations between noise or sound and the prevalence of SBS were observed, however, in sick building studies of Burge et al.[42]

Vibration, a sound-related parameter, has also been reported to cause problem-building symptoms. In a complaint investigation, Hodgson et al.[43] concluded that vibration generated from a pump room was the causal contributor to symptoms reported by office staff in an adjacent office. Significant associations were observed between vibration parameters and difficulty concentrating and ear popping in office workers.

Air Ions

Considerable research on the potential effects of air ions on the behavior, physiology, and well-being of humans and animals has been conducted over the past half-century[44] The many studies which report significant effects of air ions are not supported by others which indicate little or no effects. Because of differences in research results reported, the role of air ions and their potential effects on human health and well-being is subject to considerable controversy.

Air ions are electrically charged clusters of molecules held together by electrostatic forces and have an excess or shortage of one electron charge.[45] They are produced naturally as a result of collisions of cosmic radiation with air molecules and from the radioactive disintegration of radon. The concentration of positive and negative ions formed by these processes is exceptionally small; for every air ion there are approximately 10 quadrillion uncharged molecules.[46] In clean ambient air, the average concentration of positive and negative ions is in the range of $2–3 \times 10^3$ ions/cm^3, with positive/negative ion ratios of 1.2:1.

Both particulate-matter levels and electrostatic forces may cause a depletion of air ions from ambient and indoor environments. In the ambient air of our polluted cities, small ion concentrations may be as low as 500 positive and 300 negative ions/cm^3.[47] In buildings, electrostatic forces and other factors may reduce levels to an average of 150 positive and 50 negative ions/cm^3, and in some cases, ion concentrations may approach zero levels.[48]

The ambient atmosphere and air in buildings contain large ions produced by the attachment of small ions to the surface of aerosol particles. Aerosol particles thus serve as sinks depleting the air of small ions. Large ions are considerably less mobile than small ions and are less biologically active.[48] Stray electrical fields also cause small ion depletion by attracting and transporting them to surfaces where their electrical charge is neutralized on contact.

Several studies have reported that human exposures to either low negative ion or high positive ion concentrations diminish individual well-being with complaints of headache, nausea, lethargy, respiratory symptoms, and not feeling well.[49-51] Because of the uncertainties associated with this field of research, it is difficult to determine the significance of such results.

With this in mind and controversies involved, Charry[44] attempted to critically review the literature on the biological effects of air ions. He concluded that deficiencies in experimental design (including lack of statistical analyses), measurement of air ions, and other environmental parameters were responsible for the apparent lack of agreement which exists among scientists relative to the biological effects of small air ions.

Physiological and Behavioral Studies

A variety of experimental studies have provided evidence to support a role for small air ions in causing biological effects. These include investigations of the effects of air ion exposure on human and animal physiology and behavior and clinically observed responses in humans. Given the mixed results reported in the 57 studies reviewed, Charry[44] nevertheless concludes that certain trends are apparent when studies are viewed collectively. In human physiological and behavioral studies, biological effects appear to occur at exposures of 10^4 ions/ cm^3 (both positive and negative ion exposures). These include subtle changes in reaction time, skin conductance, and mood.

Results of physiological and behavior studies of animal exposures are more consistent. Positive ion exposures appear to depress activity, decrease arousal, impair learning and performance, and increase sensitivity to pain. Opposite effects have been generally associated with negative ions. Ion concentrations in animal (as compared to human) studies have been several orders of magnitude higher (10^5–10^6 ions/cm^3 vs. 10^3–10^4 ions/cm^3), possibly explaining the more consistent results observed in animal studies. Other trends include apparent effects on learning and performance, serotonin levels, pain and analgesia, conditioned emotional response, and a variety of physiological responses. Most notable are reports of reduction in brain serotonin levels associated with exposures to positive or negative ions at concentrations of 10^5 ions/ cm^3. Serotonin is a neurotransmitter apparently involved in sleep regulation, sexual activity, aggression, and psychotic depression in humans.

Clinical Studies

Many investigations have been conducted on humans for the purpose of treating clinical disorders such as burns and respiratory diseases, factors indicative of general health status including blood pressure, pulse, body temperature, heart activity, blood cell counts, and common symptoms such as headache, sore throat, and upper respiratory congestion. Results of these studies indicate that exposures to positive ions aggravate medical or health conditions and that negative ions have a positive effect. In general, negative ions appear to enhance respiratory function in individuals with respiratory disorders, whereas positive ions appear to impair respiratory function and produce symptoms such as sore throat, headaches, and dizziness. Clinical studies, however, viewed as a group are the least well-designed and controlled air ion studies reported in the literature. Ion levels have not been reported and presumably not measured in many studies. The significance of such results is, as a consequence, diminished by the deficiencies in methodologies employed.

Despite methodological problems which plague the science of air ions and their effects on humans, the collective weight of the various studies reported indicates that air ions can have significant biological effects and that such effects can occur in humans.

Building Studies

Low concentrations of air ions have been reported in air-conditioned buildings.[47] Consequently, several investigators have attempted to determine whether negative ionization of building spaces would reduce the prevalence of building-related complaints. In the initial studies of Hawkins,[47] negative ionization appeared to reduce significantly the complaint rate of illness symptoms. In a follow-up double-blind study, however, there was no association between negative ionization and symptom prevalence rates.[48] Similarly, Finnegan et al.[52] did not observe any significant improvement in comfort ratings or decrease in symptom prevalence in response to the addition of negative ions to building spaces. In a double-blind crossover study, Daniell et al.[53] observed no direct or residual effects of negative ion generator use on air ion levels, airborne particles, or symptom reporting. In a recent study in a Swedish hospital, the use of negative ionizers was observed to have significantly reduced SBS symptoms among hospital staff.[54]

Ion depletion has also been reported to be associated with the operation of video display terminals (VDTs). Wallach[55] observed that CRT-type VDTs exhibit a positive charge over the outer surface of the display varying from 400–12,000 volts. He suggested that this constantly positive potential quickly attracts and neutralizes negative ions and repels positive ones, causing severe ion depletion out to 50+ cm. This severe ion depletion is suggested to be the proximal cause of what Wallach[55] describes as "Video Operator's Distress Syndrome" (VODS).

In contrast to the findings of Wallach,[55] Swedish investigators[56] report that VDTs are an indirect source of air ions; that is, electrostatic charges on VDTs are a trap for air ions. They report significantly higher levels of free negative and positive ions in the air of rooms with VDTs compared to controls, with differences, evident in samples collected near the faces of building occupants.

Electrostatic Charges

The potential role of electrostatic charges in contributing to SBS symptoms has been studied by several investigators. Goethe et al.[57] conducted a study of electrostatic charges and SBS symptoms in four Swedish office buildings. Maximal peak potential differences between individuals and a grounded reference point during normal work activities varied from 1.76 kv to –3.78 kv. The mean potential differences in rooms equipped with wall-to-wall carpets was 0.41 kv and 0.07 kv in rooms with other flooring materials. There were no observed differences in the prevalence of mucous membrane symptoms among employees for whom mean voltage relative to the earth was above 0.5 kv compared to voltages below –0.5 kv. Not surprisingly, a significant correlation was observed for electrostatic potential and relative humidity.

Ancker et al.[58] theorized that electrostatic charges could cause attraction of aerosol particles which in deposition on the protruding parts of the body, such as the nose and area around the eyes, could cause irritation. They were unable, however, to demonstrate any relationship between facial symptoms and dustiness as perceived by employees.

An enhanced prevalence of SBS symptoms associated with electrostatic shock potential has been reported by Norback and Michel[59] and an enhanced prevalence of fatigue by Michel et al.[60] In the initial statistical analyses of data collected in the Danish Town Hall Study,[61] electrostatic charge was one of twelve factors observed to be related significantly to the prevalence of SBS symptoms. However, after statistical adjustments for covariation, electrostatic charges were no longer considered to be a significant factor by themselves. No significant association between the prevalence of electrostatic shocks and SBS symptoms was observed in the studies of Burge et al.[22]

Electric and Magnetic Fields

As a part of the Office Illness Project in Northern Sweden, Sandstrom et al.[62] conducted measurements of both electric and magnetic fields to determine whether they could be related to skin symptoms among VDT workers. The median value of the 50 Hz background field in the 150 offices surveyed was 0.07 μT, with approximately 5% higher than 0.5 μT. The dominating source of ELF fields was electric office equipment, not necessarily VDTs. The median value of magnetic fields in front of VDTs in the ELF-range was 0.21 μT and 0.03 μT in the VLF-range; the median value of electric fields in the VLF range was 1.5 V/m. Significant associations were observed for the highest exposed group (compared to the lowest) for both the electric background field and the magnetic field in the ELF range.

In a follow-up study of 22 office buildings, Sandstrom et al.[63] attempted to determine whether there were differences in power frequency magnetic and electric fields in offices with high and low prevalence of SBS symptoms. There were no differences in electric field strengths in high and low SBS environments. Neither were there any associations with magnetic field parameters and increased SBS prevalence. However, 30-minute recordings of magnetic fields were significantly higher in low SBS prevalence buildings. Similar associations were not observed for other measured magnetic field parameters.

Reprise

Many of the physical environmental factors discussed in the preceding pages have been implicated as risk factors for SBS symptoms. The nature of the risk (if any) may be direct or indirect. For the former to be true, there must be plausible biological mechanisms between the exposure and observed symptom or symptoms. Plausible biological responses manifested as one or more symptoms have been suggested for exposures to low relative humidity, air velocity, some lighting conditions, small air ions, and electrostatic and electromagnetic phenomena.

In cases such as temperature or thermal conditions, elevated humidity, air movement (or lack thereof), and some lighting and noise conditions, the effect on SBS may be indirect; that is, there may be no human tissue or system that is being affected in a physiological or pathological way. The effect may be one of subjective human comfort and result in either increased or decreased symptom reporting rates. Indirect effects may also include the effects of environmental factors such as temperature and relative humidity on contaminant levels. Temperature is a significant controlling variable for emissions of contaminants from a variety of source materials in indoor spaces, and both temperature and humidity significantly affect emissions and concentrations of formaldehyde.[31]

When the results of various studies are compared, it is not unusual to observe different outcomes, with some showing associations between an environmental variable and SBS symptoms, while others do not. Such mixed results may be due to the type of statistical analyses employed, outcomes associated with mass significance, study population size employed, the power of statistical analyses to detect associations, and the nature of the study design.

In some studies, statistical analyses may be relatively simple with potential associations between variables evaluated singly. Associations determined from crude odds ratios and other simple statistical tests may disappear when multivariate analyses are conducted to take covariation between two or more variables into account. Such covariation is common and confounds efforts to determine causal associations or risk factors.

Many of the cross-sectional studies conducted in systematic sick building investigations employ multivariate statistical analysis to evaluate potential associations between a variety of environmental factors, personal factors, building factors, contaminant levels, and SBS symptoms. The use of large numbers of variables in such analyses poses the risk that one or more variables will by chance be observed to be associated with the outcome variable, SBS symptoms. Because of the potential for such spurious or chance outcomes, the results of a single multivariate analysis should always be viewed with some degree of caution, particularly when there is no plausible mechanism to explain it and when it is not observed in other studies of comparable or stronger design.

Studies of buildings where the number of individuals or buildings included in the study is small may not have the statistical power to detect associations between various environmental, personal, building, and contaminant factors and SBS symptoms when such associations are in fact real. The cross-sectional studies of Burge et al.,[22] the Danish Town Hall Study,[7,8] the Dutch Office Building Study,[15] and the Office Illness Project of Northern Sweden[64] are notable because of the relatively large sample populations of both individuals and buildings included in each. Cross-sectional studies of single or several buildings may not in many cases have sufficient statistical power to evaluate associations adequately.

Some study designs are superior to others in evaluating associations among study variables. The longitudinal and/or crossover studies of Reinikainen et al.[17-19] on the effects of relative humidity and Jaakkola et al. and Reinikainen et al.[10,12] on temperature are examples of strong study designs which were implemented to answer the question of what the effects of these environmental factors were on SBS symptoms. Mendell[65] has attempted to evaluate the effect of various environmental factors on SBS symptoms by taking the strength of study design into account.

ENVIRONMENTAL SYSTEMS

In addition to heating, and in many cases cooling, environmental systems are often designed and operated to provide outdoor air and air circulation to dilute contaminants for the purpose of maintaining human comfort. Historically, comfort concerns have focused on using ventilation to reduce human bioeffluents and associated discomfort to acceptable levels. The reduction by dilution of contaminants such as tobacco smoke and other contaminants generated within building spaces has been a secondary concern.

Inadequacies in the design, operation, and maintenance of ventilation systems have long been and are still considered to be the major cause or contributing factor to occupant health complaints and dissatisfaction with IAQ. The relationship between IAQ problems and concerns and ventilation is, however, a very complex one.

Field Investigations

In reported field investigations of NIOSH,[66-68] Turner,[69,70] Woods,[71] Robertson,[72] and others, deficiencies in the design, operation, and maintenance of HVAC systems were identified as being the major contributors to illness complaints in a large percentage of buildings investigated. The belief that most building-related health complaints are due to such deficiencies is widely held. As a consequence, evaluating the performance of HVAC systems is standard practice in sick building investigations conducted by NIOSH investigators, Canadian investigators, and many private consulting firms.

Design problems have included inadequate outdoor air capacity or even absence of outdoor air intakes, poor air distribution within and among spaces, and unbalanced intake and exhaust air. Operating problems have included reduction in outdoor air supply rates to conserve energy, changes (increasing) in building occupancy, and poor HVAC system operation by inadequately trained personnel. Poor system operation may result from inadequate maintenance. Commonly encountered problems include failure to keep air intakes clean, dirty filters not replaced or serviced adequately, fouled and contaminated heating and cooling coils, and disconnected components such as damper linkages, exhaust fans, and automatic controls.[73,74]

In NIOSH field investigations[66-68] as well as those of others, determinations of ventilation as a causal or contributing factor in problem-building complaints have been based on reference to a guideline value of 1000 ppm CO_2. This 1000 ppm guideline does not reflect any health hazard associated with CO_2; rather, CO_2 is used as an indicator of human bioeffluents which have historically been of concern because of associated human odors[75] and suspected human comfort effects of human bioeffluents. Occupant discomfort has been suggested to be associated with CO_2 levels in the range of 600–1000 ppm and higher.[76,77]

Systematic Building Studies

A variety of building ventilation system factors have been suggested to be associated with the prevalence of SBS symptoms.

Type of Ventilation System

Sterling et al.,[78] noting the rapidly increasing trend away from "naturally" ventilated to "closed loop" HVAC systems and unopenable windows suggested that "sealed" or "tight" buildings were responsible for sick building phenomena over the past decade. Several early studies in the U.K. appeared to confirm such a linkage. Finnegan et al.[79] and Robertson et al.[79] in the United Kingdom (U.K.) observed significantly higher prevalence rates in mechanically as opposed to naturally ventilated noncomplaint buildings. In a larger study of 27 office buildings, Harrison et al.[81] also observed significantly higher prevalence rates in mechanically ventilated buildings.

In an expanded U.K. study which included a total of 47 office buildings, Burge et al.[22] observed significant differences in symptom prevalence rates among buildings with different types of ventilation systems, with the highest rates associated with fully air-conditioned buildings (Table 3.5). These results were confirmed by Mendell and Smith[82] who conducted a meta-analysis of data from a number of systematic European investigations. This same relationship was also observed in the Dutch Office Building Study.[49]

Ventilation Conditions — Symptom Prevalence Rates and Occupant Satisfaction/Dissatisfaction with Air Quality

A number of investigators have attempted to evaluate ventilation conditions in buildings and SBS symptom prevalence rates or satisfaction/ dissatisfaction with air quality. Because of the widespread use of indoor CO_2 levels as a guideline for ventilation adequacy, attempts have been made to determine whether there is any significant relationship between CO_2 levels and symptom prevalence. Carbon dioxide levels would, in theory, serve as an indicator of building ventilation conditions.

No significant relationship between CO_2 levels and the prevalence of SBS symptoms has been observed in the Danish Town Hall Study[8] and in 6 sick schools[16] and 11 office buildings[83] in Sweden. In U.K. studies, symptom prevalences were observed to be higher in mechanically ventilated air-conditioned buildings than in naturally ventilated buildings, despite the fact that the latter had higher CO_2 levels.[42] Broder et al.[84] were unable to observe any correlation between building-related discomfort and CO_2 levels in three Canadian office buildings. Fanger et al.[85] observed no correlation between CO_2 levels and panel-assessed dissatisfaction in 15 office spaces and 5 assembly halls in 18 noncomplaint buildings in the greater Copenhagen area. A relationship was, however, observed between perceived air quality and indoor CO_2

Table 3.5 Relationship between building/work-related symptoms and ventilation system type.

Ventilation type	Building sickness index	% With symptom									
		Dry eyes	Itchy eyes	Runny nose	Blocked nose	Dry throat	Lethargy	Headache	Flu	Difficulty breathing	Chest tightness
Natural (N = 442)	2.49	18	22	19	40	36	50	39	15	6	6
Mechanical (N = 944)	2.18	20	20	16	32	33	42	33	14	6	5
Local induction/ fan coil (N = 508)	3.81	34	33	29	58	56	66	52	28	14	11
Central induction/ fan coil (N = 1095)	3.70	31	32	28	57	54	68	47	29	12	14
"All air" (N = 1384)	3.12	31	29	21	45	46	56	43	25	9	8
Whole group (N = 4374)	3.10	27	28	23	47	46	57	43	23	9	9

From Burge, S., et al., 1987. *Ann. Occup. Hyg.*, 31:493–504. With permission from Elsevier Science Ltd.

levels, with perceived air quality decreasing with increased CO_2 levels in a study of ten school buildings.[86] Results of these studies indicate that there is apparently little or no relationship between CO_2 levels and SBS symptoms or occupant satisfaction/dissatisfaction with air quality.

The absence of an apparent association between CO_2 levels and SBS symptoms/occupant dissatisfaction with air quality is not surprising. Though a good indicator of bioeffluent levels, they are, at best, a relatively crude indicator of building ventilation conditions. Consequently, other ventilation assessment methods have been used to evaluate the effects of ventilation conditions on SBS symptom prevalence rates and/or occupant satisfaction/dissatisfaction with air quality. These have included measuring volumetric flow rates through outdoor air intakes and determining actual air exchange rates by using tracer gases. The latter is clearly superior because the former does not account for infiltration-related air exchange.

In a study of multiple floors of a problem office building, Salisbury[87] observed that symptom prevalence rates were paradoxically highest on the floor with the highest outdoor volumetric flow rate (39 L/s/person) and lowest on the one with the lowest flow rate (8.5 L/s/person). There appeared to be a trend in the seven floors investigated toward increasing symptom prevalence rates with increasing outdoor air flow. On the other hand, in a study of an office building in Alaska, Hodgson et al.[88] observed that symptom prevalence decreased with increasing outdoor air flow rates. In the 41 office-building study of Burge et al.,[22] both high and low symptom prevalence rates were observed to be independent of outdoor ventilation rates.

In a cross-sectional study of an office building in Finland, Jaakkola et al.[11] failed to observe any relationships between SBS symptom prevalence rates and ventilation in the range of 7–70 L/s/person. In another Finnish study, Ruotsalainen et al.[89] reported no consistent association between the magnitude of air flows or air exchange rates and the occurrence of SBS symptoms reported by staff in mechanically ventilated day-care centers.

In contrast, Sundell et al.[90] in the Office Illness Project of Northern Sweden were able to observe significant but weak associations between log outdoor flow rates and general symptoms ($r = -0.46$ males; $r = -0.39$ females), sensation of dryness ($r = -0.30$ males; $r = -0.39$ females), with mixed results for mucous membrane symptoms ($r = -0.05$ males; $r = 0.31$ females). These relationships can be seen in Figure 3.1. In contrast to the studies of Burge et al.,[22] none of these office buildings appeared to be fully air-conditioned.

Other investigators have attempted to evaluate the effect of experimental adjustments to the ventilation system on symptom prevalence rates or occupant satisfaction with air quality. Jaakkola et al.[11] conducted a controlled experimental study in which ventilation was shut off in one part of the study building, by 75% and 60% in two other areas, with one part of the building serving as a control. These reductions in ventilation rates resulted in a small but significant difference in symptom prevalence between test and control groups. Since symptom prevalence scores were decreasing with time (in all groups), the experimental groups were observed to show the smallest decrease. Wyon[54] conducted experimental studies in Sweden in which estimated mean high and low ventilation rates were 11.6 and 16.6 L/s/person. No differences in SBS symptom prevalence rates between these ventilation rate conditions were observed.

Nagda et al.[91] attempted to determine the effect of high (10 L/s/person) versus low (2.5 L/s/person) volumetric outdoor flow rates on symptom prevalence in a New York office building. Although they observed significantly lower symptom rates at the higher ventilation rate condition when assessed during both summer and wintertime periods, only the summertime prevalence rates were observed to be significantly higher when occupant perceptions of the relationship between symptoms and the building were taken into account. When actual air exchange measurements were made, the higher ventilation condition was only 20–25% and 38% higher than the low ventilation condition in summertime and wintertime studies, respectively. Infiltration was the major contributing factor to the limited differences in actual air exchange observed.

Hanssen[92] conducted measurements of ventilation conditions in seven school buildings in Norway. Supply air volumes were very low, varying from 1–8 L/s/person. The ventilation system was adjusted in one building to provide 8 L/s/person outdoor air with no recirculation. General or central nervous system symptoms were significantly reduced; mucous membrane symptoms were not.

Jaakkola et al.[93,94] conducted experimental studies which focused on the potential effects of recirculated air on symptom prevalence rates or occupant dissatisfaction with air quality (in contrast to ventilation rates, per se). They observed no differences in symptom prevalence and occupant dissatisfaction at recirculation percentages of 70% vs. 0%.[93] However, in a survey study of occupants in 41 randomly selected office buildings, they observed a significant increase in nasal congestion among occupants of buildings with air recirculation compared to similar buildings without air recirculation.[94] Occupants of buildings with recirculated air were, however, more satisfied with thermal climate and perceived air quality.

Canadian investigators[95,96] have attempted to evaluate the effect of experimentally adjusted ventilation conditions on SBS symptoms and environmental satisfaction in a double-blind study in a 30-story office building in the ventilation rate range of 7.5–31.5 L/s/person. No differences in symptom prevalence rates between low and high ventilation conditions were observed. Because of the potential confounding effect of time-dependent decreases in symptom prevalence rates and improved environmental satisfaction scores, Menzies et al.[97] expanded their studies to include four buildings evaluated during two seasons. Again, no differences in symptom prevalence rates under calculated (from CO_2 levels) ventilation conditions of 14 and 30 L/s/person were observed.

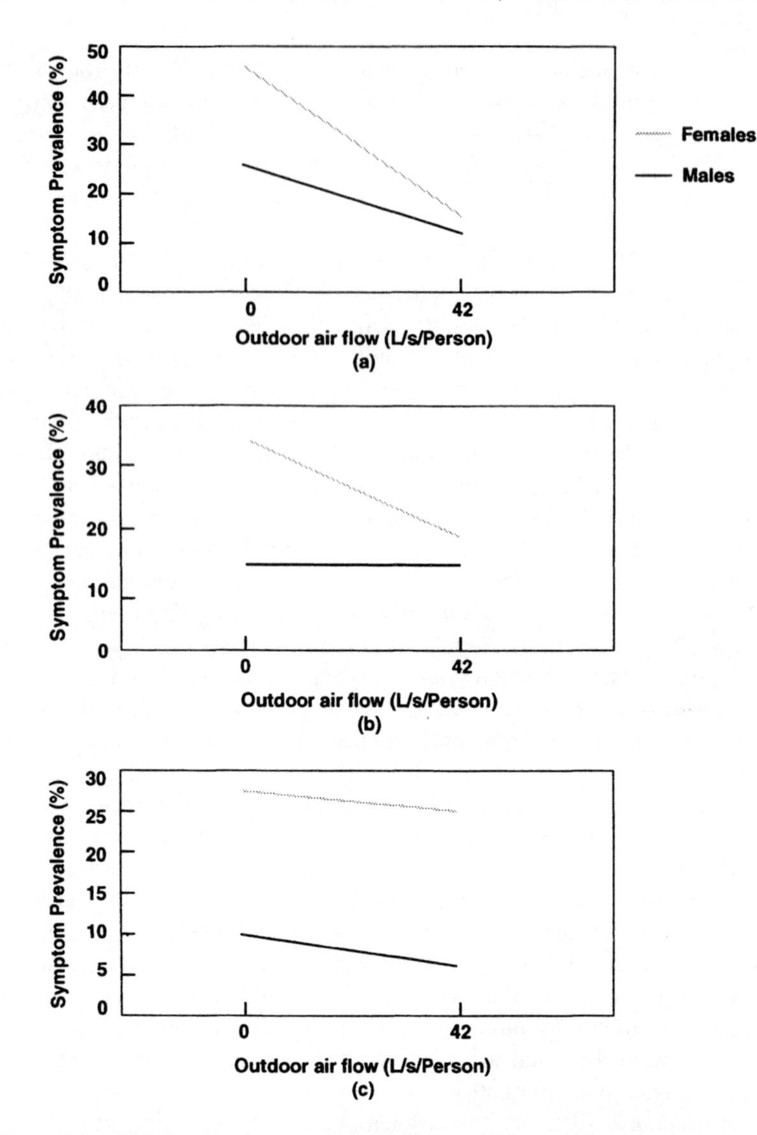

Figure 3.1 Relationships between the prevalence of (a) general symptoms, (b) mucous membrane symptoms, (c) dryness symptoms and outdoor air exchange rates in Swedish office buildings. (From Sundell, J., et al., 1991. *Post-Conference Proceedings IAQ '91: Healthy Buildings.* Atlanta. With permission.)

Investigators have also attempted to evaluate the effect of ventilation rate differences on occupant satisfaction/dissatisfaction with a variety of air or environmental quality parameters. Menzies et al.[96,97] observed no improvement in environmental satisfaction scores under the higher ventilation rate conditions used in their study. Nagda et al.[91] also reported that satisfaction/dissatisfaction with air quality parameters such as odors, tobacco smoke, and dust were

Table 3.6 Occupant dissatisfaction (%) with air quality parameters under different ventilation rate conditions (N = 580).

	Ventilation rate (L/s/person)			
Always/sometimes dissatisfied with	**<5**	**5–10**	**10–15**	**>15**
Odor	25	18	9	10
Dustiness	55	49	29	34
Stuffiness	59	70	51	48

From Palonen, J. and O. Seppanen. 1990. *Proceedings of the Fifth International Conference on Indoor Air Quality and Climate.* Vol. 4. Toronto.

relatively unaffected by ventilation rate conditions. Palonen and Seppanen,[98] on the other hand, observed general decreases in occupant dissatisfaction with odor, dustiness, and stuffiness with increasing ventilation rates from <5 to >15 L/s/person. Dissatisfaction levels, nevertheless, remained relatively high despite increased outdoor air flow rates (Table 3.6).

Workstation Characteristics

The nature of an individual's workstation relative to ventilation conditions may affect symptom prevalence rates. Menzies et al.[96] reported that individuals in open plan office areas had significantly higher SBS symptom prevalence rates and lower ventilation efficiency.

Ventilation Conditions and Contaminant Levels

Studies of the effect of increased outdoor air ventilation rates on contaminant concentrations have shown mixed results. In the previously reported studies of Nagda et al.,[91] no significant effects of outdoor ventilation rates were observed on concentrations of formaldehyde, nicotine, and total volatile organic compounds (TVOCs), with marginally significant increases in CO_2 and CO under lower ventilation conditions. This apparent ineffectiveness of ventilation system adjustments may have been due to the fact that actual ventilation rates were only marginally different. The target ventilation rates were 10 L/s/person and 2.5 L/s/person. Actual differences were only in the 20–38% range. Collett et al.[99] experienced similar problems. Despite efforts to configure HVAC systems to provide outdoor ventilation rates of 10 and 5 L/s/person, actual differences in air exchange were slight. As a consequence, no significant reduction in CO_2, formaldehyde, RSP, nicotine, and microbial concentrations were observed.

Palonen and Seppanen[98] observed no correlation between outdoor ventilation rates and measured levels of CO_2, formaldehyde, VOCs, particles, bacteria, mold and radon in their study of five office buildings. In contrast, Farant

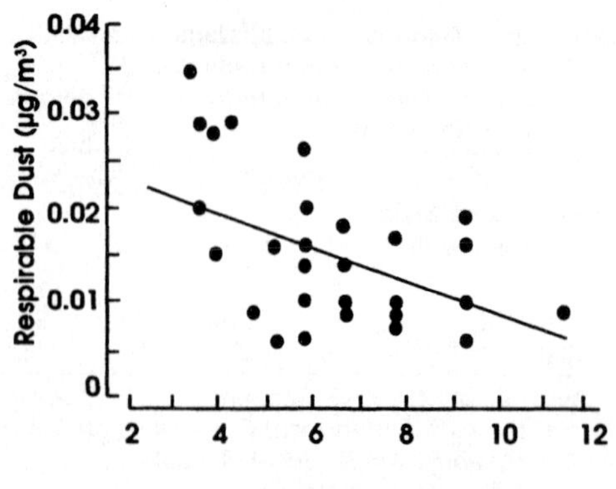

Figure 3.2 Relationship between indoor RSP levels and outdoor air flows in Swedish school buildings. (From Norback, D., et al., 1990. *Br. J. Indust. Med.*, 47:733–741. With permission.)

et al.[100] observed significant associations between outdoor ventilation rates (based on floor surface area, L/s/m^2, and the natural log of the indoor/outdoor ratios of CO_2, TVOCs, formaldehyde, and the oxides of nitrogen). In another office building study, Baldwin and Farant[101] reported a much higher correlation ($r = -0.86$) between outdoor ventilation rates and TVOC levels (a twofold increase in ventilation resulted in an approximate 75% reduction in TVOCs). In addition, significant correlations ($r = -0.50$) between ventilation rate conditions and indoor RSP levels were reported by Norback et al.[16] in a longitudinal study of 14 Swedish primary schools (Figure 3.2).

Reprise

Mixed results have been reported for studies which have attempted to determine relationships between ventilation rates, symptom prevalence, occupant satisfaction with air quality, and contaminant levels. Differences in results reflect, in part, the nature of the study conducted (cross-sectional versus experimental), how ventilation rates were determined (volumetric flow rates versus tracer gas measurement), and the ability of investigators to control ventilation rates in experimental studies.

In general, it is more difficult to observe differences in cross-sectional studies because of the variability that other factors contribute to the statistical analysis. The significant associations observed in the Office Illness Project in Northern Sweden[90] are, therefore, notable.

In theory, experimental studies are better suited to determine whether associations between ventilation rates and SBS symptom prevalence exist. However, as can be seen in the studies reported above, results have been quite mixed. Differences are likely due, in part, to methods used in determining ventilation rates and the inability of investigators to control ventilation conditions to target levels because of the significant effect of infiltration.

Many investigators attempt to determine ventilation rates by measuring the flow of outdoor air through air intakes in HVAC systems and report these in L/s/person. Such measurements are at best "guesstimates" because they do not include the often significant effect of infiltration. As such, they represent nominal ventilation rate conditions. As observed by Nagda et al.[91] and Collett et al.,[99] actual air exchange conditions in experimental studies may differ little even though outdoor flow rate adjustments would indicate differences on the order of three to four times.

It is, therefore, desirable to measure actual air exchange rates by using tracer gases or some technique that represents actual air exchange conditions. In the experimental studies of Menzies et al.,[95-97] building ventilation rates were calculated from measurements of CO_2 levels. Though this method is less accurate than tracer gas methods, it nevertheless has the advantage of taking infiltration-induced ventilation into account.

Even with accurate measurements of outdoor ventilation rates, it is nevertheless difficult to conduct experimental studies because minimum ventilation rates are determined by infiltration which is uncontrollable. This appeared to be a major problem in the studies of both Menzies et al.[97] and Wyon[54] whose low ventilation rate experimental conditions (16 and 11 L/s/person) were relatively high (above ASHRAE guidelines). Indeed, the Menzies et al.[97] study has received considerable criticism in the IAQ community because of the high ventilation rates that were evaluated and reported.

Mendell[102] has evaluated ventilation rate and SBS prevalence studies. He has observed that studies[11,90,91] which included ventilation rates below 10 L/s/person showed significant associations with SBS symptoms, while those with high and low ventilation rates above 10 L/s/person have not,[11,54,97] suggesting that ventilation rates up to 10 L/s/person are risk factors for SBS symptoms.

The HVAC System as a Source of Contaminants

The studies of Burge et al.,[22] Zweers et al.,[15] Mendell and Smith,[82] and Jaakkola et al.[93] have all reported that symptom prevalence rates are significantly affected by the type of ventilation system present in an office building. In most cases,[15,22,82] symptom prevalence rates were significantly higher in fully air-conditioned buildings. These results suggest that some factor or factors associated with fully air-conditioned buildings may uniquely contribute to SBS symptoms. Because systems that provide cooling also produce condensate water which may amplify the growth of microorganisms, Burge et al.[22] and Mendell and Smith[82] have suggested that the causal factor for increased SBS symptoms may be biogenic, resulting in allergic reactions or reactions to microbial toxins.

Condensate waters from air-conditioning/cooling are not the only source of water in HVAC systems. Many mechanically ventilated buildings are humidified either with cool mist evaporators (drip or spray) or by steam.

In the U.K. study of 47 office buildings,[103] a significant relationship was observed between the type of humidification system (none, evaporative spray, or steam) and the prevalence of itchy eyes, stuffy nose, lethargy, breathing difficulty, and chest tightness. Prevalence rates appeared to be higher in those buildings with evaporative spray humidification systems; buildings with steam humidification or no humidification had similar prevalence rates. In the Dutch Office Building Study,[15] humidified buildings had higher SBS symptom prevalence rates than either naturally ventilated or simple mechanically ventilated buildings with no apparent differences between spray and steam humidification. In meta-analyses of a number of sick building studies conducted in Europe (including the U.K. study described above), Mendell and Smith[82] were unable to observe any association between the use of humidification systems and SBS symptom prevalence.

Leaving aside for the moment the important problems of Hypersensitivity Pneumonitis, Humidifier Fever, and Legionnaires' Disease (see Chapter 6), a variety of investigations have implicated the HVAC system as a potential contributor to problem building complaints and/or occupant dissatisfaction with air quality. These include reports of NIOSH health hazard evaluations[104] which associated corrosion inhibitors (added to boiler water subsequently used for steam humidification) or their decomposition products with outbreaks of illness. In addition, Godish[105] reported that the prevalence rate of facial erythema experienced by students and staff in a primary school building was significantly reduced when first hot water flow and then operation of the HVAC system was discontinued. Fanger et al.,[85] Thorstensen et al.,[86] and Pejtersen et al.[106] reported results of studies of perceived air quality in office buildings, assembly halls, schools, and kindergartens. In all cases, operating ventilation systems were observed to be the major contributor (approximately 40%) to dissatisfaction with air quality assessed by a trained panel.

These studies and others indicate that HVAC systems and their components may serve as sources of indoor contaminants including a variety of gas-/vapor-phase chemicals, man-made mineral fibers (MMMF), and other particulate-phase materials, and particulate or gas-phase materials of biological origin.

Chemical Contaminants

Molhave and Thorsen[107] observed that ventilation systems could be a significant source of VOCs. The source strength of the ventilation system investigated was fourfold larger than all other sources including occupants and their activities. A 16-fold increase in VOC levels occurred on reactivation of the HVAC system after night-time shutdown. Major identified compounds included heptene, hexane and other alkanes. Though a source of VOC emissions, HVAC system filters accounted for only a fraction of the total VOCs apparently produced within the ventilation system. The VOC source was postulated as dust on internal system surfaces which may have served as a sink and therefore a secondary source of VOC emissions. Valbjorn et al.,[108] however, reported that VOC and odor emissions from HVAC system dusts were negligible. Pasanen et al.[109] observed that oil-treated HVAC system filters were significant sources of toluene and xylene and a variety of odorous aldehydes (e.g., hexanal, heptanal, octonal, and nonal) and light organic acids once they had become infested with microorganisms.

The HVAC system as a source of VOCs and possibly CO is suggested from the studies of Berglund et al.[110] In a study of a sick building, they observed that a larger quantity of VOCs and CO appeared to be transferred from return to supply air than would have been expected based on ventilation-associated reductions in CO_2 levels and calculated air flows. Lundin[20] reported that eye irritation in occupants of a sick library building was associated with VOC levels which were, in turn, associated with high ventilation air flows. The HVAC system was suggested to be the source of the VOC "surplus."

Godish[105] reported results of emission testing of duct liner materials taken from various HVAC system components in a problem school building. Emission testing results using NASA Test 7[111] and the purge and trap method[112] are reported in Table 3.7. Most surprising were the relatively high emissions of CO in four of the five samples tested, with the highest emissions from duct liner materials used to insulate the interior of two large AHUs. It is not immediately apparent how such materials could be a source of CO. They may have been due to some decomposition or sink process. These CO emissions may, in part, explain the results of Berglund et al.[110] who observed that ventilation did not significantly reduce CO levels in a sick building.

Significant emissions of methyl chloroform (1,1,1-trichloroethane) were observed from two duct liner samples taken from different HVAC system components. Methyl chloroform is one of the most prevalent and abundant VOCs observed in monitoring studies conducted in public access buildings.[113] Whether these materials were primary or secondary (sink) sources of methyl chloroform is not known. Methyl chloroform was one of six prominent VOCs characterizing a sick building (as compared to a healthy building) in the pattern analyses of Berglund.[114]

Sheldon et al.[115] conducted headspace analyses of a single unidentified type of duct liner as a part of a larger study of building materials. They observed only trace emissions of trimethyl benzenes. Chemical contaminants associated with water used for steam humidification of several problem buildings investigated by NIOSH field staff[104] were suggested to include boiler water corrosion-inhibiting additives diethylamine ethanol (DEAE) and cyclohexamine or some chemically altered form of DEAE.

Table 3.7 Emissions from duct liner materials collected from a problem school building.

Sample	Source	Compound	Emission concentration (µm/gm)
1	VAV box	Methyl chloroform	14.46
		Carbon monoxide	7.68
2	Reheat box	Acetone	0.66
		Ethylene chloride	0.71
		Carbon monoxide	6.55
3	Main AHU	Carbon monoxide	93.48
4	Auxiliary AHU	Carbon monoxide	101.46
5	Perimeter Heating unit	Methyl chloroform	5.76

From Godish, T., 1992. *Proceedings Eleventh International Conference Clean Air Society of Australia and New Zealand.* Brisbane. With permission.

Man-Made Mineral Fibers (MMMF) and Other Particles

HVAC systems are a potential source of MMMF, such as fibrous glass and mineral wool, as well as other particles. MMMF may be released into building air from the erosion of fibrous-glass duct materials commonly used to line the interiors of air handling units (AHUs), reheat, variable air volume (VAV), mixing and diffuser boxes, and various lengths of supply and return air ducts. Such materials serve as both thermal and acoustical insulation. In many residential and, to a lesser extent, mechanically ventilated buildings, fibrous-glass ductboard is used as supply and return air transmission systems (ducts). Ceiling tiles which serve as a base for return air plenums in many large mechanically ventilated buildings are another potential source of MMMF as well as other fibers. Ceiling tiles are composed of a variety of materials including fibrous glass, mineral wool, asbestos, and cellulose.

Fibrous-glass duct liner and ductboard, as well as ceiling tiles, are products which, in theory, should shed few fibers into supply or return airstreams since they are bound by an adhesive matrix which typically has considerable chemical and thermal stability. Phenol-formaldehyde resins are used as adhesives for fibrous-glass duct materials. Fiber shedding is further limited by latex coatings which are applied on the airstream surface of many of these materials by manufacturers and secondary fabricators.[116]

Despite use of resin binders and vinyl coatings which are putatively stable both thermally and chemically, the author's field experiences and anecdotal reports of other investigators indicate that fibrous-glass duct materials undergo physical deterioration with time with the resultant release of fibrous materials into supply airstreams. In addition to the gross damage to duct-insulating materials observed by field investigators, the potential exists for exposed surfaces to undergo physical/chemical erosion with the release of fibers and other particulate-phase materials. Such erosion may result from the decomposition of resin binders and protective emulsion coatings associated with elevated

HVAC system temperatures, high velocity air flows, reactions with air contaminants such as ozone, aging processes, and microbial growth.

The shedding potential of fibrous-glass duct materials has been studied by a number of investigators. Cholak and Shafer[117] collected air samples in the outlet of supply air ducts in six different buildings. Analyses revealed no evidence of fiber shedding. Surface wipe samples did, however, contain fibrous glass which was attributed to dust collection over several years from "unknown sources". Balzer et al.[118] studied fiber levels in supply air outlets of 13 fibrous-glass air transmission systems. A decrease in glass fiber levels was observed as air passed through these systems. Glass fiber concentrations in building air correlated closely with concentrations in outdoor air. Concentrations in ducts were consistently lower than in ambient air of San Francisco which varied from <0.05 to 10 fibers/liter (f/L) with average concentrations of 0.2 f/L. These results did not preclude the occurrence of fiber erosion. If it did occur, it may have been offset by deposition and entrapment of fibers in filters and elsewhere within the air transmission systems tested.

Gamboa et al.[119] conducted emission testing of glass fibers from duct liners and ductboard materials under controlled laboratory conditions. At test velocities of 2400 and 3000 feet/min (FPM) (731 MPM and 914 MPM), glass fiber levels increased from background values of 0.17 f/L to 0.39 and 0.51 f/L, respectively, from ductboard and duct liners with treated and untreated edges at a duct velocity of 5000 FPM (1524 MPM). Referring to unpublished studies conducted by Owens-Corning in 1980, Hays[120] indicated that the discharge of fibers from fibrous-glass duct materials was less than 0.01 f/L. In more recent studies,[120] glass fiber concentrations in the discharge airstream of ductboard systems (evaluated under laboratory conditions) were reported to be in the range of 0.0024 to 0.0061 f/L, depending on the counting technique employed.

Schneider[121] determined MMMF concentrations in air samples and settled dust in a group of kindergartens, a school, and an office building in which MMMF was suspected to be a problem and compared these to samples collected in 11 mechanically-ventilated school buildings. The problem buildings had respirable MMMF concentrations which were significantly higher than buildings used as a control. A wide range of MMMF concentrations were observed in settled dust, suggesting that transfer from surfaces by human fingers was responsible for complaints of eye irritation. Significant fibrous-glass concentrations in wipe samples of office and internal HVAC system surfaces were observed in studies conducted by Samimi.[122] These samples also contained other particulate-phase materials which appeared to have been associated with HVAC system components including cellulose fibers from both pre- and after-filters, carbon from carbon filter modules, and aluminum oxide dust from the corrosion/oxidation of the aluminum fans of reheat coils.

Studies of Cholak and Shafer,[117] Balzer et al.,[118] Gamboa et al.,[119] and Hays[120] indicate that erosion of MMMF from fibrous-glass duct materials may be relatively limited, resulting in either no increase in airborne fiber concentrations above background levels to slight increases in the airstream passing through these materials. Shedding of fibers from duct materials was suggested to be too low to be biologically significant. However, the results of Cholak and Shafer,[117] Schneider,[121] and Samimi[122] which show MMMF in surface samples indicate that measurements of airborne MMMF may not be a good indicator of fiber erosion, building contamination, or human exposure. Fibers in settled dust can be disturbed and resuspended, resulting in potential respiratory, ocular, and dermal exposure. Exposure may also result from direct dermal contact with fibrous glass in settled dust.

Several studies have implicated settled dust as a potential causal or contributing factor to SBS symptom prevalence rates.[123,124] Most notable for this discussion is the study of Hedge et al.[124] who observed significant correlations between MMMF counts and settled dust and SBS symptoms. Similar relationships were not observed for airborne MMMF levels.

Biological Contaminants

A variety of HVAC system components have been reported to be sources of contaminants produced as a consequence of biological growth. These include cooling coil condensate sinks, humidifiers, porous thermal system insulation, and HVAC filters. The nature of contamination problems, contaminants, and potential health problems associated with such biological contamination are discussed in detail in Chapter 6.

Reprise

Though ventilation systems are designed to provide outdoor air to building spaces to improve air quality, there is increasing evidence that HVAC systems may be sources of contaminants. As a consequence, ventilation systems may themselves be risk factors for outbreaks of building-related illness, SBS symptoms, and/or dissatisfaction with air quality.

Ventilation systems which are contaminant sources may confound efforts designed to determine the effect of ventilation rates on SBS symptom prevalence. Ventilation is often deemed to be the solution to "the problem". In some cases, contaminants generated within ventilation systems may be "the problem". To date, this aspect of ventilation has received too little scientific attention.

Entrainment/Re-Entry/Cross-Contamination

Ventilation or HVAC systems were previously addressed as potential risk factors for SBS symptoms or occupant dissatisfaction with air quality from the perspective of inadequate outdoor ventilation rates or that the HVAC system and its components may be a source of contaminants which affect IAQ adversely. Ventilation systems may, in addition, serve to transmit and disperse contaminants produced (1) outdoors but entrained in building air supplies, (2) indoors but exhausted to the outdoor environment from where they may re-enter the building, and (3) in different parts of a building. These problems can be described as entrainment, re-entry, and cross-contamination.

Though representing only about 10% or so of cases reported by NIOSH investigators,[67-69] entrainment/re-entry/cross-contamination problems are, nevertheless, relatively common. They are among the easiest building-related health, odor, and nuisance problems to both identify and resolve. In general, they are not associated with what has been described as SBS.

A variety of entrainment, re-entry, and cross-contamination problems have been reported by NIOSH and other investigators.[125] These have included (1) motor vehicle exhausts drawn into outdoor air intakes or through elevator shafts from lower level parking garages, (2) local exhaust ducts or boiler chimneys vented nearby and upwind of outdoor air intakes, (3) failure of flue gases to rise due to strong negative pressures in the building, (4) construction and road work contamination of intake air, (5) gasoline fumes from leaking underground storage tanks and sewer gases, and (6) cross-contaminating cafeteria odors and laboratory solvents.

Entrainment/Re-Entry

Entrainment occurs whenever contaminants are drawn into and mix with building air. The contaminants may be produced externally or may be exhausted gases, vapors, or particulate matter which re-enter the building. Such contamination may occur from air drawn through outdoor air intakes as well as by infiltration, particularly when the building is under strong negative pressure.

Building air contamination by entrainment/re-entry occurs because of a variety of design and building operating problems. One of these is the placement of outdoor air intakes. Building designers often appear to be ignorant of the need to locate intakes properly to minimize contamination of outdoor air supplies. Common design deficiencies include locating outdoor air intakes at ground level close to loading docks, on roof tops nearby and directly upwind of boiler or laboratory fume hood exhausts, and nearby and upwind of cooling tower mist drifts. Building designers, unfortunately, place too high a priority on what they perceive as an efficient use of space and building appearance. In the latter case, it is not uncommon for boiler chimneys to be not much higher or even flush with the roofline.

Significant contamination of indoor air can occur when local ventilation is used to exhaust toxic gases or vapors in multipurpose laboratory, hospital, and industrial buildings.[126,127] Re-entry of exhaust gases results as a consequence of the close proximity of exhaust ports and outdoor air intakes, the low height of exhaust vents, low exhaust stack velocities, and ventilation system imbalances. Re-entry of contaminants may occur as a consequence of interactions between exhaust gases and the building wake, the air flow resulting from the physical character of the building, and the direction and speed of the wind. Since the building wake is not coupled with the free-flowing air above, contaminants entrained in it may receive little dilution before being drawn into the building through outdoor air intakes or by infiltration. The latter can be significant when there are imbalances between exhaust and intake air, a common situation where mechanical exhaust systems are used.[126] When such imbalances occur, buildings may be under strong negative pressure, increasing the likelihood of contaminant re-entry by infiltration. For both practical and aesthetic reasons, exhaust vents/stacks on large buildings are of short height (less than a meter). Exhaust vents covered by rain shields reduce the velocity of exhaust gases which, because of the design of these outlets, are virtually released flush with the roof surface. Both factors increase the probability that contaminants will re-enter a building.

Contaminants can also become entrained in building air as a result of negative pressures which are produced by the stack effect.[128] The stack effect is responsible for the entrainment of motor vehicle exhaust gases produced in parking decks located on the lower stories of large buildings. The elevator shaft is a major conduit for the upward flows of air induced by the stack effect.

The stack effect and system imbalances can also cause below-ground contaminants, such as gasoline fumes from leaking underground storage tanks and sewer gases, to be drawn up through dry drain traps. Sewer gas vapors may be limited to ammonia, methane, hydrogen sulfide, and other vapors common to human sewage, and/or they may include vapors of a variety of solvents which have been disposed of through the building's waste drainage system. The entry of sewer gases often results when changes in building space use patterns are made and drain traps are left unmaintained. Significant illness may result from the entrainment of sewer gas heavily contaminated with toxic solvents.

Cross-Contamination

One or more AHUs may be used in the conditioning of air in large buildings. Where multiple AHUs are employed, each unit is designed and operated to serve the needs of discrete building spaces. In many instances, a single AHU serves a building space which is physically separated from spaces served by other AHUs (e.g., AHUs serving different floors). In other instances, AHUs may serve building spaces which are not physically separated, e.g., three AHUs serve, respectively, the east, middle, and west sections of a four-story building. In such instances, communication between the air supplies of each of the three AHUs readily occurs. Though physically separated, communication may occur between air supplies of different AHUs as a consequence of air transmission through passageways, such as elevators and stairwells, as well as penetrations for utility connections. Communication between AHUs may become significant when differences in pressure occur between spaces served by different AHUs. The greater the pressure imbalance, the greater the potential for air flow to the space under the greatest negative pressure. When spaces differ significantly in their contaminant-generating potential (e.g., a printing shop and a clerical office), significant cross-contamination, and therefore complaints, may occur. This is readily apparent in buildings where cooking odors associated with cafeterias are sensed in many other building spaces.

RECAPITULATION

A variety of environmental factors such as temperature, relative humidity, air movement, lighting conditions, noise, vibration, air ions, and electrical/magnetic phenomena have the potential for affecting humans either directly or indirectly. Direct effects would most probably be of a physiological nature. Indirect effects may, on the other hand, include perceptions of thermal comfort and satisfaction/dissatisfaction with building environmental conditions which may influence the reporting of SBS symptom prevalence. They may also affect contaminant concentrations.

Results of building investigations, cross-sectional epidemiological studies, and experimental studies have implicated most of the environmental factors listed above as potential risk factors for SBS symptoms and/or satisfaction/dissatisfaction with air quality. Because of differences in study designs, populations surveyed, and statistical analyses employed, results of these studies are mixed, with some showing significant associations with specific environmental factors and others not.

Significant attention has been given to both temperature and relative humidity. Investigations with strong study designs indicate that increasing temperature from 21–25°C and humidity levels less than 30% are risk factors for SBS symptoms. The studies of air ions have been mixed, and effects of ionization on SBS symptoms are not conclusive. Studies of the effects of air movement, lighting, noise, vibration, electrostatic charges, and electric and magnetic phenomena have been very limited. The uncertainties, as a consequence, are too great to make inferences relative to their contributions to SBS prevalence rates. Reported associations between SBS symptoms and seasonal mood changes are quite interesting and are in need of further study.

Environmental control systems which regulate building thermal and ventilation conditions may be risk factors for problem-building complaints and SBS symptoms. Concerns have included (1) the inadequacy of outdoor air ventilation rates, (2) ventilation systems as a source of contaminants which may cause a deterioration in air quality and contribute to health complaints and dissatisfaction with air quality, and (3) ventilation systems serving as a means of contaminant transmission from sources located outdoors, exhausted contaminants which re-enter buildings, and sources located in different parts of buildings.

The adequacy of ventilation or the relationship between outdoor ventilation rates and SBS symptoms and occupant satisfaction/dissatisfaction with air quality has received considerable attention. With the exception of the Office Illness Project of Northern Sweden, large cross-sectional epidemiological studies have failed to show an association between SBS symptoms and outdoor ventilation rates. Experimental studies have, on the other hand, reported mixed results. An evaluation of investigations with strong study designs indicate that low outdoor air ventilation rates (<10 L/s/person) are risk factors for SBS symptoms and that ventilation rates greater than 10 L/s/person apparently have no effect on SBS symptom prevalence rates.

Both field investigations and research studies have implicated HVAC systems and their components as contributing factors to illness complaints, dissatisfaction with perceived air quality, and contamination of indoor air. Contaminants generated within HVAC systems include gas-/vapor-phase substances such as VOCs, MMMF and other particles, and a variety of contaminants of biological origin. The role of the HVAC system as a source of building contaminants has to date received relatively little attention.

Contamination problems and occupant health and nuisance odor complaints associated with entrainment of contaminants from outdoor sources, reentry of exhausted contaminants through air intakes and by infiltration, and the contamination of building spaces by substances generated in another space (cross-contamination) commonly occur in buildings and are easily identified and resolved. Ventilation systems represent a risk factor for such problems because they serve to transmit contaminants from their source or sources to building spaces where they may produce air quality problems.

REFERENCES

1. Fanger, P.O. 1989. "A New Comfort Equation for Indoor Air Quality." 251–254. In: *Proceedings of IAQ '89: The Human Equation: Health and Comfort.* American Society of Heating, Refrigerating and Air-Conditioning Engineers. Atlanta.
2. Clausen, G.H. 1988. Comfort and Environmental Tobacco Smoke." 267–274. In: *Proceedings of IAQ '88: Engineering Solutions to Indoor Air Problems.* American Society of Heating, Refrigerating and Air-Conditioning Engineers. Atlanta.
3. Woods, J.E. et al. 1987. "Office Workers' Perceptions of Indoor Air Quality Effects on Discomfort and Performance." 464–468. In: *Proceedings of the Fourth International Conference on Indoor Air Quality and Climate.* Vol. 2. West Berlin.
4. Fanger, P.O. 1984. "The Philosophy Behind a Comfort Standard." 91–98. In: *Proceedings of the Third International Conference on Indoor Air Quality and Climate.* Vol. 1. Stockholm.
5. Selfridge, O. J. et al. 1990. "Thermal Comfort Dissatisfaction Responses in the Library of Congress and Environmental Protection Agency Indoor Air Quality and Work Environment Study." 665–670. In: *Proceedings of the Fifth International Conference on Indoor Air Quality and Climate.* Vol. 4. Toronto.

6. Nelson, C.J. et al. 1990. "Environmental Protection Agency Indoor Air Quality and Work Environment Study: Health Symptoms and Comfort Concerns." 615–620. In: *Proceedings of the Fifth International Conference on Indoor Air Quality and Climate*. Vol. 4. Toronto.
7. Skov, P. and O. Valbjorn. 1987. "The Sick Building Syndrome in the Office Environment: The Danish Town Hall Study." *Environ. Int.* 13:339–349.
8. Skov, P. et al. 1990. "Influence of Indoor Climate on the Sick Building Syndrome in an Office Environment." *Scand. J. Work Environ. Health.* 16:363–371.
9. Skov, P. et al. 1990. "The Danish Town Hall Study—A One-Year Follow-Up." 787–791. In: *Proceedings of the Fifth International Conference on Indoor Air Quality and Climate*. Vol. 1. Toronto.
10. Jaakkola, J.J.K. et al. 1989. "Sick Building Syndrome, Sensation of Dryness and Thermal Comfort in Relation to Room Temperature in an Office Building: Need for Individual Control of Temperature." *Environ. Int.* 15:163–168.
11. Jaakkola, J.J.K. et al. 1991. "Mechanical Ventilation in Office Buildings and the Sick Building Syndrome. An Experimental and Epidemiological Study." *Indoor Air*. 2:111–121.
12. Reinikainen, L.M. and J.J. Jaakkola. 1993. "The Effect of Room Temperature on Symptoms and Perceived Indoor Air Quality in Office Workers. A Six-Week Longitudinal Study." 47–52. In: *Proceedings of the Sixth International Conference on Indoor Air Quality and Climate*. Vol. 1. Helsinki.
13. Tamblyn, R.M. et al. 1993. "Big Air Quality Complainers—Are Their Office Environments Different From Workers with No Complaints?" 133–138. In: *Proceedings of the Sixth International Conference on Indoor Air Quality and Climate*. Vol. 1. Helsinki.
14. Hawkins, L.H. and T. Wang. 1991. "The Office Environment and the Sick Building Syndrome." 365–370. In: *Proceedings of IAQ '91: Healthy Buildings*. American Society of Heating, Refrigerating and Air-Conditioning Engineers. Atlanta.
15. Zweers, T. et al. 1992. "Health and Indoor Climate Complaints of 7043 Office Workers in 61 Buildings in the Netherlands." *Indoor Air*. 2:127–136.
16. Norback, D. et al. 1990. "Volatile Organic Compounds, Respirable Dust, and Personal Factors Related to Prevalence and Incidence of Sick Building Syndrome in Primary Schools." *Br. J. Med.* 47:733–741.
17. Reinikainen, L.M. et al. 1991. "The Effect of Air Humidification on Different Symptoms in Office Workers—An Epidemiologic Study." *Environ. Int.* 17:243–250.
18. Reinikainen, L.M. et al. 1992. "The Effect of Air Humidification on Symptoms and Perception of Indoor Air Quality in Office Workers: A Six-Period Cross-over Trial." *Arch. Environ. Health.* 47:8–15.
19. Lundin, L. 1993. "Symptom Patterns and Air Quality in a Sick Library." 127–132. In: *Proceedings of the Sixth International Conference on Indoor Air Quality and Climate*. Vol. 1. Helsinki.
20. Reinikainen, L.M. 1993. "The Effect of Humidification on Perceived Indoor Air Quality Assessed by an Untrained Odor Panel." 101–106. In: *Proceedings of the Sixth International Conference on Indoor Air Quality and Climate*. Vol. 1. Helsinki.

21. Berglund, L. and W. Cain. 1989. "Perceived Air Quality and the Thermal Environment." 93–99. In: *Proceedings of IAQ '89: The Human Equation: Health and Comfort*. American Society of Heating, Refrigerating and Air-Conditioning Engineers. Atlanta.
22. Burge, S. et al. 1987. "Sick Building Syndrome: A Study of 4373 Office Workers." *Ann. Occup. Hyg.* 31:493–504.
23. Sundell, J. and T. Lindvall. 1993. "Indoor Air Humidity and the Sensation of Dryness as Risk Indicators of SBS." 405–410. In: *Proceedings of the Sixth International Conference on Indoor Air Quality and Climate*. Vol. 1. Helsinki.
24. Hodgson, M.J. et al. 1991. "Symptoms and Microenvironmental Measures in Nonproblem Buildings." *J. Occup. Med.* 33:527–533.
25. Green, G.H. 1984. "The Health Implications of the Level of Indoor Air Humidity." 71–78. In: *Proceedings of the Third International Conference Indoor Air Quality and Climate*. Vol. 1. Stockholm.
26. McIntyre, D.A. 1978. "Response to Atmospheric Humidity at Comfortable Air Temperature: A Comparison of Three Experiments." *Ann. Occup. Hyg.* 21:177–190.
27. Lubart, J. 1962. "The Common Cold and Humidity Imbalance." *N.Y. State J. Med.* 62:817–819.
28. Sale, C.S. 1971. "Humidification During the Cold Weather to Assist Perennial Allergic Rhinitis Patients." *Ann. Allergy*. 29:356–357.
29. Kay, D.L.C. et al. 1990. "Effects of Relative Humidity on Nonsmoker Response to Environmental Tobacco Smoke." 275–280. In: *Proceedings of the Fifth International Conference on Indoor Air Quality and Climate*. Vol. 1. Toronto.
30. Anderson, I.B. et al. 1974. "Human Response to 78 Hour Exposure to Dry Air." *Arch. Environ. Health*. 29:319–324.
31. Godish, T. and J. Rouch. 1986. "Mitigation of Residential Formaldehyde by Indoor Climate Control." *Am. Ind. Hyg. Assoc. J.* 47:792–797.
32. World Health Organization. 1987. "Air Quality Guidelines for Europe." WHO Regional Publications. European Series No. 23. Copenhagen.
33. Plog, B.A. (Ed.) 1988. *Fundamentals of Industrial Hygiene*. 3rd ed. National Safety Council, Chicago.
34. Rask, D.R. et al. 1990. "Environmental Stressors and System Deficiencies Identified in Problem Office Buildings." Presented at the 83rd Annual Meeting Air and Waste Management Association, Pittsburgh, PA.
35. Abbritti, G. et al. 1990. "Sick Building Syndrome: High Prevalence in a New Air-Conditioned Building." 513–518. In: *Proceedings of the Fifth International Conference on Indoor Air Quality and Climate*. Vol. 1. Toronto.
36. Landrus, G. and J. Larkin. 1990. "Cool-White, Warm-White, and Full-Spectrum Fluorescent Light Effects on Learning, Health, and Attitudes." 705–709. In: *Proceedings of the Fifth International Conference on Indoor Air Quality and Climate*. Vol. 1. Toronto.
37. Hodgson, M.J. et al. 1990. "Symptoms and the Microenvironment in Non-Problem Buildings." 549–554. In: *Proceedings of the Fifth International Conference on Indoor Air Quality and Climate*. Vol. 1. Toronto.
38. Robertson, A.S. et al. 1989. "Building Sickness, Are Symptoms Related to the Office Lighting." *Ann. Occup. Hyg.* 33:47–59.

39. Sterling, E. and T. Sterling. 1983. "The Impact of Different Ventilation Levels and Fluorescent Lighting Types on Building Illness: An Experimental Study." *Can. J. Pub. Health.* 74:385–392.
40. Alderfer, R. et al. 1993. "Seasonal Mood Changes and Building-Related Symptoms." 309–314. In: *Proceedings of the Sixth International Conference on Indoor Air Quality and Climate.* Vol. 1. Helsinki.
41. Sverdrup, C. et al. 1990. "A Comparative Study of Indoor Climate and Human Health in 74 Day Care Centers in Malmo, Sweden." 651–655. In: *Proceedings of the Fifth International Conference on Indoor Air Quality and Climate.* Vol. 1. Toronto.
42. Burge, P.S. et al. 1990. "Sick Building Syndrome—Environmental Comparisons of Sick and Healthy Buildings." 479–484. In: *Proceedings of the Fifth International Conference on Indoor Air Quality and Climate.* Vol. 1. Toronto.
43. Hodgson, M.J. et al. 1987. "Vibration as a Cause of 'Tight Building Syndrome' Symptoms." 449–453. In: *Proceedings of the Fourth International Conference on Indoor Air Quality and Climate.* Vol. 2. West Berlin.
44. Charry, J.M. 1984. "Biological Effects of Small Air Ions: A Review of Findings and Methods." *Environ. Res.* 34:351– 389.
45. Lehtimaki, M. and G. Graeffe. 1984. "Measurement of Air Ions." 187–191. In: *Proceedings of the Third International Conference on Indoor Air Quality and Climate.* Vol. 3. Stockholm.
46. Kreuger, A.P. 1980. "On Air Ions—And Your Health, Moods and Efficiency." *Executive Health.* XVII(2) November.
47. Hawkins, L. 1981. "The Influence of Air Ions, Temperature and Humidity on Subjective Well-Being and Comfort." *J. Environ. Psychol.* 1:279–292.
48. Hawkins, L.H. and L. Morriss. 1984. "Air Ions and the Sick Building Syndrome." 197–200. In: *Proceedings of the Third International Conference on Indoor Air Quality and Climate.* Vol. 3. Stockholm.
49. Davis, J.B. 1963. "Review of Scientific Information on the Effects of Ionized Air on Humans and Animals." *Aerospace Med.* 34:35–42.
50. Kreuger, A. 1973. "Are Negative Ions Good for You?." *New Scientist.* 58:668–670.
51. Kreuger, A. and E.J. Reed. 1976. "Biological Impact of Small Air Ions." *Science.* 193:1209–1213.
52. Finnegan, M.J. et al. 1987. "Negative Ions and the Sick Building Syndrome." 547–551. In: *Proceedings of the Fourth International Conference on Indoor Air Quality and Climate.* Vol. 2. West Berlin.
53. Daniell, W. et al. 1991. "Trial of a Negative Ion Generator Device in Remediating Problems Related to Indoor Air Quality." *J. Occup. Med.* 33:683–687.
54. Wyon, D.P. 1992. "Sick Buildings and the Experimental Approach." *Environ. Tech.* 13:313–322.
55. Wallach, C. 1984. "Video Display Health Hazard Safeguards." 169–174. In: *Proceedings of the Third International Conference on Indoor Air Quality and Climate.* Vol. 3. Stockholm.
56. Nylen, P. et al. 1984. "Physical and Chemical Environment at VDT Work Stations: Air Ions, Electrostatic Fields, and PCBs." 163–167. In: *Proceedings of the Third International Conference on Indoor Air Quality and Climate.* Vol. 3. Stockholm.

57. Goethe, C.J. et al. 1989. "Electrical Potential Differences Against the Surroundings and Discomforts in Indoor Environments." *Ann. Occup. Hyg.* 33:263–267.
58. Ancker, K. et al. 1984. "Electrostatic Charge in Office Environments." 157–162. In: *Proceedings of the Third International Conference on Indoor Air Quality and Climate*. Vol. 3. Stockholm.
59. Norback, D. and I. Michel. 1987. PI "Vilka Falstorer Kan Forklara Upplevelsen Av Besvare i 'Sjuka Hus'?" (Which Factors Can Explain the Experience of Discomforts in 'Sick Buildings'?). *Hygiea*. 96:140.
60. Michel, I. et al. 1989. "An Epidemiological Study of the Relation Between Fatigue, Dental Amalgam and Other Factors." *Swedish Dent. J.* 13:33–38.
61. Valbjorn, O. and P. Skov. 1987. "Influence of Indoor Climate on the Sick Building Syndrome Prevalence." 593–597. In: *Proceedings of the Fourth International Conference on Indoor Air Quality and Climate*. Vol. 2. West Berlin.
62. Sandstrom, M. et al. 1991. "The Office Illness Project in Northern Sweden. Electric and Magnetic Fields: A Case Referent Study Among VDT Workers." Investigation Report 1991:12. National Institute of Occupational Health. Umea, Sweden. (In Swedish-English Summary)
63. Sandstrom, M. et al. 1993. "The Office Illness Project in Northern Sweden—A Study of Offices with High and Low Prevalences of SBS: Electromagnetic Fields in our Indoor Environment." 303–307. In: *Proceedings of the Sixth International Conference on Indoor Air Quality and Climate*. Vol. 1. Helsinki.
64. Sundell, J. et al. 1990. "The Office Illness Project in Northern Sweden Part III. A Case-Referent Study of SBS in Relation to Building Characteristics and Ventilation." 633– 638. In: *Proceedings of the Fifth International Conference on Indoor Air Quality and Climate*. Vol. 4. Toronto.
65. Mendell, M.J. 1993. "Optimizing Research on Office Worker Symptoms: Recommendations from a Critical Review of the Literature." 713–718. In: *Proceedings of the Sixth International Conference on Indoor Air Quality and Climate*. Vol. 1. Helsinki.
66. Melius, J. et al. 1984. "Indoor Air Quality—The NIOSH Experience." *Ann. ACGIH: Evaluating Office Environmental Problems.* 10:3–8.
67. Wallingford, K.M. and J. Carpenter. 1986. "Field Experience Overview: Investigating Sources of Indoor Air Quality Problems in Office Buildings." 448–453. In: *Proceedings of IAQ '86: Managing Indoor Air for Health and Energy Conservation*. American Society of Heating, Refrigerating and Air-Conditioning Engineers. Atlanta.
68. Seitz, T. A. 1990. "NIOSH Indoor Air Quality Investigations 1971–1988." 163–171. In: Weekes, D.M. and R.B. Gammage (Eds.). *Proceedings of the Indoor Air Quality International Symposium: The Practitioner's Approach to Indoor Air Quality Investigations*. American Industrial Hygiene Association. Akron, OH.
69. Turner, S. and P.W.H. Binnie. 1990. "An Indoor Air Quality Investigation of Twenty-Six Swiss Office Buildings." *Env. Technol.* 11:303–314.
70. Turner, S. 1990. "Two Indoor Air Quality Investigations—Oceans Apart." Presented at the 83rd Annual Meeting of the Air and Waste Management Association, Pittsburgh.

71. Woods, J.E. 1988. "Recent Developments for Heating, Cooling and Ventilating Buildings: Trends for Assuring Healthy Buildings." 99–107. In: Berglund, B. and T. Lindvall (Eds.). *Proceedings of Healthy Buildings '88*. Swedish Building Research Institute. Stockholm.
72. Robertson, G. 1988. "Sources, Nature and Symptomatology of Indoor Air Pollutants." 507–516. In: Berglund, B. and T. Lindvall (Eds.). *Proceedings of Healthy Buildings '88*. Swedish Building Research Institute. Stockholm.
73. Woods, J.E. et al. 1988. "Indoor Air Quality and the Sick Building Syndrome: A View from the United States." 15–25. In: *Proceedings of Chartered Institute of Building Services Engineers. Advances in Air-Conditioning Conference*, London, Oct. 7.
74. Morey, P.R. and D.E. Shattuck. 1989. "Role of Ventilation in the Causation of Building-Associated Illnesses." 625–642. In: Cone, J.E. and M.J. Hodgson (Eds.). *Problem Buildings: Building-Associated Illness and the Sick Building Syndrome. Occupational Medicine: State of the Art Reviews*. Hanley and Belfus, Inc. Philadelphia.
75. National Research Council. 1981. *Indoor Air Pollutants*. National Academy Press. Washington, D.C.
76. Bell, S. and B, Khati. 1983. "Indoor Air Quality in Office Buildings." *Occup. Health in Ontario*. 4:103–118.
77. Rajhans, G. 1983. "Indoor Air Quality and CO_2 Levels." *Occup. Health in Ontario*. 4:160–167.
78. Sterling, T.D. et al. 1983. "Building Illness in the White Collar Workplace." *Int. J. Health Sci.* 13:277–287.
79. Finnegan, M.J. et al. 1984. "The Sick Building Syndrome: Prevalence Studies." *Br. Med. J.* 289:1573–1575.
80. Robertson, A.S. et al. 1985. "Comparison of Health Problems Related to Work and Environmental Measurements in Two Office Buildings with Different Ventilation Systems." *Br. Med. J.* 291: 373–376.
81. Harrison, J. et al. 1987. "The Sick Building Syndrome. Further Prevalence Studies and Investigations of Possible Causes." 487–491. In: *Proceedings of the Fourth International Conference on Indoor Air Quality and Climate*. Vol. 2. West Berlin.
82. Mendell, M.J. and A.H. Smith. 1990. "Consistent Patterns of Elevated Symptoms in Air-Conditioned Office Buildings: A Re-analysis of Epidemiologic Studies." *Am. J. Pub. Health*. 80:1193–1199.
83. Norback, D. et al. 1990. "Indoor Air Quality and Personal Factors Related to the Sick Building Syndrome." *Scand. J. Work Environ. Health*. 16:121–128.
84. Broder, I. et al. 1990. "Building-Related Discomfort is Associated With Perceived Rather than Measured Levels of Indoor Environmental Variables." 221–226 In: *Proceedings of the Fifth International Conference on Indoor Air Quality and Climate*. Vol. 1. Toronto.
85. Fanger, P.O. et al. 1988. "Air Pollution Sources in Offices and Assembly Halls Quantified by the Olf Unit." *Energy and Buildings*. 12:7–19.
86. Thorstensen, E. et al. 1990. "Air Pollution Sources and Indoor Air Quality in Schools." 531–536. In: *Proceedings of the Fifth International Conference on Indoor Air Quality and Climate*. Vol. 1. Toronto.

87. Salisbury, S.A. 1984. "A Typically Frustrating Building Investigation." *Ann. ACGIH: Evaluating Office Environmental Problems.* 10:129–136.
88. Hodgson, M.J. et al. 1990. "Allergic Tracheobronchitis in Alaska." 197–202. In: *Proceedings of the Fifth International Conference on Indoor Air Quality and Climate.* Vol. 1. Toronto.
89. Ruotsalainen, R. et al. 1993. "Ventilation Rate as a Determinant of Symptoms and Unpleasant Odors among Workers in Day Care Centers." 127–130. In: *Proceedings of the Sixth International Conference on Indoor Air Quality and Climate.* Vol. 5. Helsinki.
90. Sundell, J. et al. 1991. "Influence of the Type of Ventilation and Outdoor Air Flow Rate on the Prevalence Rate of SBS Symptoms." 85–89. In: *Post-Conference Proceedings of IAQ '91: Healthy Buildings.* American Society of Heating, Refrigerating and Air–Conditioning Engineers. Atlanta.
91. Nagda, N.L. et al. 1991. "Effect of Ventilation in a Healthy Building." 101–107. In: *Proceedings of IAQ '91: Healthy Buildings.* American Society of Heating, Refrigerating and Air-Conditioning Engineers. Atlanta.
92. Hanssen, S.F. 1993. "Increased Ventilation Reduces General Symptoms But Not Sensory Reactions." 33–38. In: *Proceedings of the Sixth International Conference on Indoor Air Quality and Climate.* Vol. 5. Helsinki.
93. Jaakkola, J.J.K. et al. 1993. "The Helsinki Office Environment Study: The Type of Ventilation System and 'Sick Building Syndrome'." 285–290. In: *Proceedings of the Sixth International Conference on Indoor Air Quality and Climate.* Vol. 1. Helsinki.
94. Jaakkola, J.J.K. et al. 1990. "The Effect of Air Recirculation on Symptoms and Environmental Complaints in Office Workers. A Double Blind, Four Period Cross-Over Study." 281–286. In: *Proceedings of the Fifth International Conference on Indoor Air Quality and Climate.* Vol. 1. Toronto.
95. Menzies, R.I. et al. 1990. "Sick Building Syndrome: The Effect of Changes in Ventilation on Symptom Prevalence: The Evaluation of a Double Blind Experimental Approach." 519– 524. In: *Proceedings of the Fifth International Conference on Indoor Air Quality and Climate.* Vol. 1. Toronto.
96. Menzies, R.I. et al. 1991. "The Effect of Varying Levels of Outdoor Ventilation on Symptoms of Sick Building Syndrome." 90–96. In: *Proceedings IAQ '91: Healthy Buildings.* American Society of Heating, Refrigerating and Air-Conditioning Engineers. Atlanta.
97. Menzies, R.I. et al. 1993. "The Effect of Varying Levels of Outdoor Air Supply on the Symptoms of Sick Building Syndrome." *New Eng. J. Med.* 328:821–827.
98. Palonen, J. and O. Seppanen. 1990. "Design Criteria for Central Ventilation and Air-Conditioning Systems of Offices in Cold Climate." 299–304. In: *Proceedings of the Fifth International Conference on Indoor Air Quality and Climate.* Vol. 4. Toronto.
99. Collett, C.W. et al. 1991. "The Impact of Increased Ventilation on Indoor Air Quality." 97–100. In: *Proceedings of IAQ '91: Healthy Buildings.* American Society of Heating, Refrigerating and Air-Conditioning Engineers. Atlanta.
100. Farant, J.P. et al. 1990. "Effect of Changes in the Operation of a Building's Ventilation System on Environmental Conditions at Individual Workstations in an Office Complex." 581–585. In: *Proceedings of the Fifth International Conference on Indoor Air Quality and Climate.* Vol. 1. Toronto.

101. Baldwin, M.E. and J.P. Farant. 1990. "Study of Volatile Organic Compounds in an Office Building at Different Stages of Occupancy." 665–670. In: *Proceedings of the Fifth International Conference on Indoor Air Quality and Climate.* Vol. 2. Toronto.
102. Mendell, M.J. 1993. "Nonspecific Symptoms in Office Workers: A Review and Summary of the Epidemiological Literature." National Institute of Occupational Safety and Health. *Indoor Air* 3:227–236.
103. Hedge, A. et al. 1989. "Work-Related Illness in Offices: A Proposed Model for the 'Sick Building Syndrome'." *Environ. Int.* 15:143–158.
104. Hills, B. et al. 1990. "Workplace Exposures to the Corrosion-Inhibiting Chemicals from a Steam Humidification System." *Appl. Occup. Environ. Hyg.* 5:672–673.
105. Godish, T. 1992. "HVAC System Materials as a Potential Source of Contaminants and Cause of Symptoms in a School Building." 208–217. In: *Proceedings of the Eleventh International Conference of the Clean Air Society of Australia and New Zealand.* Brisbane.
106. Pejtersen, J. et al. 1991. "Air Pollution Sources in Kindergartens." 221–224. In: *Proceedings of IAQ '91: Healthy Buildings.* American Society of Heating, Refrigerating and Air-Conditioning Engineers. Atlanta.
107. Molhave, L. and M. Thorsen. 1991. "A Model for Investigations of Ventilation Systems as Sources for Volatile Organic Compounds in Indoor Climate." *Atmos. Environ.* 25A:241–249.
108. Valbjorn, O. et al. 1990. "Dust in Ventilation Ducts." 361–364. In: *Proceedings of the Fifth International Conference on Indoor Air Quality and Climate.* Vol. 3. Toronto.
109. Pasanen, P. et al. 1990. "Emissions of Volatile Organic Compounds from Air-Conditioning Filters of Office Buildings." 183–186. In: *Proceedings of the Fifth International Conference on Indoor Air Quality and Climate.* Vol. 3. Toronto.
110. Berglund, B. et al. 1982. "The Influence of Ventilation on Indoor/Outdoor Air Contaminants in an Office Building." *Environ. Int.* 8:395–399.
111. National Aeronautics and Space Administration. 1991. "Flammability, Odor, Offgassing, and Compatibility Requirements and Test Procedures for Materials in Environments That Support Combustion." NHB 8060.1C.
112. U.S. Environmental Protection Agency. Method 5030A.
113. Wallace, L. et al. 1987. "Volatile Organic Chemicals in 10 Public Access Buildings." 188–192. In: *Proceedings of the Fourth International Conference on Indoor Air Quality and Climate.* Vol. 1. West Berlin.
114. Berglund, B. 1990. "The Role of Sensory Reactions as Guides for Nonindustrial Indoor Air Quality." 113–130. In: Weekes, D.M. and R.B. Gammage (Eds.). In: *Proceedings of the Indoor Air Quality International Symposium: The Practitioner's Approach to Indoor Air Quality Investigations.* American Industrial Hygiene Association. Akron, OH.
115. Sheldon, L.S. et al. 1988. "Indoor Air Quality in Public Buildings." Vol. 2. EPA 600/6–88/009b.
116. Godish, T. 1987. "Heating/Cooling System Odors." *Air-Conditioning, Heating and Refrigerating News*, May 11.
117. Cholak, J. and J. Shafer. 1971. "Erosion of Fibers from Installed Fibrous Glass Ducts." *Arch. Environ. Health* 22:220–225.

118. Balzer, J.L. et al. 1971. "Fibrous Glass-Lined Air Transmission Systems: An Assessment of Their Environmental Effects." *Am. Ind. Hyg. Assoc. J.* 32:512–518.
119. Gamboa, R.R. et al. 1988. "Data on Glass Fiber Contribution to the Supply Airstream from Fiberglass Ductliner and Fiberglass Ductboard." 25–33. In: *Proceedings of IAQ '88: Engineering Solutions to Indoor Air Problems*. American Society of Heating, Refrigerating and Air-Conditioning Engineers. Atlanta.
120. Hays, P.F. 1990. "Detection of Air Stream Fibers in Fibrous Glass Ductboard Systems." 635–640. In: *Proceedings of the Fifth International Conference on Indoor Air Quality and Climate*. Vol. 3. Toronto.
121. Schneider, T. 1986. "Man-Made Fibers and Other Fibers in the Air and in Settled Dust." *Environ. Int.* 12: 60–65.
122. Samimi, B.S. 1990. "Contaminated Air in a Multi-story Research Building Equipped with 100% Fresh Air Supply Ventilation Systems." 571–574. In: *Proceedings of the Fifth International Conference on Indoor Air Quality and Climate*. Vol. 4. Toronto.
123. Raw, G.J. 1993. "Indoor Surface Pollution: A Cause of Sick Building Syndrome." Building Research Establishment. United Kingdom.
124. Hedge, A. et al. 1993. "Effects of Man-Made Mineral Fibers in Settled Dust on Sick Building Syndrome in Air-Conditioned Offices." 291–296. In: *Proceedings of the Sixth International Conference on Indoor Air Quality and Climate*. Vol. 1. Helsinki.
125. Gorman, R.W. 1984. "Cross Contamination and Entrainment." In: *Ann. ACGIH: Evaluating Office Environmental Problems*. 10:115–120.
126. Reible, D.D. et al. 1985. "The Effect of the Return of Exhausted Building Air on Indoor Air Quality." 62–71. In: Walkinshaw, D.S. (Ed.). In: *Transactions: Indoor Air Quality in Cold Climates*. Air Pollution Control Association. Pittsburgh.
127. Wilson, D. J. 1985. "Ventilation Intake Contamination by Nearby Exhausts." 335–347. In: Walkinshaw, D.S. (Ed.). In: *Transactions: Indoor Air Quality in Cold Climates*. Air Pollution Control Association. Pittsburgh.
128. Tamura, G.T. and A.G. Wilson. 1967. "Pressure Differences Caused by Chimney Effect in Three High-Rise Buildings." *ASHRAE Trans.* 73. Pt. 2.

4 OFFICE MATERIALS, EQUIPMENT, AND FURNISHINGS

A variety of materials and equipment used in the work environment have been implicated as potential risk factors for illness symptoms. These include carbonless copy paper (CCP), other papers, office copy machines, video-display terminals and potentially other computer equipment.

In addition, materials used to furnish building interiors may contribute to SBS symptoms. Although emissions from a variety of building materials and furnishings may be responsible for sick building complaints, much research has focused on floor coverings, particularly textile materials, and their potential for contributing to SBS symptoms and indoor air contamination.

OFFICE MATERIALS AND EQUIPMENT

Carbonless Copy Paper

Carbonless copy paper has been extensively studied as a potential causal factor in office/work-related health complaints. These have included complaint investigations, human and animal exposure evaluations, cross-sectional epidemiological studies, and studies of CCP constituents potentially responsible for symptoms reportedly associated with its use by office workers.

Product Characterization

Carbonless copy paper is a pressure-sensitive material developed by the National Cash Register (NCR) Corporation of Dayton, Ohio, in the 1950s. It is often referred to as NCR paper, which both describes the company which invented it and its characteristic "no carbon required". The principle of CCP is to cause color formation on a copy sheet from chemical reactions between colorless color formers (contained in microcapsules) and a developer. In three-part CCP forms, the top sheet (CB) is coated on the backside with color formers contained within microcapsules; the middle sheet (CFB) is coated on the front

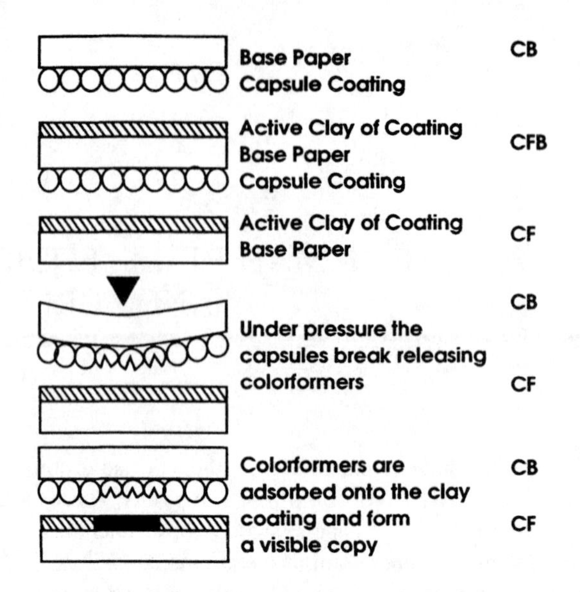

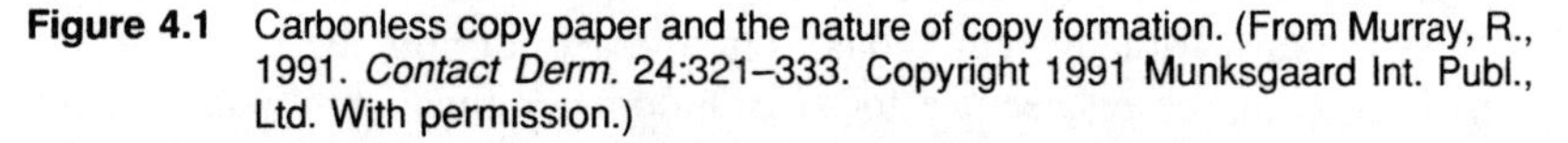

Figure 4.1 Carbonless copy paper and the nature of copy formation. (From Murray, R., 1991. *Contact Derm.* 24:321–333. Copyright 1991 Munksgaard Int. Publ., Ltd. With permission.)

with the color developer and on the back with color formers; and the third sheet (CF) is coated on the front with color developer (Figure 4.1). When CCP is struck by type or by the pressure of a pen or pencil, the microcapsules break, releasing color formers which come in contact with developing compounds on the pages beneath, producing a visible copy.[1]

The microcapsules are typically a few micrometers in diameter. They consist of an outer layer made of gelatin, gum arabic, or carboxymethyl cellulose which encapsulates oils which transfer a dye to the developer on the underlying sheet. A variety of oils and dilutents are used for the capsule oils including hydrogenated terphenyls mixed with aliphatic hydrocarbons, diaryl ethanes, alkyl napthalenes, chlorinated paraffins, and alkyl benzenes. Starch or other binders with additives such as fine cellulose powder are used to adhere the microcapsules to the base paper.

Color developers are based on clays, phenolic resins, or salts of aromatic carboxylic acids. Clay developers predominate in Europe; phenolic resins, in the United States and Japan. Clay types are made by activating bentonite, montmorillonite, attapulgite, zeolite, or synthetic products such as magnesium or aluminum silicate with a mineral acid. The developer coating contains one or more binders which are not reactive with the clay or other binders. Commonly used binding substances include styrene-butadiene rubber, acrylic latex, and water-soluble polymers such as carboxymethyl cellulose, polyvinyl acetate, and polyvinyl alcohol.

A number of color-forming chemicals have been used since the early development of CCP. These include crystal violet lactone, benzoyl leuco methylene blue, rhodamine blue lactom, paratoluene sulfonate (Michler's hydrol), indolyl red, phthalide red, phthalide violet, spirodipyranes, and fluorans. In printing business forms, commercial printers use conventional inks on the top form. Desensitizing inks are used to deactivate the developer coating to prevent the print on the top form from transferring to the lower sheets.

Complaint Investigations

A number of complaint investigations associated with CCP have been reported over the past decade and a half. Calnan[2] in the late 1970s and early 1980s investigated a number of episodes of symptom outbreaks in office workers which were apparently associated with the handling of large numbers of CCP forms in relatively small office environments. Reported symptoms included itchy rashes on the hands, swollen eyelids, headaches, burning throat and tongue, fatigue, excessive thirst, burning sensation on the face and forehead, backache, nausea, eye irritation, sore and dry burning lips, facial rash, and dry throat. Because of the pattern of symptoms in each episode, Calnan[2] concluded that some component of CCP was responsible. Since complaints were associated with used paper, the rupture of capsules containing color-forming chemicals was believed to be a significant factor in contributing to worker complaints. He concluded, however, that the color formers could not have caused the eye, nose, and throat symptoms and suggested that the responsible agent had to be associated with the encapsulated solvents.

Menne et al.[3] conducted investigations in response to complaints from employees of the Copenhagen Telephone Company and intensively evaluated 38 of 70 complaints reported in a 6-month period. Twenty-six individuals had only skin symptoms, nine skin and mucous membrane irritation, and three mucous membrane symptoms only. Skin symptoms were characterized as temporary redness with burning and itching that occurred within 2 to 3 hours after commencing work and resolving at night or over weekends. Skin symptoms were typically first observed on the hands. Fifty-five percent of individuals with skin symptoms had active eczema. Mucous membrane symptoms included itching of the eyes or nose, hoarseness and burning sensation in the mouth.

Menne and his co-workers[3] conducted a questionnaire study in response to employee reports of an apparent relationship between skin and mucous membrane complaints and intensity of work with CCP. Of 1855 employees surveyed, 208 or 10% reported skin and/or mucous membrane symptoms. A significant relationship between number of contacts with CCP and symptoms of rash, itching, and mucous membrane irritation was observed (Figure 4.2).

Exposure Investigations

A number of studies have been conducted in which investigators attempted to determine whether individuals would respond to CCP materials or

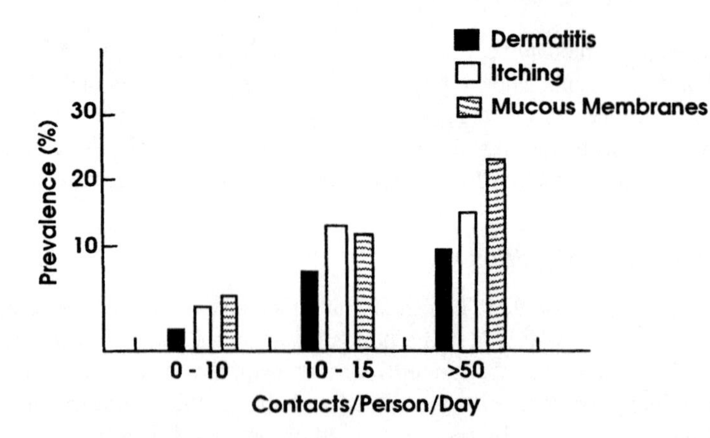

Figure 4.2 Relationship between carbonless copy paper exposure and employee symptom prevalence. (From Menne, T. et al., 1981. *Contact Derm.* 7:72–76. Copyright 1981 Munksgaard Int. Publ., Ltd. With permission.)

components in patch testing or in controlled dermal or inhalation exposures. The irritancy potential of CCP vapors has also been evaluated using mouse bioassays.

In addition to questionnaire studies, Menne et al.[3] conducted extensive dermatological testing including standard patch tests of 17 paper substances and with the paper itself. No positive reactions were observed for any CCP components. Relatively weak positive reactions were observed among 6 of 26 individuals pin-prick tested with one type of CCP and 5 of 26 with another. However, since other investigators conducted similar tests without positive results, Menne et al.[3] were uncertain as to the significance of their results.

Kleinman and Horstman[4] conducted a survey of employees who used CCP in 61 offices at the University of Washington. Most frequently reported complaints were headaches; dry, cracking, peeling skin; shortness of breath; burning eyes; dizziness; and sore irritated throat. A significant correlation was observed between the amount of CCP used and symptom prevalence. The rate of complaints was on the order of 10–15%.

Health complaints associated with CCP have been intensively studied in Sweden. As a consequence of these investigations, Goethe et al.[5] concluded that handling large quantities of CCP may induce dose-related and relatively benign nonallergic irritative symptoms particularly of the mucous membranes of the nose and mouth. However, this response was not specific to CCP; it could also be elicited by handling large quantities of ordinary bond paper albeit with much lower prevalence rates.

Jeansson et al., also in Sweden,[6] conducted patch-test evaluations of office workers whose symptoms were suspected to be caused by CCP exposure. One hundred and thirty-four individuals were patch tested with standard allergens and the CCP they used. An additional 50 individuals were tested with approximately 60 chemicals used in the production of CCP. There was no increase in the prevalence of formaldehyde allergy. Two individuals, however, developed weak reactions to two solvents (isoparrafins and kerosene) used in CCP. Two other individuals had probable allergic reactions to resin components (melamine formaldehyde and resorcin). Individuals who tested positive to either the solvents or resin components did not, however, test positive to the CCP they used. Two other individuals reacted to ordinary carbon paper. Because of patch-test results and histories of the complainants in this study, Jeansson et al.[6] concluded that there was no pattern of symptoms or complaints that could have linked them with working with CCP.

Marks[7] in the United States conducted an investigation involving a young female office worker who had a history of intermittent skin symptoms of the face and neck. Patch testing revealed a positive reaction to the color former, a key ingredient of which was paratoluene sulfonate of Michler's hydrol (PTSMH). The reaction was further identified to be associated with the Michler's hydrol component of the PTSMH molecule. Subsequent avoidance of CCP resulted in the complete resolution of symptoms. Murray[1] reports that PTSMH is no longer used in the manufacture of CCP.

Marks et al.[8] conducted a blind controlled challenge study of a young woman who experienced pruritus, eye and throat irritation, hoarseness, shortness of breath, and fatigue within a half hour of exposure to CCP. She developed contact urticaria of the hand that held the paper and changes in pulmonary function characteristic of upper airway obstruction on CCP challenge. Symptoms were associated with increased plasma prostaglandin levels.

La Marte et al.[9] have reported that alkylphenol novalac resin, a major ingredient in some CCP formulations, caused recurrent episodes of hoarseness, cough, flushing, pruritus, and rash within 30 minutes of a cutaneous challenge. Marked edema of the larynx and a sixfold increase in plasma histamine levels were observed. Exposed subjects experienced mast cell/basophil-mediated systemic adverse reactions to the alkylphenol novalac resin component of CCP, and because of the laryngeal edema, La Marte et al.[9] concluded that such reactions could be life- threatening.

Morgan and Camp[10] exposed 30 workers who had complained of prior sensitivity to vapors generated from CCP forms and to bond paper in a random single-blind fashion. A significant increase in nasal impedance (a measure of nasal congestion) on exposure to CCP vapors was observed. However, there were no differences in symptom reporting rates for exposures to CCP and plain bond paper.

Wolkoff et al.[11] conducted mouse bioassays of irritation effects associated with CCP emissions. The greatest degree of sensory irritation was associated with the CF layer of a CCP which had been previously reported to cause complaints. The CF layer contained the highest concentrations of free formaldehyde (219 mg/kg paper) as compared to the CB (17 mg/kg) and CFB layers (30 mg/kg). Sensory irritation was reported to be caused by trigeminal nerve stimulation.

Cross-Sectional Epidemiological Studies

Handling carbonless copy paper has been identified as a risk factor for SBS symptoms in the Danish Town Hall Study.[12] Danish investigators observed a significant relationship between the number of pages of CCP handled by office workers per day and the prevalence of mucous membrane and general symptoms. The use of CCP more than 1 hour per day was also observed to be a significant risk factor for eye, nose, and throat irritation; tight chest/difficulty breathing; and fatigue or sleepiness in the California Healthy Building Study.[13] Handling CCP without consideration for the level of exposure did not appear to be a risk factor for SBS symptoms in the Dutch Office Building Study.[14] There did appear to be a trend to increased oronasal and fever symptoms with CCP use.

Potential Causal Factors

Morgan and Camp[10] suggested that formaldehyde and a variety of volatile organic compounds (VOCs) were responsible for the significant nasal congestion associated with CCP exposures in their studies. Goeckel et al.[15] observed significant quantities of formaldehyde in CCP forms as determined by aqueous extraction and in samples collected by passing air through a burette packed with a standard quantity of CCP. Formaldehyde levels in 15 liter air samples varied from 0.33–0.72 ppm in previously unopened packages of CCP and 0.04–0.25 ppm from forms which had been removed from their original packages for some time. Air concentrations as high as 0.5 ppm were measured in filing cabinet drawers where CCP forms were being stored. They suggested that formaldehyde may have been one of the components of CCP responsible for causing eye, skin, and respiratory symptoms.

A potential causal role for formaldehyde has been suggested by NIOSH investigators[16] in a health hazard investigation conducted at the municipal court office in Englewood, Colorado. Eight employees reported complaints of eye, skin, and respiratory irritation. The pattern of symptoms reported by employees suggested that a cabinet containing CCP forms was the source of the problem. The various forms were analyzed and formaldehyde determined to be a significant component of each. Air concentrations varied from nondetectable to 0.04 ppm. Investigators felt that the close proximity of the CCP storage cabinet to a series of hot water boiler pipes would increase emissions from the stored CCP. Slightly higher formaldehyde levels were found in the cabinet than in area samples. In another investigation involving CCP, NIOSH investigators[17] were unable to detect formaldehyde in qualitative analyses of bulk samples. However, they were able to detect such substances as dibutyl phthalate, diethyl phthalate, and dioctyl adipate in both bulk samples and on the white cotton gloves employees were asked to wear while using CCP. In a study conducted by Norback and Goethe[18] in Sweden, no emission of formaldehyde could be demonstrated even during intense handling. In another study, "fresh"

CCP was observed to emit small detectable levels of formaldehyde; "old" CCP handled by individuals with irritative symptoms did not.[19]

In addition to formaldehyde, Wolkoff and Nielsen[20] reported that CCP emitted a variety of VOCs. These included monoalkylbenzenes (C_{16}-C_{18}), chlorinated paraffins (C_{13}-C_{15}), tetradecane, and dialkyl napthalenes. In the studies of La Marte et al.,[9] human reactions to CCP appeared to be associated with alkylphenol novolac which can be best described as a mixture of relatively small molecules of phenol-formaldehyde.

Headspace analyses of vapors emitted at 50°C from CCP were conducted by Molhave and Grunnet.[21] Over 42 different chemicals were identified with approximately 90% (C_6-C_{14}) alkanes and alkenes. Terphenyls in the capsule oils were suggested as the most likely cause of reported eye, skin, and respiratory irritation associated with CCP. Because of the low vapor pressure of terphenyl compounds, they postulated that exposure may occur as a consequence of adsorption on dust particles.

Other potential causes of symptoms reportedly associated with CCP exposure have been suggested. Menne and Hjorth[22] reported three cases of dermatitis of the palms and fingertips which appeared to be caused by frictional trauma associated with the handling of large quantities of CCP. Skin lesions were restricted to areas where maximum job-related friction occurred. These were characterized by redness, scaling, vesicles, and occasionally pustules. Conclusions were, in part, based on negative patch test results.

Representatives of manufacturers have claimed that the positive patch tests associated with CCP reported by Menne et al. [3] were not due to the paper itself but to desensitizing inks used by printers.[23] As previously described, commercial printers use desensitizing inks to prevent print placed on the front page of CCP forms from transferring to other pages. The studies of Norback et al.[24] also indicate that desensitizing inks may play a role in symptoms reported to be associated with CCP. They conducted studies which focused on what was described as the exposure factor (nature of exposure from CCP) as compared to those not exposed. The 276 samples of paper provided them by individuals complaining of symptoms were chemically analyzed and separated into 14 different brands and types. The greatest degree of complaints was associated with a CCP in which mono-iso-propyl biphenyl (MIPB) was used as a color former. They concluded, however, that the irritating factor was not likely to be MIPB but some other component which was also present. Skin symptoms were observed to be correlated with the type of desensitizing ink used; mucous membrane symptoms were not.

Reprise

A variety of studies have implicated CCP or some component of CCP as a causal factor in skin and/or mucous membrane irritation associated with handling CCP. A number of components have been suggested to cause either skin symptoms, mucous membrane symptoms, or both. These include a dye (PTSMH), formaldehyde, alkylphenol novolac, terphenyls, and desensitizing inks. Dermatological testing employing CCP and its components has, for the most part, resulted in negative results and as a consequence has not been particularly illuminating. One of the major difficulties in attempting to identify causal agents associated with CCP is that the product differs from manufacturer-to-manufacturer relative to its chemical components. This heterogeneity makes the problem more difficult to assess. In many cases, office workers may handle CCP forms from a variety of sources, particularly if copies of invoices, etc. originate from outside the building.

Adding to the uncertainty associated with identifying components of CCP which may be causal factors in CCP-related skin and upper respiratory symptoms are claims that such symptoms are at relatively low frequency in manufacturing and printing plants where exposures to airborne substances emanating from CCP would be expected to be higher. For example, representatives of CCP manufacturers[1,23] claim that similar symptoms are not reported among manufacturing employees. Unfortunately, no supporting data were provided. Norback and Goethe[18] have, however, reported significantly higher symptom prevalence in office workers who handled CCP as compared to workers in CCP printing plants.

Other Papers

Carbonless copy paper is not the only type of paper which has been reported to cause symptoms on exposure. Hjorth[25] described five patients with positive reactions to carbon papers, and Calnan and Connor[26] described a patient with dermatitis from nigrosine which was present in a special carbon paper used for computer work.

In studies of CCP, Morgan and Camp[8] exposed subjects to the emissions of both CCP and plain bond paper. Though they reported significant differences in nasal function in CCP-exposed subjects as compared to those exposed to plain bond paper, 30% of the latter group complained of symptoms which included congestion, dryness, cough, itching nose, dryness around the eyes, headache, and light-headedness. As described above, Goethe et al.[5] observed that handling large quantities of bond paper could induce symptoms associated with CCP but at lower prevalence rates.

Brooks and Davis[27] report that "green bar" computer paper and preprinted paper forms subjected to thermal fusing in laser printers and copiers are potentially significant sources of VOCs. Emissions from the latter have included acetaldehyde, acetic acid, acetone, acrolein, benzaldehyde, butanal, 1,5-dimethylcyclopentene, 2-ethyl furan, heptane, hexamethyl cyclosiloxane, hexanal, 4-hydroxy-4-methyl pentanone, isopropanol, pentanal, 2-pentyl furan, propionaldehyde, 1,1,1 trichloroethane, and paper dust. Emission rates of VOCs depended on the design, preparation, and curing of the preprinted forms, as well as the paper itself.

Extensive testing of emissions from electrostatically copied, laser-printed and matrix-printed paper were conducted by Wolkoff et al.[28] Results are summarized in Table 4.1. Since many of reported compounds were also found in the toner powder, they may have originated from both the paper and the toner. The most commonly observed compounds were benzene, 1-butanal, toluene, hexanal, 1-butyl ether, ethyl benzene, *m*-, *p*-, and *o*-xylene, styrene, 2-phenylpropane, 1-phenylpropane, ethyl toluene isomers, benzaldehyde, diethylbenzene isomers, and 2-ethylhexyl acrylate.

Office Equipment

Significant exposures to a variety of potentially irritating substances may occur in many buildings from commonly used office equipment. These include a variety of duplicators, electrophotographic printers, laser printers and copiers, microfiche and blueprint machines, signature machines, computers and video-display terminals (VDTs). Individual pieces of office equipment may represent significant point sources with exposures occurring to those working with the equipment and those nearby.

Evidence to implicate individual types of office equipment as potential contributing or risk factors for work/building-related health symptoms is extensive in some cases and limited in others. Such evidence includes complaint investigations, source characterization studies, cross-sectional and, in some instances, case-control epidemiological studies, and studies conducted in exposure chambers. Office equipment of concern includes wet-process photocopiers, electrostatic photocopiers, laser printers, diazo-photocopiers, microfilm copiers, spirit duplicating machines, and VDTs and computer equipment.

Wet-Process Photocopiers

Wet-process photocopy (WPC) machines have been reported to be a significant source of VOCs in Canadian office buildings.[29-31] In a study of 28 buildings, Tsuchiya and Stewart[31] observed that 18 had a VOCs fingerprint characteristic of WPC. In eight buildings, copier vapor accounted for 90% of TVOC concentrations. Hodgson et al.[32] reported that the dominant source of VOCs in a newly constructed Portland, Oregon, office building was liquid-process photocopiers and plotters which emitted a characteristic mixture of C_{10}-C_{11} isoparaffinic hydrocarbons.

Wet-process photocopiers typically employ two liquids, a dispersant and a toner. Tsuchiya and Stewart[31] have reported that the dispersant consists primarily of *n*-nonane and *n*-undecane, with the toner primarily *n*-decane and *n*-undecane.

Kerr and Sauer[33] studied emissions of VOCs from WPCs. Emission rates of 0.222–0.258 g/copy were similar to that reported by the manufacturer. Maximum TVOC concentrations of 3 g/m^3 in the exhausts of WPCs have been reported.[31]

Table 4.1 Volatile organic compounds emitted from photocopied (A-F), laser printed (G-I) and matrix printed (J,K) paper.

	Paper type										
Compound	**A**	**B**	**C**	**D**	**E**	**F**	**G**	**H**	**I**	**J**	**K**
Hexane	X										
1,1-Dicloro-1-nitroethane	X										
Benzene	X	X+	X+	X+	X+	X+	X	X+	X	X	X
Octene (isomer)	X										
Pentanal	X				+						
Trichloroethene	X										
1-Butanol	X+	X+		X	X+	X+	X+	X	X+		X
Toluene	X+	X+	X+	X+	X+		X+	X+	X		X
Pyridine					X+						
4-Methyl-2-pentanone		X+		X+							
Hexanal	X	X	X	X	X	X	X	X	X		X
C_4-Cyclohexane isomers			+	X+			+		+		
1-Butyl ether		X+		+	X	+	X+	X+	X+	X	X
m- and *p*-Xylene	X+	X+	X+	X+	X+	X+	X+	X+	X+	X	X
o-Xylene	X+	X+	X+	X+	X+	X+	X+	X+	X+	X	X
Styrene	X+	X+	X+	X+	X+	X+	X+	X+	X+	X	X
1-Butyl acrylate		X+			X+		X+				
2-Phenylpropane	X	X+	X	X+	X+	X+	X+	X+	X+	X	X
3-Heptanol			X+								
1-Phenylpropane	X	X+	X+	X+	X+	X+	X+	X+	X+	X	X
Ethyl toluene (isomers)	X	X+	X+	X+	X+	X	X+	X+	X+	X	
3-Ethoxy-3-ethyl-4,4, dimethylpentane		X+									
1-Butyl methacrylate	X+					+					
Benzaldehyde	X			X+	X+	X+		X+	X+	X	X
Diethylbenzene isomers			X	X			X+	X+	X		X
2-Ethyl-1-hexanol				X+	+		+	X+	X+		X
2-Ethylhexyl acetate			+	+				X+	+		
2,2-Azo-bis-isobutyronitrile	+	X+		X+					X+		
2-Ethylhexyl acrylate			X+	X+				X+	X+	X	X
Methylbiphenyl		+								X	

Note: x = detected in paper; + = detected in toner powder.

From Wolkoff, P. et al., 1993. *Indoor Air,* 3:118–123. With permission.

The high VOC emission rates of WPCs and the significant contribution of such copiers to TVOC concentrations in Canadian office buildings have led to suggestions that they are contributors to SBS.[30] There are, however, no published reports which link employee symptoms/health complaints to direct exposure to WPC emissions. The health concerns lie in the fact that VOC emissions may cause SBS symptoms because they contribute significantly to building TVOC levels, and in such buildings they may exceed those reported by Molhave and his co-workers[34] to cause mucous membrane and central nervous system effects in exposed individuals in the range of 5 and 25 mg/m^3, and symptoms at exposure concentrations less than 1 mg/m^3 proposed by Molhave.[35]

Greenwood[36] has suggested that since C_{10}-C_{15} isoparaffinic hydrocarbons used in WPCs have a very low order of acute toxicity, and office exposures are substantially below average industry occupational exposure standards (range

100–400 ppm), that emissions from WPCs are a relatively minor health concern. A guideline exposure value of 20 ppm was recommended based on the ASHRAE practice of setting indoor nonindustrial exposure guidelines at 1/10 the TLV and unpublished data which indicated that consumer complaints did not occur when equipment emissions were less than 50 ppm. This suggested guideline is also based on studies conducted by RICOH Corporation in which a panel of six individuals were exposed to isoparaffinic hydrocarbon emissions from their machines. Slight discomfort at 54 ppm and significant discomfort at 143 ppm were reported. Odor was imperceptible at 22 ppm, weak at 48 ppm, and clearly discernable at 108 ppm.[37]

Health complaints associated with WPC machines have been reported by Jensen and Rold-Petersen in Denmark.[38] In one case, several workers employed in a large room of a post office complained of itching erythema of the face, sore eyes, and headache. Similar complaints were reported in another post office building. Such complaints were typical in offices where similar copying machines were used.

Electrostatic Copying Machines

Electrostatic or xerographic copy machines have been used for over 40 years and are the most common reproduction machines found in office environments.

Because of voltages (5–10 kv) used to produce electrical charges and discharges of photoconductive material, electrostatic copiers produce ozone (O_3), a relatively toxic gas. Ozone emissions from photocopying devices have been reported by a number of investigators.[39-41]

Allen et al.[39] tested two electrostatic photocopy machines for O_3 emissions. Rates were observed to be in the range of 48–158 μg O_3/copy. Equilibrium O_3 concentrations in small, poorly ventilated rooms were in the range of 0.105–0.204 ppm. Ozone concentrations were reduced to background levels when photocopiers were serviced.

Selway et al.[40] measured O_3 emissions from ten electrostatic photocopying machines. Emission rates varied from <1 to 54 μg O_3/copy. Emissions were insufficient to cause O_3 levels in a small room to rise above background levels in over half the machines tested. Four machines produced sufficient O_3 to cause room air concentrations to rise to a range of 0.055–0.153 ppm.

Braun-Hansen and Andersen[41] studied O_3 emissions from 69 photocopying machines. Emission rates averaged 259 ± 302 μg/min with a range of 0–1350 μg/min. The maximum O_3 concentration in the breathing zone of 19 operators varied from 0.001–0.15 ppm. The TLV for O_3 is 0.10 ppm.[42]

Ozone emissions varied from copier to copier. Increases in voltage and electrical power (watts) were observed to result in increased O_3 production. Machines using alternating current (AC) were observed to produce twice the amount of O_3 as those using positive direct current (DC), with negative DC producing 10 times as much O_3 as positive DC.[41]

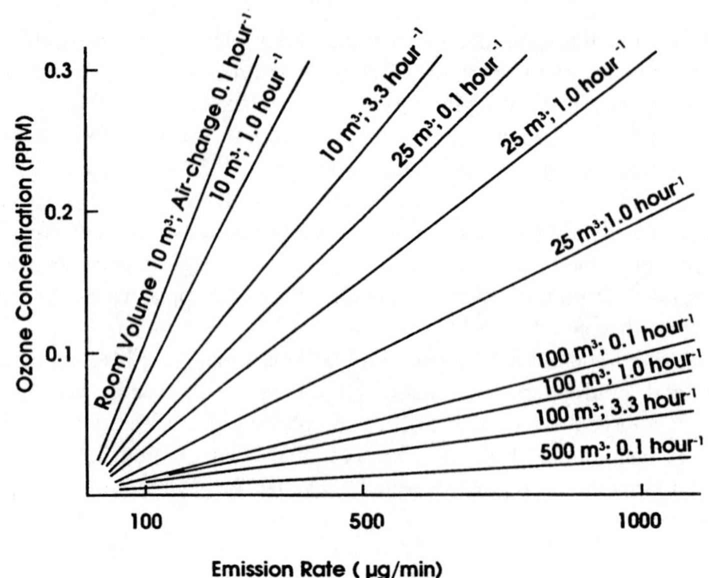

Figure 4.3 Relationship between room volume, air exchange rates, and ozone concentrations associated with electrostatic copier use. (From Braun-Hansen, T.B. and B. Andersen, 1986. *Am. Ind. Hyg. Assoc. J.*, 47:659–665. With permission.)

The actual exposures experienced by an operator depend on emission rates, O_3 decomposition rates, room volume, room air exchange, and the use and effectiveness of machine O_3 filters. The common placement of photocopiers in poorly ventilated rooms is a potentially significant factor in operator O_3 exposures. The effect of room volume and air exchange rate on O_3 concentrations can be seen in Figure 4.3. In general, the larger the machine, the higher the emission rate, and the greater the potential for human exposure.

Electrostatic copiers release other contaminants as well. Selenium, the most commonly used photoconductive material in electrostatic photocopy machines, has been measured in exhaust concentrations in the range of 0.01–0.6 μg/m^3.[43,44] In addition to selenium, cadmium sulfide, zinc oxide, and organic polymers are used as photoconductive material. Cadmium concentrations of 0.2 μg/m^3 have been reported from data obtained from Xerox Corporation and Minolta by Braun-Hansen and Andersen.[41] Studies by Braun-Hansen and Andersen,[41] however, revealed no detectable levels of either selenium or cadmium associated with photocopier use.

Emissions of toner powder can also occur from electrostatic photocopiers. The toner powder consists of carbon black which contains a resin (80–90% of the mixture) which adheres carbon black to paper. Toner powder particles are typically in the size range of 10–20 μm. Braun-Hansen and Andersen[41] have measured toner dust levels in exhaust channels of five different photocopiers. Dust concentrations were in the range of 0.09–0.46 mg/m^3 compared to normal office concentrations in the range of 0.05–0.5 mg/m^3.

Schnell et al.[45] conducted studies of black carbon emissions from a photocopier in a 20' × 15' workroom. They observed that the operating photocopier could raise carbon black levels from background values of 200–500 ng/m³ to 600-2000 ng/m³ within 10–15 minutes of use. Black carbon aerosol values of 2000 ng/m³ are similar to those observed in urban areas. Carbon black particles are deposited in the immediate vicinity of the photocopier because of their electrostatic charged nature.

A potential role of office photocopiers in contributing to SBS symptoms is reported in the Danish Town Hall Study.[12] Office workers who did photocopying work were at increased risk for work-related mucosal irritation and general symptoms. Photocopying more than 25 sheets of paper per week appeared to be a significant risk factor for mucous membrane symptoms. The type of photocopying machines were not identified but are here presumed to be electrostatic devices because these are the most commonly and widely used in Denmark.[41]

Laser Printers

Laser or electrophotographic printers also appear to have some potential for emitting contaminants and causing health effects. Laser printers employ a high voltage generator which charges the surface of a continuously rotating photoconducting drum.[45] The surface is exposed to a scanning laser beam which discharges the photoconductor surface selectively in accordance with the character matrix pattern to be printed. The toner, a black thermoplastic powder, is charged with an electrostatic potential and is attracted to the drum surface in the developer station. A negative image is produced on the drum surface. The paper is brought into contact with the drum surface, and the toner is transferred to it by means of a second high voltage generator. The image produced is made fast by means of heat or chemical agents. The toner has been described as consisting of 10% carbon black, 5% charge-control agent derived from diphenyl hydrazones, and 85% printing binder, a styrene-acrylate copolymer.[46] The source of air contaminants is suggested to be toner compounds subjected to thermal fusing. Laser printers which do not use heat fusion may employ a mixture of Freon and acetone. Such printers can produce elevated Freon and acetone levels in the breathing zone of operators. Sonnino and Povan[46] report acetone levels of 1–2 and 10 ppm in ventilated and poorly ventilated spaces, respectively, and Freon levels of 10–18 and 53 ppm, respectively, under the same conditions.

In addition to photocopier emissions, Schnell et al.[45] attempted to measure black carbon emissions from an operating laser printer. They were unable to detect any black carbon emissions, presumably because toner particles were so highly charged that they deposited out before they could be sampled and measured.

Brooks and Davis[27] describe case histories in which operators of electrophotographic printers complained of health symptoms. One case involved an outbreak of a facial rash. The problem appeared to be due to fugitive solvent emissions from vapor-fused printers. In the second case, employees complained of eye, nose, and throat irritation. Investigators concluded that symptoms were due to VOC emissions from preprinted forms.

Skoner et al.[47] reported a case history of an individual who experienced nasal congestion, headaches, diffuse retrosternal and epigastric discomfort after a computer system with laser printer was installed in his workstation. Challenge studies showed that symptom onset occurred after handling printer paper for 10 minutes and sitting next to an operating unit.

Toners

Braun-Hansen and Andersen[41] report that vapors associated with electrostatic photocopier toners produced both an unpleasant odor and a feeling of discomfort. These vapors were suggested to consist of residual monomers from toner resin and decomposition products including unstable ozonides, diperoxides, and epoxides, together with stable oxygenated hydrocarbons (lower aldehydes, ketones, and carboxylic acids).

Wolkoff et al.[28] studied emissions from six photocopier and three laser printer toners. VOCs identified included solvent residues (benzene, toluene, xylene, octene, C_4-cychohexanes, 1-butanol, butyl acetate, 2-ethoxy-ethanol), monomers (styrene and acrylate esters), monomer impurities (ethyl, propyl, and isopropyl benzenes and diphenyl butane isomers), coalescent agent (2,2,4 trimethyl-1,3 pentandiol monoisobutyrate, also known as Texanol), monomer or polymer oxidation products (benzaldehyde, acetophone). 4-phenycyclohexene (4-PC), a compound usually associated with styrene-butadiene latex bonded carpet, was observed in the volatiles of one toner. A total of 61 VOCs were identified in the nine toners evaluated.

Diazo-Photocopiers

Paper used in conjunction with a diazo-photocopying process was reported to cause eczematous skin symptoms on the face, neck, arms, and hands of an individual handling such paper routinely.[48] The contact skin sensitivity was determined to be a reaction to thiourea used as an antioxidant to prevent yellow discoloration in the diazo-photocopying process.

Microfilm Copiers

Skin symptoms have also been reported to be associated with microfilm copy paper. Tencati and Novey[49] reported a case of palpable purpura on the lower extremities of a librarian apparently associated with emissions of behemic acid from heat-activated photocopy paper associated with three machines designed to make copies of microfilm.

Spirit Duplicating Machines

A spirit duplicator is a device used to reproduce printed material by transferring an alcohol-based purple dye to duplicating paper. The duplicating process consists of placing a master (with a reverse image printed on it with an alcohol-soluble dye) on a drum. A thin layer of alcohol is applied to each duplicating sheet by means of an alcohol-saturated wick. As paper comes in contact with the master copy, the alcohol dissolves a small amount of the dye, transferring it to the finished paper. The "spirit" typically used is methyl alcohol.

Spirit duplicators do not provide a high-quality reproduction. As a consequence, they are not used in office environments. They do, however, receive considerable use in elementary and secondary schools.

NIOSH,[50] in 1980, conducted a health hazard evaluation of potential adverse effects of spirit duplicator use among teacher aides in a school district in the state of Washington. Teacher aides reportedly experienced significantly higher prevalence rates of blurred vision, headache, dizziness, and nausea compared to a less/nonexposed group of teachers. Concentrations of methyl alcohol in the breathing zone of exposed individuals varied from 365–3080 ppm. Fifteen of the 21 measurements of methyl alcohol exceeded the NIOSH-recommended exposure limit of 800 ppm. Based on symptom characteristics of methyl alcohol toxicity, 45% of 84 teacher aides appeared to have been adversely affected by exposure to spirit duplicator vapors. In another study, Kingley and Hirsch[51] reported headaches among spirit duplicator operators at methyl alcohol concentrations ranging from 200–375 ppm.

Video-Display Terminals/Computers

Working with VDTs or computers appears to pose a variety of health-related concerns. These include (1) SBS-type symptoms which may be related to chemical emissions or the electrostatic nature of such devices, (2) complaints of skin symptoms, (3) complaints related to job stress and the ergonomics of VDT work, (4) exposure to extra-low frequency (ELF) electromagnetic radiation, and (5) potential reproductive hazards. Issues related to the electrostatic nature of such devices have been discussed in Chapter 3.

SBS-Type Complaints. Brooks and Davis[27] have reported a case history in which health complaints appeared to be associated with a VDT. An employee reported experiencing eye, nose, and throat irritation shortly after the installation of a new computer. When the computer terminal was removed from the office, symptom severity diminished. The complaints were apparently associated with VOC emissions from the volatilization of residual solvents and adhesives from electronic components.

A large number of cross-sectional epidemiological studies have implicated working with VDTs or computers as a risk factor for SBS symptoms. Skov et al.,[12] in the Danish Town Hall Study, reported a significant association between working at a VDT (weekly or on a daily basis) and mucous membrane irritation and general symptoms. In a study of two office buildings in Great Britain, Harrison et al.[52] were not able to observe any difference in symptom complaint rates between users and non-users of VDTs. They observed, however, an excess of respiratory tract symptoms among workers in a room housing computers in one of the buildings compared to the rest of the building. Hedge et al.,[53] in a study of 4373 workers in 47 office buildings, observed a significant association between SBS symptoms and length of time employees worked with VDTs; those working with VDTs more than 6 hours/day reported more symptoms. They also reported more job stress, a factor highly correlated with SBS symptoms.

A significant association between hours worked at a VDT and sick building symptoms has been reported in the Dutch Office Building Study,[14] the USEPA Headquarters Building Study,[54] and a Canadian office building study.[55] No associations between VDT usage and SBS symptoms were observed in the California Healthy Building Study.[13]

VDT work has been shown to be significantly associated with increased inflammatory cells (polymorphonuclear neutrophils) in eye tear fluid of office workers in sick and reference buildings.[56] This is a significant observation because it is an objective measure of eye irritation. Computers and VDTs have been reported to emit a variety of substances. Because of high internal operating temperatures (circa 60°C), VOC emissions are very probable, particularly when products are new.[27] Emissions may occur from electronic components, adhesives, and plastic covers. Reported emissions from computers/VDTs are summarized in Table 4.2.

Skin Symptoms. A variety of skin symptoms have reportedly been associated with VDT work. Such reports have been primarily, but not exclusively, from the Scandinavian countries. Linden and Rolfsen[57] observed an itchy dermatitis of the face in ten patients coinciding with exposures to VDTs. Exposure to an electrostatic field provoked the rash in four patients. Nilsen[58] reported a transient itchy face rash in six healthy female operators of VDTs who associated their rash with VDT usage. They were free of the rash on weekends and holidays. Though circumstantial evidence implicated VDT use, provocation tests could not produce rash in patients or controls.

Liden and Wahlberg,[59] in an epidemiological study of office workers in Sweden, observed that individuals with rosacea, seborrhoeic dermatitis, and acne were over-represented in the VDT-exposed group. As a result, a pilot study was conducted to determine whether individuals with rosacea (a reddening of the cheek areas of the face) experienced any symptom aggravation as a result of VDT work. Those reporting skin symptom aggravation worked more than 4 hours/day at a VDT, while those reporting no aggravation had a lower VDT exposure (1–4 hours/day). Berg and Langlet[60] also reported a rosacea-like dermatitis in patients who attributed skin symptoms to working with VDTs.

Table 4.2 VOC emissions from VDTs/computers.

2,6-*Bis* (1,1 dimethyl)-4-methyl phenol	Heptadecane
n-Butanol	Hexanedioc acid
2-Butanone	4-Hydroxy benzalde hyde
2-Butoxyethanol	3-Methylene-2-pentanone
Butyl 2-methylpropyl phthalate	2-Methyl-2-propenoic acid
Caproloactum	Ozone
Cresol	Phenol
Decamethyl cyclopentasiloxane	Phosphoric acid
Diisoctyl phthalate	2-tert-Butylazo-2-methoxy-4-methyl-pentane
Dimethylbenzene	
Dodecamethyl cyclohexasiloxane	
2-Ethoxyethyl acetate	Toluene
Ethylbenzene	Xylene

From Brooks, B.O. and W.F. Davis, 1991. *Understanding Indoor Air Quality*. CRC Press, Boca Raton, FL. With permission.

Feldman et al.,[61] in the United States, reported a case of a white male who had a 15-month history of redness, burning, and itching of the dorsal areas of the hands and distal areas of the forearm. Skin symptoms developed after 2 weeks of working at a new job at a computer terminal. Skin symptoms developed on challenge response to the VDT with which he worked. A case of dermatitis in a 36-year-old woman who complained of tingling and reddening of her upper cheeks after 2–3 hours of exposure to her computer terminal has been reported by Fisher.[62]

Berg[63] examined 201 patients who claimed that their skin problems were associated with working with VDTs. In 18% of these patients,the condition was reported to improve overnight, with another 21% improving over weekends. In 25 patients, skin symptoms occurred primarily on the cheek turned toward the VDT. Fifty percent had relatively severe rosacea with pain, burning, and itching. Other patients were observed to have seborrhoeic eczema, acne vulgaris, or atopic dermatitis. An unusually high prevalence of migraine headache (40%) was observed among these individuals. After an 8-month follow-up period, 67% of the patients reported fewer skin complaints. An attempt to correlate skin problems with electrostatic fields created by the VDTs was not successful.

Knave et al.[64] conducted a study of 400 VDT workers and 150 selected control subjects. Females in the VDT-exposed group were observed to have had a significantly higher frequency of skin symptoms.

If skin symptoms are associated with VDT use, what are the likely causal factors? Initial speculation has focused on emissions of ultraviolet light (UV-A and UV-B) and the production of electrostatic fields. Provocation tests for both UV-A and UV-B conducted by Nilsen[58] were negative. Liden and Wahlberg[59] suggested that VDT emissions of UV-A and UV-B are too low to cause dermatological symptoms with exposures being less than 1% of Swedish occupational standards.

A potential causal role of UV emissions was initially considered because of the apparent phototoxic nature of some of the reported skin symptoms (rosacea) among VDT workers. Attention has also been focused on electrostatic charges because VDTs are known to create significant electrostatic fields. Linden and Rolfsen[57] reported that they were able to provoke characteristic skin symptoms by exposing patients to electrostatic fields. Other investigators,[60,63] however, were unable to reproduce the effect of electrostatic fields. This may have been due in part to the fact that shields used in their studies were not effective in reducing electrostatic fields.

Only very limited evidence is available to implicate chemical emissions from computers/VDTs in causing skin problems in VDT operators. Emission of skin irritants from computers and VDTs is probable. For example, Brooks and Davis[27] report emissions of cresol, a potent phototoxic substance. Reported rosacea-like symptoms are suggestive of a phototoxic effect. Epoxy resins have also been associated with phototoxic effects.[65-67] Because of their wide applications as adhesives and their bonding power, epoxy resins may be present in some computer equipment. Under the internal operating conditions of some computer components (circa 60°C), volatilization of residual substances from epoxy resins used as adhesives or in other components may occur.

Ergonomic Problems/Job Stress. The ergonomic nature and stressful aspects of VDT work have received considerable public health attention. Eyestrain, for example, is the most frequently reported health complaint among VDT users. Studies of visual discomfort associated with VDT use indicate that more than half (and up to 90% or more) of VDT operators experience some visual discomfort[68,69] which may include burning, itching, tiredness, aching, soreness, watering, blurred vision, and altered color perception.[70] Though eye-related complaints are common among clerical workers, there appear to be significantly more complaints among VDT users, particularly those using VDTs more than 4 hours/day, with intensive data entry generating more complaints than other VDT tasks.[71] Significantly higher frequencies of eye discomfort were reported among VDT workers in the case-control study of Knave et al.[64]

VDT operators report high frequencies of back, neck, shoulder, and general muscle discomfort. In one of the largest studies of its kind, NIOSH investigators observed that 64% of VDT operators experienced neck discomfort, 62% upper back discomfort, and 71% lower back discomfort, with a frequency of a few times a week or daily.[72] Causes of physical discomfort and muscular strain associated with VDT work have been suggested to include (1) repetitive work with static postures, (2) concentrated mental and visual effort, and (3) workstation design.[70]

Stress can be described as the response of the human body to the various physical, emotional, and environmental demands made on it. Continued or chronic stress has been linked with coronary heart disease, ulcers, psychological disorders, as well as accidents, substance abuse, and work absenteeism. Excessive stress in the workplace is common, with clerical workers experiencing high levels of stress compared to others in the same work environment. Stress levels among VDT operators may be high because of the repetitive nature of the work, job insecurity, negative feelings about computer monitoring, shiftwork, machine pacing, incentive pay schemes, increased workload, role ambiguities, reduced social interaction, and lack of opportunities for advancement.[70]

Rossignol et al.[73] conducted a study of 1545 Massachusetts clerical workers. Those who used VDTs 4 or more hours/day reported increased prevalence of vision problems compared to individuals not working with VDTs. The prevalence rate was dose-dependent, with those working with VDTs 7 or more hours/day having the most vision problems. Vision complaints included eyestrain/sore eyes, blurred vision, tearing/itching, burning/irritation, redness, and vision-associated absenteeism.

The prevalence rate for musculoskeletal conditions were observed to increase for high VDT usage (>4 hours/day). Musculoskeletal complaints included neck pain (stiffness, soreness), shoulder pain (stiffness, soreness), lower back pain, upper back pain, arm pain (stiffness, soreness), pain in hands, fingers, and wrists, numbness and tingling in hands, and missed work because of musculoskeletal problems. Rossignol et al.[73] concluded that the increased prevalence rates of musculoskeletal problems among VDT users probably resulted from the constrained posture and static loading of muscles associated with such work.

Several studies have attempted to evaluate whether VDT work increases the prevalence of headache. No significant relationship between VDT use and headache was observed by Knave et al.[64] A significant relationship was, however, reported by Rossignol et al.[73] Though these are mixed results, let us assume for the moment that VDT work does increase the prevalence of headaches in office environments. In such cases, it would be a potentially significant confounding variable in evaluating sick building complaints. In addition, the stressful nature of VDT work may increase symptom reporting rates.

Electromagnetic Radiation. Questions have been raised from time to time about the potential for VDTs to emit harmful electromagnetic radiation. The ionizing radiation potential of VDTs has received some attention because of reproductive health concerns. Several studies indicate that, with the exception of equipment malfunction, X-ray emissions from computer terminals are not above ambient radiation levels in the general area of the device being measured.[75,76]

VDTs emit non-ionizing radiation of various types. These include ultraviolet, visible light, infrared, microwaves, radio frequencies, and very-low frequency (VLF) and extremely-low frequency (ELF) electromagnetic radiation.

All electrical devices on activation produce electric and magnetic fields. VDTs produce characteristic fields associated with their cathode-ray tube electron beam acceleration and deflection circuitry. The strongest field produced is an electrostatic field that exists between the charged screen and nearby objects. Following this in strength is a magnetic field with a fundamental frequency in the ELF band of 60–70 Hz. Third is a weaker magnetic field in the frequency range of 15–22 Hz. The weakest field is a pulsating electric field in the VLF band.

Studies of electrical and magnetic field exposures at VDT operators' normal working position (approximately 50 cm from the screen) indicate that field strengths are well within the range of magnetic and electric fields in residential and workplace environments and well within occupational guidelines.[70,75,76] Nevertheless, there is a sense of unease about such exposures as increasing concern continues to develop that exposures to ELF electric and/or magnetic fields may have significant health effects since several studies have suggested a linkage between leukemia[77,78] and exposure to electromagnetic fields.

Reproductive Hazards. Several clusters of reproductive failure have been reported among VDT operators in the United States, Canada, and Western Europe in the past decade.[70] These clusters have given rise to concerns that VDT work poses unique reproductive hazards. A number of epidemiological studies have attempted to evaluate potential associations between VDT work during pregnancy and untoward reproductive outcomes such as spontaneous abortion and birth defects. Goldhaber et al.,[79] in a case-control study of 1583 pregnant women, observed a significantly higher risk of spontaneous abortion for women who reported using VDTs more than 20 hours/week during their first trimester of pregnancy; no significant elevation of risk for birth defects was observed. In contrast, other case-control studies have failed to observe any significant association between VDT usage during pregnancy and spontaneous abortion or birth defects.[80-82]

Reprise. Health and safety concerns associated with VDT usage in office environments are a subject of considerable controversy. While there is widespread agreement that working with VDTs causes discomfort problems, there is little agreement on other issues, particularly those associated with exposures to ELF and VLF electric and magnetic fields and with potential reproductive hazards. In the United States and Europe, VDT health and safety issues are a subject of union activism and, in the United States, political activism as well. In contrast to potential reproductive hazards, potential hazards associated with exposures to electric and magnetic fields, and ergonomic problems, the role of VDTs in SBS complaints is at present a minor concern. Nevertheless, a number of studies have indicated that increased hours of VDT usage are associated with higher SBS symptom reporting rates.

Human Exposure Studies

Wolkoff et al.[83] conducted studies in which humans were exposed to a simulated environment in a climatic chamber. The simulated office environment included three personal computers, a photocopying machine, and a laser printer. The photocopier and laser printer were selected for high O_3 and TVOC emissions and the largest range of VOCs.

Thirty female volunteers were exposed in groups of five in the simulated work environment and control chamber. In the former case, volunteers conducted routine clerical work using office equipment. Though air was circulated at a high rate, (60 ACH), the actual fresh air exchange rate was 0.5 ACH or about 5 L/s/person. The CO_2 level was a relatively high 3100 ppm. The O_3 concentration reached peak levels of 52 $\mu g/m^3$ (27 ppb). Formaldehyde levels were relatively high (95 ± 11 $\mu g/m^3$, 0.08 ± 0.008 ppm) and three times higher than the control chamber. TVOC concentrations adjusted for fragrances were relatively low, 82 $\mu g/m^3$ and 71 $\mu g/m^3$ in the simulated office and control chamber, respectively. There was a tendency toward higher total particle (50 ± 43 $\mu g/m^3$) and respirable particle concentrations (44 ± 27 $\mu g/m^3$) in the office climatic chamber compared to 29 ± 16 $\mu g/m^3$ and 27 ± 14 $\mu g/m^3$, respectively, in the control chamber without machines. The significant increase in formaldehyde concentrations in the simulated office environment had no known cause. Wolkoff et al.[83] speculated that it may have been associated with ozonolysis reactions of VOCs emitted from the toner powder, VOCs in air, or thermal decomposition of photocopy/printer paper.

Subjective assessments based on the use of questionnaires showed significantly increased perception of headache, mucous membrane irritation, and dryness of the eyes, nose, and throat and dry and tight facial skin in the simulated office chamber. Significant epithelial damage of eye conjunctiva was observed after exposure. This corresponded with perceived significant eye irritation and increase in decipol values. Though other objective (clinical) measurements were made, such as tear film quality index, nasal volumina, and nasal lavage fluid, no significant changes were observed. Wolkoff et al.[83] concluded that both subjective and objective symptoms could only be ascribed to the presence and use of office machines. Potential causes of observed eye irritation were suggested to be the integrated exposure of VOCs, O_3, formaldehyde, other irritants, the physical/chemical effect of exposure to particles, and the effect of working with VDTs.

A relationship between O_3, VOCs, and formaldehyde has been reported by Weschler et al.[84] who observed that even for moderate O_3 levels (55–86 $\mu g/m^3$; 28–44 ppb), selected VOCs were reduced by as much as 90% when exposed to O_3, with formaldehyde levels increasing by as much as threefold. In a study conducted in 29 buildings in Northern Sweden, Sundell et al.[85] observed what appeared to be an anomalous phenomenon; SBS symptom prevalence increased as TVOC concentrations decreased from intake to room air. Formaldehyde levels though low ($\bar{x}$ 31 μg/m^3, 25 ppb; range 11–59 μg/m^3, 9–48 ppb) were observed to be moderately but positively associated with SBS symptoms. Room formaldehyde levels showed a significant association with TVOC levels in intake and supply air but not with room air. This indicated that a decrease in TVOCs was associated with an increase in formaldehyde levels.

FLOOR COVERING

A variety of investigations have implicated floor covering as a potential cause of health complaints in specific problem building environments or risk factors for SBS symptoms in general. Floor coverings used to furnish office and institutional building environments typically are of two types, fabric carpeting and/or vinyl flooring. Both (1) are used in large quantities and have a large surface-to-volume ratio; (2) are for the most part composite materials, that is, they are made from a variety of substances which are responsible for their physical and chemical properties; (3) require fastening to a substratum by some physical means or by the use of a bonding agent; and (4) require periodic cleaning. Because of these factors, floor coverings can have a potentially significant effect on IAQ and possibly human health and comfort as well.

Effects may be either direct or indirect. In the former case, floor coverings may themselves release toxic contaminants which may affect the health and well-being of building occupants. In the latter case, a variety of air contamination problems can be anticipated to occur. These include toxic emissions from bonding agents used to apply floor coverings, shampoos and waxes used in cleaning, and reservoir effects involving VOCs and/or immunogenic macromolecular organic dust (MOD). Soiled carpeting may serve as a medium for microbiological growth and a source for air contamination by bacteria, fungi, and organic dust.

Carpeting

Wall-to-wall textile carpeting is a widely used and popular floor covering in North America and western Europe. Its popularity stems primarily from its aesthetic appeal. It is attractive and conveys a sense of warmth. Part of its appeal is acoustical; carpeting absorbs sound and reduces undesirable reverberation.

Carpeting is produced in a large variety of types or grades depending on desired applications including residential, office, retail establishments, and institutional. Non-residential types are described as commercial or industrial grades. They differ from residential carpets in terms of thickness of the pile (carpet fabric), fabric density, nature of the weave, backing materials, and means of application. Typically, commercial/industrial grade carpets have shorter piles, are more dense, and are applied to floor substrata by bonding agents. Commercial/industrial grade carpeting is, in general, subject to more intense wear and soiling than residential carpeting.

Though carpeting can be described as a generic product consisting of fibers of various length and density bound to backing materials, its composition varies widely. It may differ relative to the fiber-type employed (synthetic vs. natural fiber), backing materials, dyes and mordants, solvents for dyes, biocides, bonding agents, and the various formulations used for purposes of providing stain and soil resistance. Consequently, carpeting can be expected to vary considerably in terms of its potential emissions of chemical substances including residual monomers associated with synthetic fabric fibers and backing materials, solvents used in product manufacture, etc.

Complaint Investigations

Complaints of ill health associated with the installation of wall-to-wall carpeting have been relatively common in the United States. Most complaint reports have, however, been associated with residential dwellings. In the Federal Republic of Germany, general complaint rates associated with new carpeting increased from 56 in 1984 to 650 in 1989.[86] Reported symptoms included mucous membrane irritation, fatigue, runny eyes, and headache.

Illness symptoms associated with recarpeting of a Canadian office building has been reported by Kerr.[87] In a questionnaire survey, 60% of 124 respondents indicated that they experienced slight, moderate, or strong symptom responses. Major symptoms included headache, tiredness, eye, nose, and throat problems, and nausea. Symptoms were experienced for 1 to 3 days for lightly affected individuals; for 3 to 5 days for moderately affected individuals. Symptoms mostly resolved by the end of 7 days.

The best-known apparent outbreak of illness symptoms associated with the installation of new carpeting occurred in the USEPA headquarters building.[88,89] Numerous studies have been conducted on the building and carpets as a potential cause of building-related symptoms. The USEPA carpet initiative and dialogue[90] with carpet manufacturers were a direct result of that episode. Because of continued air quality complaints in the USEPA headquarters office complex, an extensive employee questionnaire and environmental monitoring study were conducted. In one statistical analysis, throat symptoms (sore throat, dry throat, hoarseness) were observed to be significantly associated with employees in areas where new carpet had been installed in the year-long period covered by the survey.[91] A suggestion of an apparent effect on dizziness in males was also noted. In another statistical analysis of the same data, Wallace et al.[92] reported significant associations between a number of symptom clusters including nasal symptoms (sinus congestion, runny nose, etc.); headache/nausea; chest tightness, wheezing, and shortness of breath; sore throat, dry throat, hoarseness; dizziness; and eye irritation and reports of new carpet odor.

No significant associations between symptoms and new carpet odor were observed for the Library of Congress building.

In another complaint-related investigation,[93] fatigue and malaise among 5 of 12 building occupants were reported to be associated with poorly cleaned wall-to-wall carpeting. After carpet removal, symptoms reportedly resolved.

Epidemiological Studies

Significant associations between SBS symptom prevalence and textile carpeting have been reported for five of six cross-sectional epidemiological studies. In the Danish Town Hall Study,[94] textile floor covering was observed to be related to mucous membrane but not general symptoms. In addition, Danish investigators observed significant associations between the area of fleecy surfaces (fleece factor) and the prevalence of both mucous membrane irritation and general symptoms. Carpeting was one of the major fleecy materials present.

In Swedish studies of primary schools,[95] an increased prevalence of eye and upper respiratory symptoms, face rashes, headache, abnormal tiredness, and a sense of being electrically charged was reported in buildings with wall-to-wall carpeting (compared to those with hard floor covering). On removal of carpeting, most of the reported symptoms (except airway symptoms) decreased to the level of buildings without carpeting. Additional studies[96] showed a relationship between carpeting and SBS symptoms, with carpet removal associated with a reduction in the number of symptoms.

A significant association between symptom prevalence and carpeting was reported in the Canadian studies of Menzies et al.[97] and in the California Healthy Building Study.[13] In the latter case, Fisk et al.[13] reported that the presence of carpet in study spaces was associated with eye, nose, and throat symptoms, tight chest or difficulty breathing, fatigue/sleepiness, and headache.

Hansen et al.[98] conducted a survey study of Danish asthmatic school children. Asthmatic children in schools with wall-to-wall carpeting were reported to have had significantly more severe asthmatic symptoms.

In contrast to the cross-sectional and survey studies reported above, no apparent association between SBS symptoms and carpeting was observed in the very large office building study (69 buildings) conducted in the Netherlands.[14]

Exposure Studies

In addition to complaint investigations and epidemiological studies, human and animal studies have indicated that exposure to carpeting materials may cause a variety of SBS symptoms.

Danish investigators[99] studied the effect of exposures to carpet emissions on human subjects under controlled laboratory chamber conditions. Significant

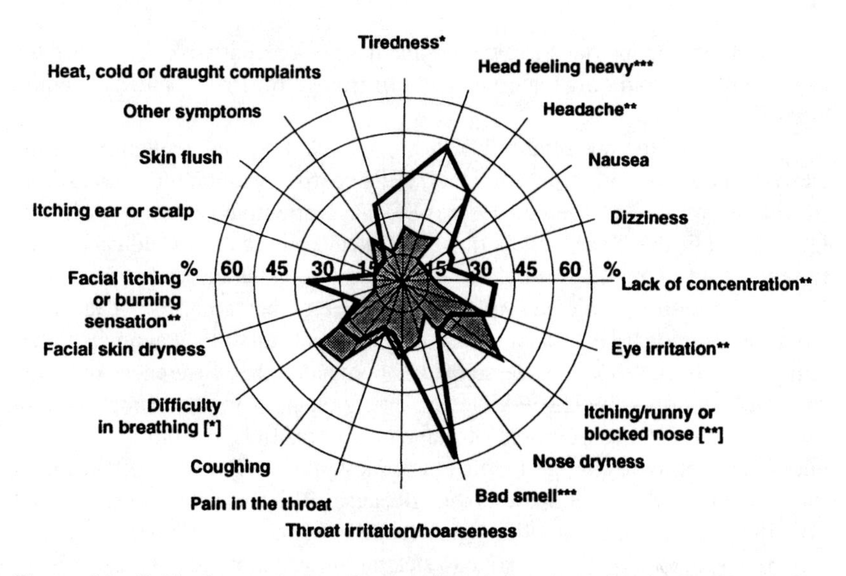

Figure 4.4 Subjective responses of test subjects before (cross-hatched area) and after exposure to carpet emissions. (From Johnsen, C.R. et al., 1991. *Indoor Air,* 1:377–388. Copyright 1991 Munksgaard Int. Publ., Ltd. With permission.)

symptom responses were reported by 20 asthmatic and 5 non-asthmatic subjects after a 6-hour exposure to nylon carpet with rubber backing. These included head feeling heavy, tiredness, eye irritation, itchy/runny or blocked nose, bad smell, and facial itching/burning sensation (Figure 4.4). Eye symptoms were observed in objective testing. In addition, subjects reported significant changes in general comfort and increased decipol values (a measure of satisfaction/dissatisfaction with IAQ). Carpet emission exposures did not provoke asthmatic attacks in individuals with hyperresponsive airways.

Anderson[100] reported significant irritant, pulmonary, and neurotoxic effects of carpet emissions using mouse bioassays. In a study of twelve different "complaint" carpet samples, Anderson[101] reported that half produced moderate to severe sensory effects in multiple 1-hour exposures, with two samples producing no sensory irritation. All 12 complaint carpet samples appeared to cause pulmonary irritation. Clinical observations included swollen faces, subcutaneous hemorrhages, altered posture, loss of balance, hypo- or hyperactivity, tremors, paralysis of one or more limbs, convulsions, and/or death. Apparently half of the animals died after the third or fourth 1-hour exposure.

Mouse bioassays were used by Wolkoff et al.[102] to test the irritancy of a variety of materials including carpeting. No significant sensory irritation was elicited in response to exposure to carpet samples tested.

Mouse bioassay studies of material emissions are based on the use of ASTM Method E 981-84[103] which determines the irritancy of chemicals by changes in animal respiratory rates. The technique has been widely accepted as a scientifically valid means of testing the sensory irritancy potential of a variety of toxic industrial chemicals[104] and has been recently recommended for use in screening building materials and furnishings for irritant chemical emissions.[105]

Because of the potential public health, regulatory, and economic significance of the Anderson carpet studies, USEPA, carpet manufacturers and Alarie (developer of ASTM Method E 981–84) have attempted to replicate them. USEPA studies[106-106a] as well as those conducted by the carpet industry failed to replicate Anderson's results. Bioassay studies conducted by Alarie[106a] confirmed the results reported by Anderson that carpet emissions can cause irritation, pulmonary effects, neurotoxicity and mortality albeit at much higher exposure concentrations. Anderson[107] has contended that differences in bioassay results reported by USEPA[106] were due to changes in the testing protocol made by USEPA for purposes of standardization. Indeed significant differences in bioassay testing protocols exist among the studies of USEPA, carpet manufacturers, Alarie and Anderson. Because of these differences, it is not possible to determine what the significance is of the initial bioassay studies reported by Anderson[100,101] relative to sick building symptoms. The disparity in results indicate that additional studies are necessary.

USEPA Investigations

The USEPA has conducted a number of studies to identify and quantify contaminants associated with textile carpeting in building air and in product emissions. These studies were stimulated in good measure by concerns associated with the USEPA headquarters building saga.

As an immediate consequence of that episode, USEPA technical staff conducted intensive investigations to identify carpet-related contaminants which may have been responsible for the outbreak of irritation-type symptoms among USEPA office employees. In addition, USEPA took the relatively unusual step of calling upon the American rug and carpet industry to undertake a voluntary effort to conduct periodic total VOCs (TVOCs) analyses on a company-by-company and product-by-product basis for the purposes of providing the public with comparative information on TVOC emissions.[90] USEPA also called for a public dialogue with the rug and carpet industry, consumer groups, federal agencies, etc., with the goal of characterizing emissions and identifying feasible VOC controls. Specifically excluded were any attempts to characterize the health effects of chemicals emitted from carpeting.

Singhvi et al.,[88] in response to the outbreaks of illness in the USEPA headquarters building in 1987, conducted a monitoring study of VOCs in the affected areas of the USEPA headquarters building. Sampled VOCs were all in the ppb range and were fairly typical of those reported in other environments. In response to the studies of Van Ert et al. at the University of Arizona[108] who reported that 4-phenylcyclohexene (4-PC), an incidental by-product of styrene-butadiene rubber (SBR) manufacture, was responsible for the noxious odor associated with carpets, Singhvi et al.[109] conducted a variety of 4-PC monitoring studies. These included sampling of air inside plastic wrapping around a new roll of carpeting stored in a warehouse, headspace analyses of carpet samples equilibrated in a chamber, solvent extraction of organic compounds from carpet samples, and monitoring of air concentrations of 4-PC in affected areas and in other buildings where carpeting was recently installed. SBR is a latex product used to bind fibers to carpet backing materials.

Air samples collected from inside the plastic wrapping of a new roll of carpeting (similar to that installed in the USEPA building) were observed to have a 4-PC concentration of 70 ppb. This carpet was made of nylon and had a secondary jute backing. In addition to 4-PC, a variety of other compounds were identified by headspace and solvent extraction analyses. These included 2,6 *bis*-(1,1-dimethylethyl)-4-methylphenol, phenol, xylenes, toluene, 1,1,1 trichlorethane, styrene, cumene, dichlorobenzene, alkylbenzenes, and isomers of ethyltoluene.[109]

Indoor air monitoring for 4-PC was conducted in USEPA headquarters offices at several different times between March and November, 1988. Those renovated within a six month period were observed to have 4-PC concentrations in the range of 0.03 to 6.65 ppb.[88] Three months later (in response to carpet aging and ventilation), the concentration of 4-PC decreased to 0.22 ppb in the space which previously had the highest concentration.

Carpet taken from the same batch was unrolled and aired in a warehouse.[88] This carpet produced 4-PC air concentrations of 4.9 ppb. After a month of airing, it was installed in office spaces in Edison, New Jersey. The initial concentration of 5.2 ppb 4-PC decreased to 0.31 ppb over a 6-month period. Seven additional carpeting materials were selected for analysis and air monitoring. Only those with jute secondary backings (as compared to foam rubber backings) were observed to emit 4-PC. Room 4-PC concentrations associated with the four carpet types ranged from 0.9 to 11.8 ppb. The 4-PC source was determined to be the SBR latex used to bind the jute backing to the primary backing of the carpet.

USEPA staff[89] conducted a risk assessment based on the known toxicity of 4-PC. They indicated that 4-PC was an irritant which, because of its similarity to other chemicals, could be expected to cause neurotoxic effects and possibly cancer. The toxicity of 4-PC was suggested to be similar to that of O_3. Based on a comparison of 4-PC levels measured in the USEPA headquarters building to occupational O_3 standards and ASHRAE IAQ guidelines, USEPA staff concluded that the reported levels posed minimal risks to employees. Because little scientific information was available on the toxicity of 4-PC, the risk analysis had a high degree of uncertainty. A comparison of 4-PC levels and O_3 standards/guidelines can be seen in Figure 4.5.

A variety of contaminant emissions were reported for five of the seven other carpet samples evaluated by Singhvi et al.[88] These included derivatives of cyclopropane, propanol and phthalates, styrene, acetic acid, benzaldehyde, trichloroethylene, tetrachloroethylene, xylenes, cyclohexane, naphthalene, methylethylketone, cresol, and a variety of alkanes and alkenes.

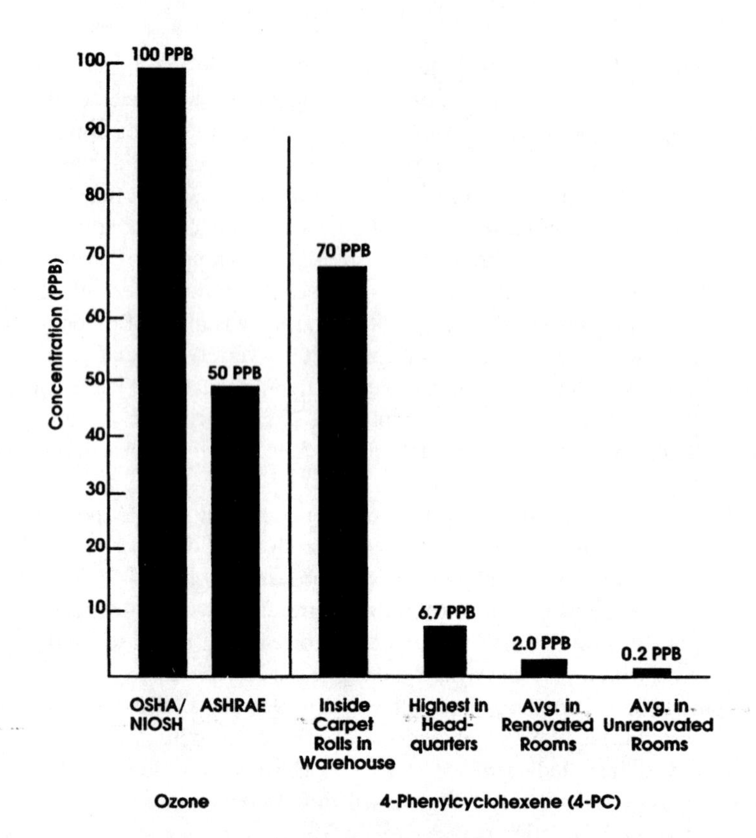

Figure 4.5 Comparison of 4-PC levels sssociated with USEPA headquarters carpeting and ozone standards/guidelines. (From Weitzman, D. and R. Singhvi, 1990. *Proceedings Indoor Air Quality International Symposium: The Practitioner's Approach to Indoor Air Quality Investigations.* With permission.)

Emission Studies — Carpet Materials

Molhave[110] conducted emission determinations of 42 commonly used building materials, including two types of felt carpet and two other types of textile floor covering. A total of 24 and 28 different VOCs were identified in felt carpet emissions and 28 and 61 VOCs from textile floor covering. Emissions of VOCs varied in the range of 1.95–3.15 mg/m^3 for three of the samples and was relatively high for the fourth (39.6 mg/m^3). Specific compounds were not reported.

Bayer and Papanicolopolous[111] conducted extensive environmental chamber testing of carpeting and other textile products to identify and quantify VOCs emitted. Those most frequently detected are summarized in Table 4.3. Other detected compounds included heterocyclic compounds such as furans and pyridines, sulfur-containing compounds such as dimethyldisulfide, and nitrogen-containing compounds such as ethanamines. Concentrations ranged

Table 4.3 VOCs detected in carpet emissions.

Benzene	Ethylmethylbenzenes
4-Phenylcyclohexene	Trimethylbenzenes
Ethanol	Chlorobenzene
Carbon disulfide	Chloroform
Acetone	Benzaldehyde
Ethyl acetate	Styrene
Ethylbenzene	Undecanes
Methylene chloride	Xylenes
Tetrachloroethene	Trichloroethene
Toluene	Phenol
1,1,1-Trichloroethane	Dimethylheptanes
1,2 Dichloroethane	Butyl benzyl phthalate
Hexanes	1,4-Dioxane
Octanal	Pentanal
Acetaldehyde	Methylcyclopentane
Methylcyclopentanol	Hexene

From Bayer, C.W. and C.D. Papanicolopolous, 1990. *Proceedings Fifth International Conference on Indoor Air Quality and Climate.* Vol. 3. Toronto.

from 10 ppt to 50 ppm, depending on the individual carpet tested. Measurements were made under dynamic laboratory chamber conditions of 75°F (24°C), 50% RH, and an air exchange rate of 0.5 ACH.

Intensive studies of TVOC and 4-PC emissions from carpet samples were conducted by Black.[112] TVOC emission rates of 0.071–0.315 mg/m^2 hour and 4-PC 0.04–0.152 mg/m^2 hour (at a chamber air exchange rate of 1.0 ACH) were reported for eight products tested. Emissions of TVOCs and 4-PC were observed in most cases to decrease exponentially with time. Average half-lives of 24 hours and 3 days were observed for TVOCs and 4-PC, respectively. Black,[112] as a consequence, concluded that emissions of VOCs from SBR latex-backed nylon carpets were very low and decreased rapidly with time. As such, carpets were determined to be significantly less potent emitters than other known VOC sources.

Black et al.[113] conducted more extensive studies of VOC and 4-PC emissions from SBR-backed carpets taken from manufacturers' finishing line. TVOC emissions from the 19 products tested under controlled chamber conditions decreased exponentially with an average decrease of 58 and 91% within 24 hours and 1 week, respectively. A typical decay pattern can be seen in Figure 4.6. Emissions of 4-PC were observed to peak within the first 24 hours, after which they were observed to decrease with time but more slowly than TVOC levels (Figure 4.7). Emissions of 4-PC decreased by 63% at the end of one week.

Emission Studies — Adhesives

In nonresidential environments, carpeting is almost always installed to the floor substratum by means of adhesives which vary in formulation relative to

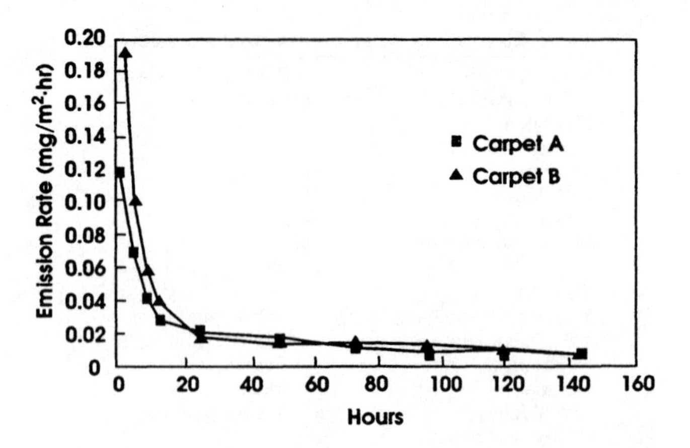

Figure 4.6 Typical time-dependent decrease of total volatile organic compound (TVOCs) emissions from carpet products. (From Black, M.S. et al., *Proceedings IAQ '91: Healthy Buildings.* With permission.)

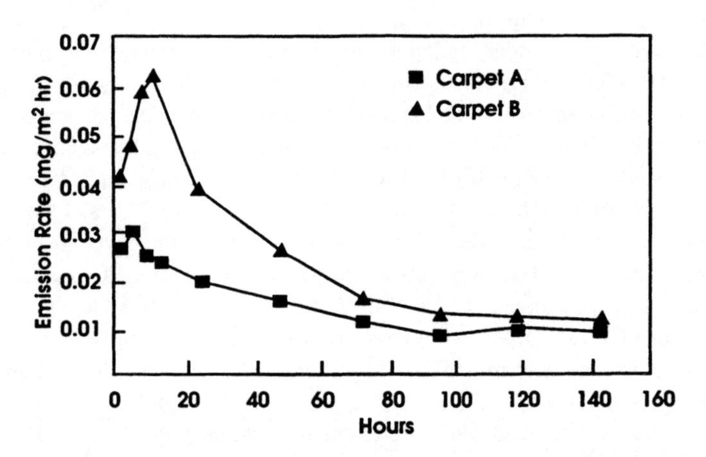

Figure 4.7 Time-dependent decreases in 4-PC emissions from carpet products. (From Black, M.S. et al., 1991. *Proceedings IAQ '91: Healthy Buildings.* With permission.)

the binding agents and solvents which vaporize from them on curing. These adhesives can be significant sources of VOCs. Black et al.[112] conducted emission testing of four latex carpet adhesives. Three of the four adhesives emitted TVOCs at levels which were several orders of magnitude higher than carpeting materials. Though TVOCs decreased exponentially with time, it required 35 days before adhesive TVOC levels were similar to those of SBR-backed carpeting after a period of 24 hours. The fourth, or low emission adhesive, emitted TVOCs in the same range of SBR-backed carpet with a similar rate of decay with time.

Emission Studies — Carpet Systems

Black et al.[113] conducted studies of TVOC emissions from carpet systems. Highest emission rates were observed with a single "glue down" with carpet and adhesive and a double "glue down" with carpet adhesive, cushion and adhesive. TVOC emissions (1000 times higher than non-adhesive systems) were primarily associated with the adhesive. High adhesive VOC emissions were also observed in the recarpeting study of Kerr.[87]

Davidson et al.[114] studied emissions from a new carpet installation using a special "low-emitting" VOCs adhesive in both laboratory chamber tests and an office complex. Both carpet and adhesive emitted very low VOC levels. The adhesive was observed to cure more rapidly when evaluated independently than when it was a component of the carpet assembly. The majority of VOCs detected were emitted by the adhesive. Ninety percent reduction in VOCs emission from the carpet assembly was predicted to occur in about 250 hours. The primary VOC present during the installation was 1,1,1 trichloroethane, a major component of the seam sealant. One week after carpet installation, TVOC levels were less in the carpet installation area than in the building control site.

Carpeting/Substrate Reactions

In addition to VOCs which appear to be associated with carpeting materials themselves, there is evidence that contaminants can be produced as a consequence of chemical reactions involving carpet components. McLauglin and Aigner[115] reported that calcium oxide in carpeting combined with water caused a hydrolysis of phthalate plasticizers producing higher alcohols such as 2-ethyl-1-hexanol and heptanol. Hydrolysis can also occur when plasticizers migrate to alkaline concrete surfaces. Hydrolysis reactions in polyvinyl chloride (PVC) backed modular-type carpets produced an intense odor on four floors of a Portland, Oregon, office building with the 2-ethyl-1-hexanol concentration ranging from 34–138 $\mu g/m^3$.

Sink Effects

VOC emissions decrease rapidly after carpeting is installed.[109,110] As a consequence, health complaints associated with carpet TVOC exposures are not expected to persist for more than a week or so.[87] In the building studies of Norback and Torgen in Sweden, carpeting had been in place for years, 8–10 years.[95] In such cases and in other reported associations between carpeting and symptoms in cross-sectional epidemiological studies, it is unlikely that emissions of VOCs initially reported for carpeting are the major contributing cause. In the Danish Town Hall Study,[116] the significance of the relationship between the prevalence of SBS and fleecy surfaces was suggested to be due to fleecy surfaces serving as a reservoir for the adsorption and release of VOCs. Fleecy surfaces such as carpeting have a large specific (total surface per weight or volume) surface area and, as such, have a large capacity for adsorbing different volatile substances.

Carpet-sink studies have been conducted by several investigators. Tichenor et al.[117] reported that carpet was a much stronger sink for tetrachloroethylene than other materials tested (e.g., wallboard, ceiling tile, and pillow). Borazzo et al.[118] observed that VOCs adsorption on carpet fibers depended on VOC volatility; the less volatile the species, the more strongly adsorbed.

MOD Reservoir and Microbial Amplification

In addition to VOCs, carpeting may serve as a reservoir for organic matter which provides a medium for microbial growth. A significant association has been reported for MOD extracted from carpeting and other horizontal surfaces and mucous membrane irritation and general symptoms in the Danish Town Hall Study.[119]

Because carpeting is subject to foot traffic, it has the potential for being soiled by various kinds of materials. Such soiling, and the deposition associated with it, can result in the accumulation of organic matter and growth of microorganisms. Several investigators have reported significant concentrations of bacteria, molds, and total proteins in wall-to-wall carpets as compared to hard floor covering.[119-123] In addition to the microbiological content of carpeting, somewhat higher levels of bacteria and mold have been reported in air samples in building spaces with carpeting as compared to those with hard floor covering.[122]

Electrostatic Shock Potential

A significant association between the presence of carpeting and the inconvenience of being electrostatically charged was observed in the studies of Norback and Torgen.[95] The phenomenon of electrostatic shock is well known to individuals who work in carpeted environments, particularly on cold dry days. What effect do electrostatic charges and other electrostatic fields have on human comfort and well-being? A variety of studies have sought to investigate the potential relationship between static electricity and SBS. These were previously discussed in Chapter 3.

Carpet Cleaning Agents

In addition to factors directly related to carpeting such as VOCs, potential VOC reservoir effects, MOD and microorganisms, other carpet-related factors may cause or contribute to building-related health complaints. Among these are carpet cleaning agents. Shampoos used in normal or excessive quantities have been reported to cause allergic contact dermatitis[124] and respiratory irritation.[125] Freshly cleaned carpets were reported to be associated with Kawasaki Syndrome, a disease affecting children.[126] This initial association, however, could not be confirmed by other investigators.[127]

Vinyl Floor Covering

Vinyl floor covering has, in comparison to textile carpeting, received little attention relative to its potential to cause both air contamination and building-related complaints. It has not been assessed as a potential risk factor in cross-sectional epidemiological studies. Interest in potential health effects has resulted primarily from complaint investigations in the Nordic countries.

Complaint Investigations

Rittfeldt and colleagues[128] in Sweden have proposed that vinyl floor coverings may contribute to "temporary sick buildings" due to emissions of benzyl and benzal chloride associated with the plasticizer butyl benzyl phthalate.

Gustafsson[129] reports that a large number of cases of unpleasant odor have been associated with vinyl floor covering in Sweden. Odor was associated with the formation of higher alcohols (e.g., 2-ethylhexanol) produced from the hydrolysis of phthalate plasticizers in contact with alkaline concrete in the presence of moisture. The use of self-leveling cement was apparently associated with a number of these cases of plasticizer degradation.[130-132] In one case, decomposition produced not only 2-ethylhexanol but phenol and cresol as well.[133] Decomposition of vinyl plasticizers to odorous ethylhexyl acrylate[134] and octonol has also been reported.

Exposure Studies

Johnsen et al.[99] and Wolkoff et al.[102] exposed human subjects to emissions from rubber (vinyl) floor covering under controlled chamber conditions. They reported significant increases in general discomfort, decipol ratings, and symptom prevalence including tiredness, head feeling heavy, headache, lack of concentration, eye irritation, itchy/runny or blocked nose, bad smell, difficulty breathing, and facial itching or burning sensation.[99] Significant objective symptoms of eye irritation were also observed. Exposure chamber TVOC levels were 1.923 mg/m^3. No effects on sensory irritation were observed in mouse bioassays.[102]

Indoor Air Contaminants

A variety of substances emitted by vinyl floor covering have been measured in building environments. These have included, in addition to those previously described in complaint investigations, styrene,[135] 2,2,4-trimethyl-1,3 pentanediol diisobutyrate, "TXIB",[136] dodecylbenzene,[137] dodecene,[137] and methylphenol.[137] Compounds such as dodecene,[137] TXIB,[136] and ethylhexyl acrylate[132] have been observed at levels in excess of 50 μg/m^3 several years after installation. Dodecene appears to be a decomposition product of dodecylbenzene formed in the manufacture of vinyl flooring.[137] Ethylhexyl acrylate is produced when a surface lacquer layer is incompletely cured. Emissions of higher alcohols from vinyl floor covering occurs when the underlying concrete has not been allowed to dry properly.[138] Phenol and methylphenol (*p*-cresol) are emitted from vinyl flooring containing phosphor plasticizers as a result of moisture degradation.[137]

Linoleum, though technically not a vinyl product, is nevertheless a source of emissions. Because of its content of cured linseed oil, linoleum can produce maladors as a result of fatty acid emissions.[137] Linoleic and unsaturated fatty acids can form low molecular weight aldehydes. Floor tiles based on cork and phenolic resin can emit phenol to indoor air.[137]

Plasticizers

Vinyl floor covering often contains a significant concentration (range 25–50%) of plasticizers which makes them soft and pliable. The most common plasticizers are phthalate esters such as diethylhexylphthalate (DEHP). As described previously, DEHP may hydrolyze under alkaline floor substrate and moisture conditions to produce ethylhexanol.[129,138]

Because of their relatively large molecular weights (>200 mass units) and boiling points (>250°C), phthalate ester plasticizers have very low vapor pressures and are solids at room temperatures. Nevertheless, measurable concentrations of dibutyl phthalate (DBP) and diisobutyl phthalate (DiBP) have been reported in buildings. These have been in the range of 2–16 μg/m^3 for DBP and 3–32 μg/m^3 for DiBP.[139] These exposure concentrations have been suggested to cause toxic effects on house plants known as the "white leaves" phenomenon.[140]

Rosell[136] has reported significant levels of 2,2,4-trimethyl-1,3 pentanediol diisobutyrate (TXIB) in the air of 8 so-called sick buildings (2–7 years old) in concentrations ranging from 100–1000 μg/m^3. TXIB is a common plasticizer in PVC floor covering where it comprises 7–8% of product weight. TXIB is semi-volatile with a molecular weight of 286 and a boiling point of 280°C. Based on toxicity studies conducted by its manufacturer (Eastman Kodak), it has a very low order of toxicity, is a slight skin irritant, and is not a sensitizer.[141]

Emission Testing

In addition to benzyl and benzal chloride, Rittfeldt et al.[128] observed a variety of other compounds being emitted from vinyl floor tiles determined by headspace analysis. They included significant emissions of 2-butanone, toluene, trichloroethane, cyclohexanone, and 1- (2-butoxiethoxi) ethanol with lesser concentrations of butanol, hexanol, benzaldehyde, phenol, benzyl butyl ether, and decanol.

Saarela and Sundell[142] reported results of emission testing of a variety of flooring materials including unused PVC and PVC material associated with complaint environments (Table 4.4). Emissions were observed to vary widely. In some cases, PVC floor coverings were strong emitters of substances which decreased relatively slowly with time.

RECAPITULATION

Various office materials and equipment have been implicated as potential contributing factors to work- or building-related health complaints. Both complaint investigations and cross-sectional epidemiological studies have contributed to our understanding of the health-affecting or contaminant-generating potential of CCP, other papers, copying machines, and VDTs/computer equipment.

Handling CCP appears to be a risk factor for mucous membrane and, in some cases, skin symptoms. Human patch-testing studies have, for the most part, failed to identify causal substances. Exposure studies indicate that CCP contains potent sensory irritants.

Working with office copy machines such as wet-process photocopiers, electrostatic photocopiers, laser printers, diazo-copiers, microfilm copiers, and spirit duplicators has been implicated as a causal/contributing factor to worker symptoms in a number of complaint investigations. Attention has been focused on wet-process photocopiers because of their significant emissions of VOCs and electrostatic and laser printers because of their widespread use and potential to emit O_3 and a variety of emissions from toner powders. Working with office copying machines has been identified as a risk factor for SBS symptoms in the Danish Town Hall Study and in controlled chamber human exposure studies.

Working with VDTs represents a number of health concerns. These include SBS symptoms, skin symptoms, ergonomic problems and stress, exposure to electromagnetic radiation, and potential reproductive hazards. Working with VDTs a minimum number of 6 hours/day has been shown to be a risk factor for SBS symptoms in most epidemiological studies conducted. However, working with VDTs may also cause a variety of ergonomic problems such as eyestrain, headache, neuromuscular problems, and stress that could

Table 4.4 TVOC emission rates from various flooring materials.

Material	Age (years)	History	Emission rate mg/m²/h
Linoleum	30.0	Subject to complaint	0.064
Cork	0.3	New material	0.805
Cork	2.0	Subject to complaint	0.007
English PVC	1.0	Unused	1.122
Finnish PVC	<0.5	Unused	1.443
Finnish PVC	0.5	Unused	2.192
Finnish PVC	1.0	Unused	1.629
Finnish PVC	2–3.0	Subject to complaint	0.273
Central Europe PVC	1–3.0	Subject to complaint	7.034
Swedish PVC	1–2.0	Subject to complaint	0.910

From Saarela, K. and J. Sundell, 1991. *Proceedings IAQ '91: Healthy Buildings.* With permission.

contribute to some of the same symptoms which characterize SBS and result in increased SBS reporting rates. The role of VDTs in causing skin symptoms has yet to be defined, and risks from exposure to electromagnetic radiation and reproductive hazards have not been established.

Floor covering has been identified as a potentially significant risk factor for building-related health complaints and source of indoor contaminants. Particular concern has focused on textile floor coverings, that is, carpeting. Five of six cross-sectional epidemiological studies have shown significant associations between SBS symptom prevalence rates and the presence of carpet, and a survey of asthmatics in schools indicates that the presence of carpeting is a risk factor for increased symptom severity. Exposure studies indicate that textile floor covering emissions can cause SBS-type symptoms in humans. Mouse bioassay studies have shown mixed results. Results reported by a private laboratory indicate that carpet emissions may cause significant sensory irritation, pulmonary irritation, and neurotoxic effects. USEPA efforts to replicate these studies by the USEPA and carpet manufacturers have not been able to confirm the results of mouse bioassay studies.

Carpeting has been implicated as an indirect contributor to building occupant health complaints. Carpeting appears to serve as a reservoir for MOD which has been shown to be associated with SBS symptoms in the Danish Town Hall Study.

Because of the outbreak of health and odor complaints associated with the installation of new carpeting in the USEPA headquarters building, a number of source characterization studies of carpeting, adhesives, and carpet system assemblies have been conducted. Emission studies have shown that these materials are initially a significant source of a large number of VOCs that decrease rapidly with time. Highest concentrations of TVOC emissions are associated with carpet adhesives. Rapid decreases in TVOC emissions from carpeting materials suggest that such emissions should not pose significant long-term contamination problems. Carpeting, however, has a significant potential to serve as a sink for a variety of VOCs and organic dust.

Significant emissions of VOCs and semivolatile VOCs have been reported for vinyl and related types of floor covering. Potential causal relationships with SBS symptoms have not been evaluated in epidemiological studies. These materials have reportedly been associated with malodor problems as a result of reactions between plasticizers and floor substrates. Additionally, persistently high plasticizer concentrations in building air have been associated with emissions from vinyl floor covering. Emission testing indicates that these materials have the potential to emit a variety of VOCs and semi-VOCs whose toxicity to humans is, for the most part, unknown.

REFERENCES

1. Murray, R. 1991. "Health Aspects of Carbonless Copy Paper." *Contact Derm.* 24:321–333.
2. Calnan, C. D. 1979. "Carbon and Carbonless Copy Paper." *Acta Dermatovener.* 59, Suppl. 85:27–32.
3. Menne, T. et al. 1981. "Skin and Mucous Membrane Problems from 'No Carbon Required' Paper." *Contact Derm.* 7:72–76.
4. Kleinman, G. and S.W. Horstman. 1982. "Health Complaints Attributed to the Use of Carbonless Copy Paper (A Preliminary Report)." *Am. Ind. Hyg. Assoc. J.* 43:432–435.
5. Goethe, C.J. et al. 1981. "Carbonless Copy Papers and Health Effects." *O Puscula Medica.* 56:1–15.
6. Jeansson, I. et al. 1984. "Complaints Related to the Handling of Carbonless Copy Paper in Sweden." *Am. Ind. Hyg. Assoc. J.* 45(B24–B27).
7. Marks, J.G. 1981. "Allergic Contact Dermatitis from Carbonless Copy Paper." *J. Am. Med. Assoc.* 245:2331–2332.
8. Marks, J.G. et al. 1984. "Contact Urticaria and Airway Obstruction with Carbonless Copy Paper." *J. Am. Med. Assoc.* 262:1038–1040.
9. La Marte, F.P. et al. 1988. "Acute Systemic Reactions to Carbonless Copy Paper Associated with Histamine Release." *J. Amer. Med. Assoc.* 260:242–243.
10. Morgan, M. and J. Camp. 1986. "Upper Respiratory Irritation from Controlled Exposure to Vapor from Carbonless Copy Forms." *J. Occup. Med.* 28:415–419.
11. Wolkoff, P. et al. 1988. "Airway-Irritating Effect of Carbonless Paper Examined by the Sensory Irritation Test in Mice." *Environ. Int.* 14:43–48.
12. Skov, P. et al. 1989. "Influence of Personal Characteristics, Job-Related Factors and Psychosocial Factors on the Sick Building Syndrome." *Scand. J. Work Environ. Health.* 15:286–295.
13. Fisk, W.J. et al. 1993. "The California Healthy Building Study, Phase 1: A Summary." 279–284. In: *Proceedings of the Sixth International Conference on Indoor Air Quality and Climate.* Vol. 1. Helsinki.
14. Zweers, T. et al. 1992. "Health and Indoor Climate Complaints of 7043 Office Workers in 61 Buildings in the Netherlands." *Indoor Air.* 2:127–136.
15. Goeckel, D.L. et al. 1981. "Formaldehyde Emissions from Carbonless Copy Paper Forms." *Am. Ind. Hyg. Assoc. J.* 42:474–476.

16. National Institute of Occupational Safety and Health. 1984. "Health Hazard Evaluation Report. Municiple Court Section, City of Englewood, CO." *HETA*. 83–313–1534.
17. National Institute of Occupational Safety and Health. 1982. "Health Hazard Evaluation Report. General Telephone Company, York, PA." *HETA*. 81–275–1122.
18. Norback, D. and C.J. Goethe. 1983. "Carbonless Copy Paper, Air Pollution and Indoor Climate." *Arbete och Halsa*. 37:37–50 [In Swedish-English Summary].
19. Norback, D. 1983. "Emissions of Chemicals from Carbonless Copy Papers." *Arbete och Halsa*. 37:51–62 [In Swedish- English Summary].
20. Wolkoff, P. and G. Nielsen. 1987. "Carbonless Copy Paper as a Potential Non-Building Indoor Irritant." 89–93. In: *Proceedings of the Fourth International Conference on Indoor Air Quality and Climate*. Vol. 1. West Berlin.
21. Molhave, L. and K. Grunnet. 1981. "Addendum: Headspace Analyses of Gases and Vapors Emitted by Carbonless Paper." *Contact Derm*. 7:76.
22. Menne, T. and N. Hjorth. 1985. "Frictional Contact Dermatitis." *Am. J. Ind. Med*. 8:401–402.
23. Thomas, N.G.H. 1981. Letters to the Editor. *Contact Derm*. 7:220–221.
24. Norback, D. et al. 1988. "A Search for Discomfort-Inducing Factors in Carbonless Copying Paper." *Am. Ind. Hyg. Assoc. J*. 49:117–120.
25. Hjorth, N. 1964. "Contact Dermatitis from Cellulose Acetate Film." *Berufsdermatosen*. 12:86.
26. Calnan, C.D. and B.L. Connor. 1975. "Carbon Paper Dermatitis Due to Nigrosine." *Berufsdermatosen*. 20:248.
27. Brooks, B.O. and W.F. Davis. 1991. *Understanding Indoor Air Quality*. CRC Press, Boca Raton, FL.
28. Wolkoff, P. et al. 1993. "Comparison of Volatile Organic Compounds from Processed Paper and Toners from Office Copiers and Printers: Methods, Emission Rates, and Modeled Concentrations." *Indoor Air*. 3:118–123.
29. Tsuchiya, Y. 1988. "Volatile Organic Compounds in Indoor Air." *Chemosphere*. 17:79–82.
30. Tsuchiya, Y. et al. 1988. "Wet Process Copying Machines: A Source of Volatile Organic Compound Emissions in Buildings." *Environ. Toxicol. Chem*. 7:15–18.
31. Tsuchiya, Y. and J.B. Stewart. 1990. "Volatile Organic Compounds in the Air of Canadian Buildings with Special Reference to Wet Process Photocopying Machines." 633–638. In: *Proceedings of the Fifth International Conference on Indoor Air Quality and Climate*. Vol. 2. Toronto.
32. Hodgson, A.T. et al. 1991. "Sources and Source Strengths of Volatile Organic Compounds in a New Office Building." *J. Air Waste Manage. Assoc*. 41:1461–1468.
33. Kerr, G. and P. Sauer. 1990. "Control Strategies for Liquid Process Photocopier Emissions." 759–764. In: *Proceedings of the Fifth International Conference on Indoor Air Quality and Climate*. Vol. 3. Toronto.
34. Molhave, L. et al. 1986. "Human Reactions to Low Concentrations of Volatile Organic Compounds." *Environ. Int*. 12:157–176.

35. Molhave, L. 1990. "The Sick Building Syndrome (SBS) Caused by Exposures to Volatile Organic Compounds (VOCs)." 1–18. In: Weekes, D.M. and R.B. Gammage (Eds.) In: *Proceedings of the Indoor Air Quality International Symposium: The Practitioner's Approach to Indoor Air Quality Investigations.* American Industrial Hygiene Association, Akron, OH.
36. Greenwood, M.R. 1990. "The Toxicity of Isoparaffinic Hydrocarbons and Current Exposure Practices in the Non-Industrial (Office) Indoor Air Environment." 169–175. In: *Proceedings of the Fifth International Conference on Indoor Air Quality and Climate.* Vol. 5. Toronto.
37. Katayangi, M. 1988. "Report on Results of Organoleptic Examination on DT Developer." Ricoh Corporation, Environmental Control Center. Tokyo.
38. Jensen, M. and J. Rold-Petersen. 1979. "Itching Erythema Among Post Office Workers Caused by a Photocopying Machine with a Wet Toner." *Contact Derm.* 5:389–391.
39. Allen, R.J. et al. 1978. "Characterization of Potential Indoor Sources of Ozone." *Am. Ind. Hyg. Assoc. J.* 39:459–466.
40. Selway, M.D. et al. 1980. "Ozone Production from Photocopying Machines." *Am. Ind. Hyg. Assoc. J.* 41:455–459.
41. Braun-Hansen, T.B. and B. Andersen. 1986. "Ozone and Other Air Pollutants from Photocopying Machines." *Am. Ind. Hyg. Assoc. J.* 47:659–665.
42. American Conference of Governmental Industrial Hygienists. 1993. *Threshold Limit Values and Biological Exposure Indices.* Cincinnati, OH.
43. Markin, J.M. et al. 1976. "Elevation of Selenium Levels in Air by Xerography." *Nature.* 259:204–205.
44. Parent, R.A. 1976. "Elevation of Selenium Levels in Air by Xerography." *Nature.* 263:708.
45. Schnell, R.C. et al. 1992. "Black Carbon Aerosol Output from a Photocopier." Presented at the Annual Meeting Air and Waste Management Association, Kansas City. Paper # 92–79.17.
46. Sonnino, A. and I. Povan. "Possible Hazards from Laser Printers." In: Grandjean, E. (Ed.). *Ergonomics and Health in Modern Offices.* Taylor and Francis, London.
47. Skoner, D.P. et al. 1990. "Laser Printer Rhinitis." *New Eng. J. Med.* 322:1323.
48. Kellett, J.K. et al. 1984. "Contact Sensitivity to Thiourea in Photocopy Paper." *Contact Derm.* 11:124.
49. Tencati, J.R. and H.S. Novey. 1983. "Hypersensitivity Angitis Caused by Fumes from Heat-Activated Photocopy Paper." *Ann. Int. Med.* 98:320–322.
50. Frederick, L.J. et al. 1984. "Investigation and Control of Occupational Hazards Associated with the Use of Spirit Duplicators." *Am. Ind. Hyg. Assoc. J.* 45:51–55.
51. Kingley, W.H. and F.G. Hirsch. (1954–55). "Toxicologic Considerations in Direct Process Spirit Duplicating Machines." *Compen. Med.* 40:7–8.
52. Harrison, J. et al. 1987. "The Sick Building Syndrome, Further Prevalence Studies and Investigation of Possible Causes." 487–491. In: *Proceedings of the Fourth International Conference on Indoor Air Quality and Climate.* Vol. 2. West Berlin.
53. Hedge, A.S. et al. 1989. "Work-Related Illness in Offices: A Proposed Model of the Sick Building Syndrome." *Envir. Int.* 15:143–158.

54. Nelson, C.J. et al. 1991. "EPA's Indoor Air Quality and Work Environment Survey: Relationships of Employees' Self- Reported Health Symptoms with Direct Indoor Air Quality Measurements." 22–32. In: *Proceedings of IAQ '91: Healthy Buildings*. American Society of Heating, Refrigerating and Air-Conditioning Engineers. Atlanta.
55. Menzies, R. et al. 1991. "The Effect of Varying Levels of Outdoor Ventilation on Symptoms of Sick Building Syndrome." 90–96. In: *Proceedings of IAQ '91: Healthy Buildings*. American Society of Heating, Refrigerating and Air-Conditioning Engineers. Atlanta.
56. Kjaergaard, S. and J. Brandt. 1993. "Objective Human Conjunctival Reactions to Dust Exposure, VDT Work and Temperature in Sick Buildings." 41–45. In: *Proceedings of the Sixth International Conference on Indoor Air Quality and Climate*. Vol. 1. Helsinki.
57. Linden, V. and S. Rolfsen. 1981. "Video Computer Terminals and Occupational Dermatitis." *Scand. J. Work Environ. Health*. 7:62–67.
58. Nilsen, A. 1982. "Facial Rash in Visual Display Unit Operators." *Contact Derm*. 8:25–28.
59. Liden, C. and J.E. Wahlberg. 1985. "Does Visual Display Work Provoke Rosacea?" *Contact Derm*. 13:235–241.
60. Berg, M. and I. Langlet. 1987. "Defective Video Displays, Shields and Skin Problems." *Lancet*. 1:800.
61. Feldman, L. et al. 1985. Letter to the Editor. "Terminal Illness." *J. Am. Acad. Dermatol*. 12:366.
62. Fisher, A.M. 1986. "Terminal Dermatitis Due to Computers (Visual Display Units)." *Cutis*. 37:153–154.
63. Berg, M. 1988. "Skin Problems in Workers Using Visual Display Terminals: A Study of 201 Patients." *Contact Derm*. 19:335–341.
64. Knave, B.G. et al. 1985. "Work with Video Display Terminals Among Office Workers. I. Subjective Symptoms and Discomfort." *Scand. J. Work Environ. Health*. 11:457–566.
65. Goranssan, K. et al. 1984. "An Outbreak of Occupational Photodermatosis of the Face in a Factory in Northern Sweden. Part I. Medical Investigation." 367–370. In: *Proceedings of the Third International Conference on Indoor Air Quality and Climate*. Vol. 3. Stockholm.
66. Goranssan, K. et al. 1984. "An Outbreak of Occupational Photodermatosis of the Face in a Factory in Northern Sweden. Part II. Chemical and Technical Investigation." 371–375. In: *Proceedings of the Third International Conference on Indoor Air Quality and Climate*. Vol. 3. Stockholm.
67. Allen, H. and K. Kaidbey. 1979. "Persistent Photosensitivity Following Occupational Exposure to Epoxy Resin." *Arch. Dermatol*. 115:1307–1310.
68. Rose, L. 1987. "Workplace Video Display Terminals & Visual Fatigue." *J. Occup. Med*. 29:321.
69. Ong, C.N. 1984. "VDT Work Place Design and Physical Fatigue: A Case Study in Singapore." 486–487. In: Grandjean, E. (Ed.). *Ergonomics and Health in Modern Offices*. Taylor and Francis, London.
70. Scalet, E.A. 1987. *VDT Health and Safety: Issues and Solutions*. Ergosyst Associates, Lawrence, KS.

71. Finn, L. and I.G. Ramberg. 1986. "Eye Fatigue Among VDU Users and Non-VDU Users." 42–52. In: *Proceedings of the International Scientific Conference: Work with Display Units*. Stockholm.
72. Sauter, S.L. 1986. "Chronic Neck-Shoulder Discomfort in VDT Use: Prevalence and Medical Observations." 154. In: *Proceedings of the International Scientific Conference: Work with Display Units*. Stockholm.
73. Rossignol, A.M. et al. 1987. "Video Display Terminal Use and Reported Health Symptoms Among Massachusetts Clerical Workers." *J. Occup. Med.* 29:112–118.
74. Hedge, A. 1989. "Environmental Conditions & Health in Offices." *Int. Rev. Ergonomics*. 2:87–110.
75. Weiss, M.M. and R.C. Petersen. 1979. "Electromagnetic Radiation Emitted from Video Computer Terminals." *Am. Ind. Hyg. Assoc. J.* 40:300–309.
76. Walsh, M.L. et al. 1991. "Hazard Assessments of Video Display Units." *Am. Ind. Hyg. Assoc. J.* 52:324–331.
77. Wertheimer, N. and E. Leeper. 1979. "Electrical Wiring Configurations and Childhood Cancer." *Am. J. Epidemiol.* 109:273–284.
78. Savitz, D.A. et al. 1988. "Case Control Study of Childhood Cancer and Exposure to 60Hz Magnetic Fields." *Am. J. Epidemiol.* 128:21–38.
79. Goldhaber, M.K. et al. 1988. "The Risk of Miscarriage and Birth Defects Among Women who use Video Display Terminals During Pregnancy." *Am. J. Ind. Med.* 13:695–706.
80. Ericson, A. and B. Kallen. 1986. "An Epidemiological Study of Work with Video Screens and Pregnancy Outcome. II. A Case-Control Study." *Am. J. Ind. Med.* 9:459–475.
81. Bryant, H.E. and E.J. Love. 1989. "Video Display Terminal Use and Spontaneous Abortion Risk." *Int. J. Epidemiol.* 18:132–138.
82. Roman, E. et al. 1992. "Spontaneous Abortion and Work with Visual Display Units." *Br. J. Ind. Med.* 49:507–512.
83. Wolkoff, P. et al. 1992. "A Study of Human Reactions to Office Machines in a Climatic Chamber." *J. Exp. Anal. Environ. Epidemiol.* Suppl. 1:71–96.
84. Weschler, C.J. et al. 1992. "Ozone Chemistry, Volatile Organic Compounds, and Carpets." *Environ. Sci. Technol.* 26:2371–2377.
85. Sundell, J. et al. 1993. "TVOC and Formaldehyde as Risk Indicators of SBS." 579–589. In: *Proceedings of the Sixth International Conference on Indoor Air Quality and Climate*. Vol. 1. Helsinki.
86. Schroder, F. 1990. "Textile Floor Coverings and Indoor Air Quality." 719–723. In: *Proceedings of the Fifth International Conference on Indoor Air Quality and Climate*. Vol. 3. Toronto.
87. Kerr, G. 1991. "Chemical Emissions During Recarpeting of a Canadian Office Building." 147–154. In: *Proceedings Jacques Cartier Conference*. Montreal.
88. Singhvi, R. et al. 1988. "A Final Summary Report on Indoor Air Monitoring Performed at EPA Headquarters, Washington, D.C. on May 24, 25 and June 6, 1988."
89. Weitzman, D. and R. Singhvi. 1990. "Indoor Air Quality at EPA Headquarters Building: A Case Study." 151–162. In: Weekes, D.M. and R.B. Gammage (Eds.). *Proceedings of the Indoor Air Quality International Symposium: The Practitioner's Approach to Indoor Air Quality Investigations*. American Industrial Hygiene Association. Akron, OH.

90. U.S. Environmental Protection Agency. 1990. "Carpet Response to Citizen's Petition." *Notice Fed. Reg.* 55(79):17404–17409.
91. Wallace, L.A. et al. 1991. "Workplace Characteristics Associated with Health and Comfort Concerns in Three Office Buildings in Washington, D.C." 56–60. In: *Proceedings of IAQ '91: Healthy Buildings.* American Society of Heating, Refrigerating and Air-Conditioning Engineers. Atlanta.
92. Wallace, L.A. et al. 1993. "Association of Personal and Workplace Characteristics with Reported Health Symptoms of 6771 Government Employees in Washington, D.C." 427–432. In: *Proceedings of the Sixth International Conference on Indoor Air Quality and Climate.* Vol. 1. Helsinki.
93. Nexo, E. et al. 1983. "Extreme Fatigue and Malaise Syndrome Caused by Badly-Cleaned Wall-to-Wall Carpets?" *Ecol. Dis.* 2:415–418.
94. Skov, P. et al. 1990. "Influence of Indoor Climate on the Sick Building Syndrome in an Office Environment." *Scand. J. Work Environ. Health.* 16:367–371.
95. Norback, D. and M. Torgen. 1989. "A Longitudinal Study Relating Carpeting with Sick Building Syndrome." *Environ. Int.* 15:129–135.
96. Norback, D. et al. 1990. "Volatile Organic Compounds, Respirable Dust, and Personal Factors Related to the Sick Building Syndrome in Primary Schools." *Br. J. Ind. Med.* 47:733–741.
97. Menzies, R. et al. 1992. "Case-Control Study of Microenvironmental Exposures to Aero-Allergens as a Cause of Respiratory Symptoms—Part of the Sick Building Syndrome (SBS) Symptom Complex." 201–210. In: *Proceedings IAQ '92: Environments for People.* American Society of Heating, Refrigerating and Air-Conditioning Engineers. Atlanta.
98. Hansen, L. et al. 1987. "Carpeting in Schools as an Indoor Pollutant." 727–731. In: *Proceedings of the Fourth International Conference on Indoor Air Quality and Climate.* Vol. 2. West Berlin.
99. Johnsen, C.R. et al. 1991. "A Study of Human Reactions to Emissions from Building Materials in Climate Chambers. Part I: Clinical Data, Performance and Comfort." *Indoor Air.* 1:377–388.
100. Anderson, R.C. 1991. "Measuring Respiratory Irritancy of Emissions." 19–23. In: *Post-Conference Proceedings of IAQ '91: Healthy Buildings.* American Society of Heating, Refrigerating and Air-Conditioning Engineers. Atlanta.
101. Anderson, R.C. 1993. "Toxic Emissions from Carpets." 651–656. In: *Proceedings of the Sixth International Conference on Indoor Air Quality and Climate.* Vol. 1. Helsinki.
102. Wolkoff, P. et al. 1991. "A Study of Human Reactions to Emissions from Building Materials in Climate Chambers. Part II: VOC Measurements, Mouse Bioassay, and Decipol Evaluation in the 1–2 mg/m^3 TVOC Range." *Indoor Air.* 4:389–403.
103. American Society for Testing and Materials. 1984. "Standard Test Method for Estimating Sensory Irritancy of Airborne Chemicals." Designation: E 981–84. American Society for Testing and Materials. Philadelphia.
104. Alarie, Y. 1981. "Dose-Response Analysis in Animal Studies: Prediction of Human Responses." *Environ. Health Perspect.* 42:9–13.
105. Nielsen, G.D. and Y. Alarie. 1992. "Animal Assays for Upper Airway Irritation: Screening of Materials and Structure-Activity Relations." *Ann. N.Y. Acad. Sci.* 641:164–175.

106. Tucker, W.G. 1993. "Bioresponse Testing of Sources of Indoor Air Contaminants." 621–626. In: *Proceedings of the Sixth International Conference on Indoor Air Quality and Climate*. Vol. 1. Helsinki.

106a. Dyer, R.S. 1994. "Report: Open Scientific Meeting on Carpet Emissions: Animal Testing." USEPA Health Effects Research Laboratory. Research Triangle Park, N.C.

107. Anderson, R.C. 1993. "Toxic Emissions from Carpets." Presented at the Sixth International Conference on Indoor Air Quality and Climate, Helsinki.

108. Van Ert, M.D. et al. 1987. "Identification and Characterization of 4-Phenylcyclohexene—An Emission Product from New Carpeting." Unpublished research report. University of Arizona.

109. Singhvi, R. et al. 1990. "4-Phenylcyclohexene from Carpet and Indoor Air Quality." 671–676. In: *Proceedings of the Fifth International Conference on Indoor Air Quality and Climate*. Vol. 4. Toronto.

110. Molhave, L. 1982. "Indoor Air Pollution Due to Organic Gases and Vapors of Solvents in Building Materials." *Environ. Int.* 8:117–127.

111. Bayer, C.W. and C.D. Papanicolopolous. 1990. "Exposure Assessments to Volatile Organic Compound Emissions from Textile Products." 725–730. In: *Proceedings of the Fifth International Conference on Indoor Air Quality and Climate*. Vol. 3. Toronto.

112. Black, M. 1990. "Environmental Chamber Technology for the Study of Volatile Organic Compounds Emissions from Manufactured Products." 713–718. In: *Proceedings of the Fifth International Conference on Indoor Air Quality and Climate*. Vol. 3. Toronto.

113. Black, M.S. et al. 1991. "A Methodology for Determining VOC Emissions from New SBR-Latex-Backed Carpet, Adhesives, Cushions, and Installed Systems and Predicting Their Impact on Indoor Air Quality." 267–272. In: *Proceedings of IAQ '91: Healthy Buildings*. American Society of Heating, Refrigerating and Air-Conditioning Engineers. Atlanta.

114. Davidson, J.L. et al. 1991. "Carpet Installation During Building Renovation and Its Impact on Indoor VOC Concentrations." 299–303. In: *Proceedings of IAQ '91: Healthy Buildings*. American Society of Heating, Refrigerating and Air-Conditioning Engineers. Atlanta.

115. McLaughlin, P. and R. Aigner. 1990. "Higher Alcohols as Indoor Air Pollutants: Source, Cause, Mitigation." 587–591. In: *Proceedings of the Fifth International Conference on Indoor Air Quality and Climate*. Vol. 3. Toronto.

116. Nielson, P.A. 1987. "Potential Pollutants—Their Importance to the Sick Building Syndrome—and Their Release Mechanism." 598–602. In: *Proceedings of the Fourth International Conference on Indoor Air Quality and Climate*. Vol. 2. West Berlin.

117. Tichenor, B. et al. 1990. "Evaluation of Indoor Air Pollutant Sinks for Vapor Phase Organic Compounds." 623–628. In: *Proceedings of the Fifth International Conference on Indoor Air Quality and Climate*. Vol. 3. Toronto.

118. Borazzo, J.E. et al. 1990. "Sorption of Organic Vapors to Indoor Surfaces of Synthetic and Natural Fibrous Materials." 617–622. In: *Proceedings of the Fifth International Conference on Indoor Air Quality and Climate*. Vol. 3. Toronto.

119. Gravesen, S. et al. 1990. "The Role of Potential Immunogenic Components of Dust (MOD) in the Sick Building Syndrome." 9–14. In: *Proceedings of the Fifth International Conference on Indoor Air Quality and Climate*. Vol. 1. Toronto.
120. Anderson, R.L. et al. 1982. "Carpeting in Hospitals: An Epidemiological Evaluation." *J. Clin. Microbiol.* 15:408–415.
121. Gravesen, S. 1987. "Microbiological Studies on Carpets Versus Hard Floors in Non-industrial Occupations." 668–672. In: *Proceedings of the Fourth International Conference on Indoor Air Quality and Climate*. Vol. 1. West Berlin.
122. Gravesen, S. et al. 1986. "Demonstration of Microorganisms and Dust in Schools and Offices—An Observational Study of Non-industrial Buildings." *Allergy*. 41:520–525.
123. Gravesen, S. et al. 1983. "Aerobiology of Schools and Public Institutions—Part of a Study." *Ecol. Dis.* 2:411–413.
124. Taylor, A.E.M. and C. Hindson. 1982. "Facial Dermatitis from Alkyl Phenoxyacetate in a Dry Carpet Shampoo." *Contact Derm.* 8:70.
125. Kreiss, K. et al. 1982. "Respiratory Irritation Due to Carpet Shampoo: Two Outbreaks." *Environ. Int.* 8:336–341.
126. Patriarca, P.A. et al. 1982. "Kawasaki Syndrome: Association with the Application of Rug Shampoo." *Lancet*. 2:578–580.
127. Rogers, M.F. et al. 1985. "Kawasaki Syndrome: Is Exposure to Rug Shampoo Important?" *Am. J. Dis. Child.* 139:777–779.
128. Rittfeldt, L. et al. 1984. "Indoor Air Pollutants Due to Vinyl Floor Tiles." 297–301. In: *Proceedings of the Third International Conference on Indoor Air Quality and Climate*. Vol. 3. Stockholm.
129. Gustafsson, H. 1992. "Building Materials Identified as Major Sources for Indoor Air Pollutants. A Critical Review of Cases." Swedish Council for Building Research. D10:1992.
130. Andersson, B. 1983. "Kartlaggning Av Inomhusluften Vid Televerkets Radiokontor i Lulia Med Avseende Pa Mogel, Ammoniak, Formaldehyd, 2-ethylhexanol." Och Andra Flytiga Organiska Foreningar, Uppdragrsrapport 117.
131. Nilsson, L.-O. 1984. "Fukt i Flyspackel. Verksamhet 1981–84." *Fuktgruppen Lunds Teknisa Hogskolo.*
132. Sodermanlands Lans Landsting. 1988. "Sammanstallnig Av Atgarder i Sambaud Med Miljoproblem." *Vardcentral i Skiflinge.*
133. Hellstrom, B. 1989. "Betraffande Rapport 1988–10–31 Om Forekomsten Av Luftforoneningor mm. Hogelidsskolan. HEPA Byggkonsulter AB.
134. Rosell, L. 1990. "Stallukt Pa Ett Sjukhas Orsakat av PVC-Malta." *SP Dnr*. 309–90–1453.
135. Wolkoff, P. 1990. "Proposal of Methods for Developing Healthy Building Materials: Laboratory and Field Experiments." *Environ. Tech.* 11:327–338.
136. Rosell, L. 1990. "High Levels of Semi-VOC in Indoor Air Due to Emission from Vinyl Floorings." 707–711. In: *Proceedings of the Fifth International Conference on Indoor Air Quality and Climate*. Vol. 3. Toronto.
137. Gustafsson, H. 1991. "Building Materials Identified as Major Emission Sources." 259–261. In: *Proceedings of IAQ '91: Healthy Buildings*. American Society of Heating, Refrigerating and Air-Conditioning Engineers. Atlanta.

138. Andersson, B. et al. 1984. "Mass Spectrophotometric Identification of 2-ethylhexanol in Indoor Air: Recovery Studies by Charcoal Sampling and Gas Chromatographic Analysis at the Micrograms Per Cubic Meter Level." *J. Chromatog*. 291: 257–263.
139. Vedel, A. and P.A. Nielsen. 1984. "Phthalate Esters in the Indoor Environment." 309–314. In: *Proceedings of the Third International Conference on Indoor Air Quality and Climate*. Vol. 3. Stockholm.
140. Virgin, H.I. et al. 1984. "Effects of Di-N-Butyl- Phthalate on the Chlorophyll Formation in Green Plants." 355–360. In: *Proceedings of the Third International Conference on Indoor Air Quality and Climate*. Vol. 3. Stockholm.
141. Astill, B.D. et al. 1972. "The Toxicology and Fate of 2,2,4-trimethyl-1,3 pentanediol diisobutyrate." *Toxicol. Appl. Pharmacol*. 22:387–399.
142. Saarela, K. and J. Sundell. 1991. "Comparative Emission Studies of Flooring Materials with Reference to Nordic Guidelines." 262–266. In: *Proceedings of IAQ '91: Healthy Buildings*. American Society of Heating, Refrigerating and Air-Conditioning Engineers. Atlanta.

5 GAS/VAPOR- AND PARTICULATE-PHASE CONTAMINANTS

It is incumbent on investigators of building/work-related health, comfort, or odor complaints that they identify the cause or causes so that effective remedial measures can be implemented. As indicated in Chapter 1 and reemphasized in Chapter 7, complaint investigations fail, in many cases, to identify causal factors that are presumed to be airborne and are gas/vapor- or particulate-phase materials (with particular emphasis on the former). Indeed, sick building syndrome (SBS) is a phenomenon that is defined by the fact that no single causal substance/agent has been identified. A number of cross-sectional epidemiological studies have attempted to determine causal associations between contaminants and illness or air quality complaints. These studies have reported mixed results with some showing associations between contaminants or potential sources and SBS symptoms with no similar associations in other studies.

It is increasingly becoming evident that sick building phenomena are multifactorial, that symptoms associated with building environments may be due to a number of causal factors and exposures. In some cases, symptoms may be due to contaminant exposures which are area- or building-wide. In other cases, symptoms may be associated with personal exposures associated with handling carbonless copy paper, working with video display terminals and office copiers.

Despite the fact that building occupants may experience symptoms which are unique to their work activities, there is nevertheless a strong sense among those in the IAQ community that area- or building-wide contamination problems are responsible for a significant portion of health and/or comfort complaints. As a consequence, we continue to seek answers to the question of what the contaminants and sources are. It is commonly assumed that causal contaminants are gas/vapor- or particulate-phase substances or materials which are emitted from various building materials, furnishings, etc. There is increasing interest in biogenic contaminants, that is, contaminants of biological origin. Biogenic contaminants are discussed in detail in Chapter 6.

Systematic building studies which have focused on a population of buildings or intensive studies of one or more problem buildings have attempted to determine whether significant relationships exist between symptom prevalence rates and a limited number of relatively easily measured or assessed gas/vapor- or particulate-phase contaminants. These have included the bioeffluent carbon dioxide (CO_2), combustion by-products such as carbon monoxide (CO), formaldehyde, volatile organic compounds (VOCs), and particulate matter (respirable suspended particles, settled dust, etc.).

BIOEFFLUENTS

Contaminants generated by the human body, described as bioeffluents, have historically been a major IAQ concern. This concern has been both of an odor and comfort nature.[1] It has been suggested that occupant discomfort is associated with bioeffluent levels in the range of 600–1000 ppm CO_2 or higher.[2,3]

Because CO_2 is the bioeffluent produced in the greatest abundance and is easily measured, it has served as an indicator of bioeffluent levels and a crude indicator of ventilation adequacy. ASHRAE Standard 62-1989,[4] which recommends a ventilation rate of 20 CFM (10 L/s)/person for office buildings and 15 CFM (7.5 L/s)/person for schools and other buildings, would be equal to guideline CO_2 values of approximately 800 and 1000 ppm CO_2. Carbon dioxide at these levels or even several thousands of ppm higher are generally not considered to be toxic.

Carbon dioxide levels in the ambient (outdoor) air averages approximately 355 ppm. Within a building, CO_2 metabolically produced and released in human respiration may cause CO_2 to vary from those in the ambient environment to levels in excess of 4500 ppm.[5] Carbon dioxide levels in a building and by implication, bioeffluent levels, are dependent on occupant density and the outdoor ventilation rate. The relationship between occupancy, ventilation, and CO_2 levels can be seen in Figures 5.1a and 5.1b.[6] Other bioeffluents reported for indoor air include acetone, acetaldehyde, acetic acid, alkyl alcohol, amyl alcohol, butyric acid, diethyl ketone, ethylacetate, ethyl alcohol, methyl alcohol, phenol, and toluene.[7]

Health and Occupant Comfort Concerns

Circumstantial Evidence

It is conventional wisdom among many IAQ investigators that there is, at least at a minimum, a relationship between bioeffluent levels and human comfort. This belief is based in good measure on the personal experience of IAQ investigators in high-density poorly ventilated building spaces. The atmosphere in such environments is typically oppressive even when no sensory irritation is perceived.

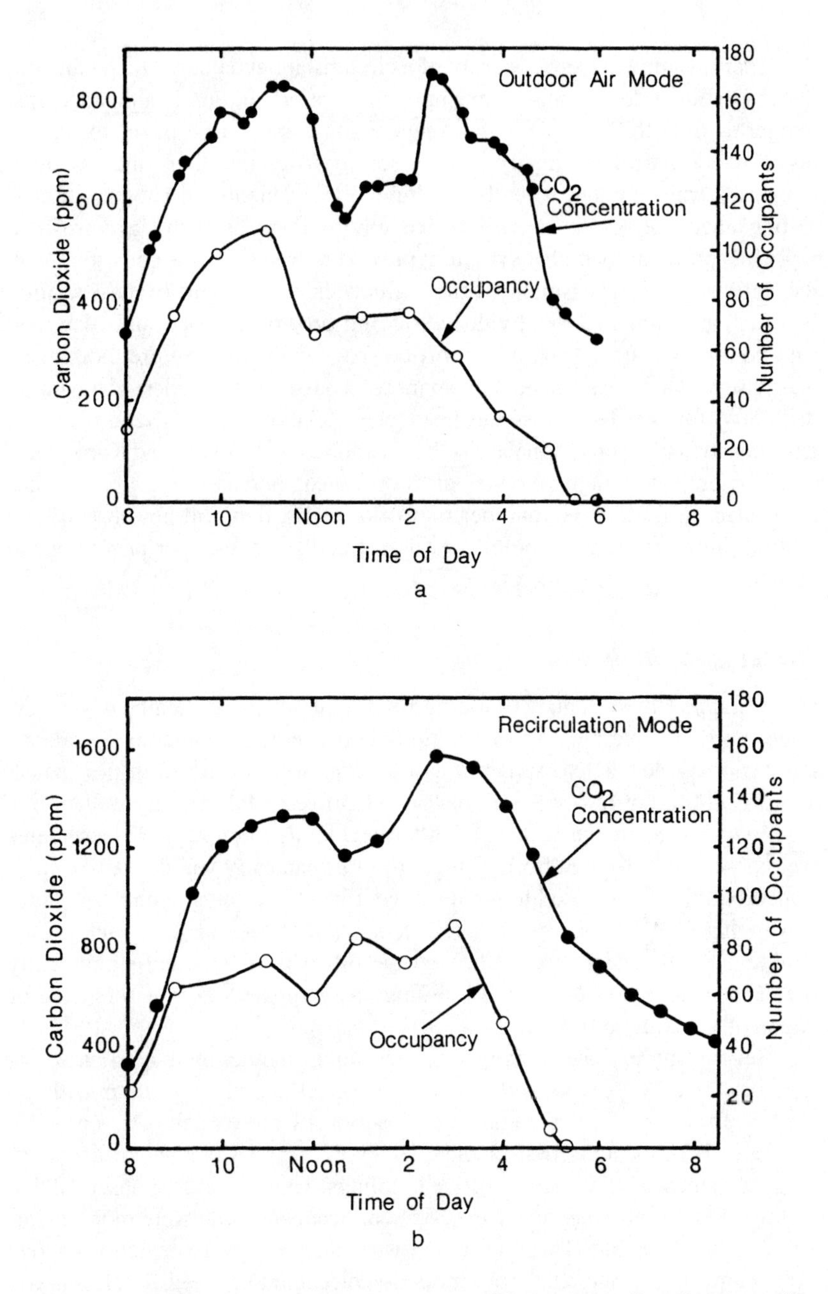

Figure 5.1 Effect of occupancy and ventilation on CO_2 levels in a San Francisco office building. (From Turiel, I. et al., 1983. *Atmos. Environ.* 17:51–64. With permission from Elsevier Science Ltd.)

There is some evidence, mostly of a circumstantial nature, which indicates that bioeffluent levels may contribute to comfort and/or health concerns. Fanger et al.,[8,9] for example, have suggested that about 13% of the expressed dissatisfaction with air quality as assessed in office buildings and assembly halls by a trained panel were due to human bioeffluents. In another study.[10] respondents who assessed their office environment as being overcrowded reported significantly higher symptom prevalence rates. This is consistent with the California Healthy Building Study which showed a significant association between the number of individuals sharing an office space and increased symptom prevalence.[11] Sharing an office space with two or more individuals was observed to be associated with an increased risk for tight chest or breathing difficulty, fatigue/sleepiness, and headache. Relationships between occupant density and SBS symptoms have also been reported by Valbjorn and Korsgaard[12] and Taylor.[13] Although Skov et al.[14] observed increased prevalence rates associated with increased number of workstations, they did not observe any relationships between symptom prevalence and floor area per person or the number of occupants in the office.[15]

Systematic Studies

A number of systematic building investigations have attempted to determine whether there are any associations between SBS symptom prevalence rates and CO_2 levels. No apparent associations were observed in the Danish Town Hall Study[14] or 6 sick schools[16] or 11 office buildings[17] in Sweden. In a study conducted in the United Kingdom (U.K.), symptom prevalence rates were observed to be significantly higher in mechanically ventilated air-conditioned buildings compared to naturally ventilated buildings, though the latter had higher CO_2 levels.[18] Similar results were reported for the Dutch Office Building Study.[14] Norback et al.,[19] on the other hand, reported significantly higher CO_2 levels in eight sick buildings as compared to four reference or apparently healthy buildings.

Several studies have attempted to determine whether there is any association between CO_2 levels and occupant comfort or dissatisfaction with air quality. Broder et al.[20] were unable to observe any correlation between building-related occupant discomfort and CO_2 levels in three Canadian office buildings. In a study of 10 Canadian office buildings, Tamblyn et al.[21] observed that average CO_2 levels were lower in offices of occupants who were more dissatisfied with air quality. The daily range was smaller as were deviations above 1000 ppm CO_2. Fanger et al.[8] observed no correlation between CO_2 levels and panel-assessed dissatisfaction in 15 noncomplaint office spaces and 5 assembly halls in 18 buildings in the greater Copenhagen area. However, in a study of ten school buildings,[22] a relationship was observed between perceived air quality and indoor CO_2 levels, with perceived air quality decreasing with higher CO_2 levels.

The studies described above have, for the most part, indicated that there is little or no association between CO_2 (and ostensibly bioeffluent levels) and SBS symptoms, occupant comfort, or dissatisfaction with air quality. One should, however, be cautious in interpreting the significance of these results. Though cross-sectional epidemiological studies are able to determine differences in symptom prevalence rates or occupant satisfaction/dissatisfaction in putatively high and low CO_2 buildings, they are a relatively crude way of determining whether dose-response relationships exist. In theory, a more suitable approach would be to conduct longitudinal studies in which both symptom prevalence and bioeffluent levels would vary over time. Because of the coupling between ventilation and human bioeffluents, as indicated by CO_2 levels, even such studies would have difficulty in establishing a direct relationship between SBS symptoms and occupant satisfaction/dissatisfaction with air quality because the concentration of other contaminants would vary as well. Separating the potential effects of bioeffluents from those of other contaminants would require experimental studies in which bioeffluent levels would be varied by occupant density changes at a standard building air exchange.

Potential Biological Mechanisms

The scientific evidence to support an association between bioeffluent levels, SBS symptoms and occupant comfort or dissatisfaction with air quality is relatively limited. There are nevertheless biologically plausible mechanisms which could contribute to such effects. Readers may find the following discussion speculative, but at the same time interesting, if not fascinating.

Human Pheromones

Though few studies of human bioeffluents have been conducted, it is quite probable that they would include human pheromones. Pheromones are volatile or semivolatile compounds produced by animals to elicit specific responses in individuals of the same species. Those responses are usually associated with sexual behavior and reproductive processes. Pheromones may become airborne and act to initiate sexually related responses by means of olfactory receptors. In mice, for example, volatile pheromones have been shown to affect sexual behavior in individuals several meters downwind of a pheromone source.[23]

Pheromonal effects have been reported to be near universal in social mammals including primates. Typically, male odors in such species serve to mark territory, assert dominance, repel rivals, attract females, and synchronize female sexual cycles.[24] Sexually related pheromonal effects have been postulated for humans and a variety of studies have been conducted with androstenol, a putative human pheromone primarily produced by males. Androstenol (5-androst-16-en-3-ol) is found in human male sweat secreted by axillary apocrine glands,[25] in human male semen,[26] and in the urine of both males and females, with much higher concentrations in the former.[27] Because androstenol has not been reported to have any known function and because it is a steroid compound, a human pheromonal role has been proposed.[24,28] Consequently, a number of investigations have been conducted to determine the effect of human pheromonal excretions on social interactions. Reported effects include synchronization of female menstrual cycles,[29-32] mood changes,[33] male and female choice performance,[34-36] social attitude changes,[37] and performance in an assessment of people test.[38]

Of potential interest in regard to IAQ concerns are those studies which focus on male and female choice selections on exposure to pheromonal preparations. It has been postulated that human pheromones, particularly androstenol, can act as an individual space regulator for human males.[33] In other mammals, scent marking communicates dominance over and intolerance of members of the same species. It functions to regulate the number of individuals within a limited area around an individual. In this way, scent marking functions to prevent intrusion into an area that is analogous to personal space. The typical response of male mammals to a scent mark is either avoidance or aggression. Gustavson et al.[34] applied androstenol and a comparison odor (androsterone) to restroom stalls to test the hypothesis that male pheromone can serve a spacing function. Males were observed to avoid androstenol-treated stalls whereas neither affected female stall selection. Kirk-Smith and Booth[35] working with androsterone observed that significantly more women (than men) used androsterone-sprayed chairs in a dental waiting room. Similar results were reported by Clark in a theater study.[36] Kirk-Smith and Booth[37] suggested that females were attracted to the odor of androsterone because they associated its odor with males, whereas it reminded males of other males and made them feel uncomfortable and even slightly aggressive. An aversion reaction of males to male odors was suggested.

Androstenol and androsterone are both steroid compounds found in human axillary sweat and urine.[28] The two compounds differ in the fact that androstenol has a hydroxyl group on the fifth position in the first benzene ring, whereas in androsterone the hydroxyl group is replaced by a ketone functional group. Androsterone has a wine-like pungent odor whereas androstenol has a musky odor. The odor threshold for androsterone is reported to range from 0.2–2 ppb. However, it appears that a large percentage of individuals cannot detect the odor of androsterone (circa 46%).

In the studies of Kirk-Smith and Booth[35] and Clark,[36] human females appeared to be attracted to androsterone, whereas males tended to avoid it. In the studies of Gustavson et al.,[34] no selective behavior was observed toward androsterone by either males or females, but avoidance behavior to androstenol was observed for males. Though the effect of androsterone was not consistent in all studies, it does appear that androgen steroids may elicit an avoidance or aversion response in males, suggesting that they may play a spacing function. It is interesting to speculate what the effect would be on individuals psychologically and possibly physiologically in poorly ventilated office spaces where avoidance is not a choice. Would the avoidance or aversion reaction result in general symptoms such as headache or fatigue as a consequence of sensory overload?

As has been described in Chapter 2, it is females who consistently report the highest prevalence rates of SBS symptoms. Could this gender-related phenomenon be related to human pheromones as well? Unfortunately, less is known of potential pheromonal effects associated with female secretions. There is evidence of menstrual synchronization associated with females in residence halls which indicates a pheromonal response.[39] Additionally, more than 30 compounds have been identified in human vaginal secretions.[40] Some of these are odoriferous and may behave as pheromones. If females also produce space-regulating pheromones of a volatile and olfactory receptive nature, then avoidance or aversion reactions may occur among females in response to female pheromones. Because of the high density of females in many office environments (in comparison to males), sensory overload to female pheromones might be expected in poorly ventilated spaces.

The case for human pheromones as a potential causal factor for at least a portion of SBS-type complaints is at best speculative. A need exists for the conduct of sampling studies to confirm the presence of human pheromones in poorly ventilated spaces followed by their quantification. Further studies would be needed to determine whether symptoms can be induced by pheromones (which may act as space regulators) when avoidance is not possible.

FORMALDEHYDE

Formaldehyde is a ubiquitous contaminant of indoor spaces. It is a substance of considerable sensory irritation potential[41,42] and apparent neurotoxic effects as well.[43] It has been reported to cause a wide variety of symptoms including most of those which characterize SBS[44-48] at concentrations in the range of 61–610 μg/m³ (0.05–0.50 ppm).

Formaldehyde has for the most part been a problem in residences where relatively high >120 μg/m³ (70.10 ppm) concentrations have been reported.[45,46,49-53] In the U.S. and Canada, residential formaldehyde contamination has been associated with mobile/modular homes, stick-built homes,[46] and homes retrofit insulated with urea-formaldehyde foam insulation.[45,51] With the exception of UFFI and acid-cured finishes, residential formaldehyde contamination has been the result of the use of urea-formaldehyde bonded wood products as construction materials and furnishings.[53]

Complaint Investigations

Formaldehyde has rarely been identified as a causal factor of illness complaints in problem building investigations. This has primarily been due to the fact that formaldehyde levels measured in large office and institutional building environments[6,14,54-58] have been relatively low (typically in the range of 12–50 μg/m^3 (10–40 ppb). These levels are below the World Health Organization's threshold guideline value of 100 μg/m^3 (82 ppb) and the target guideline value of 61 μg/m^3 (50 ppb) recommended by the California Air Resources Board.[60]

Elevated formaldehyde levels and associated sick building symptoms have been reported in modular Danish day-care centers,[47] in prefabricated Yugoslavian (Croatian) office buildings, kindergartens and school buildings,[61,62] a kindergarten building in the Slovak Republic,[63] American modular homes used as temporary offices,[64] and Swiss office and school buildings.[65] In general, these would be considered exceptional cases of formaldehyde contamination of office and other public-access buildings.

Systematic Studies

Formaldehyde levels are commonly measured in problem building investigations but have not been the focus of much attention in epidemiological assessments of single or large building populations. Potential relationships between formaldehyde levels and SBS symptoms have been evaluated in a few studies. No association between formaldehyde exposures and SBS symptoms were observed in the Danish Town Hall Study,[14] the sick building study of Norback et al.,[17] and the sick building study of Hodgson et al.[66] On the other hand, Sundell et al.[58] in the Office Illness Project of Northern Sweden observed that formaldehyde levels above 31 μg/m^3 (25 ppb) were a significant indicator of mucous membrane symptoms, skin symptoms, and at least one symptom reported each week of their study. The prevalence of at least one symptom each week was observed to show a very modest but significant ($r = 0.28$) dose-response relationship with formaldehyde concentrations. An association was also observed between the depletion of VOCs from intake to room air and SBS symptoms and between formaldehyde levels and "lost" VOCs, suggesting that formaldehyde was being generated as VOC levels decreased. Since these covary, it would be difficult to assign a causal relationship. Rather, Sundell et al.[58] suggested that both formaldehyde and depleted or "lost" VOCs should be viewed as potential risk factors for SBS symptoms.

Though the levels of formaldehyde reported by Sundell et al.[58] are relatively low, they do not differ substantially from levels reported by Falk et al.[67] who observed that exposure concentrations of 73 μg/m^3 (60 ppb) for 120 minutes in controlled chamber studies resulted in significant objectively measured swelling of nasal mucosa in a test group of sensitive individuals.

Formaldehyde levels in both the Sundell et al.[58] and the Falk studies were below the European threshold guideline[59] of 100 μg/m^3 (82 ppb), and the former were below California[60] air quality guidelines.

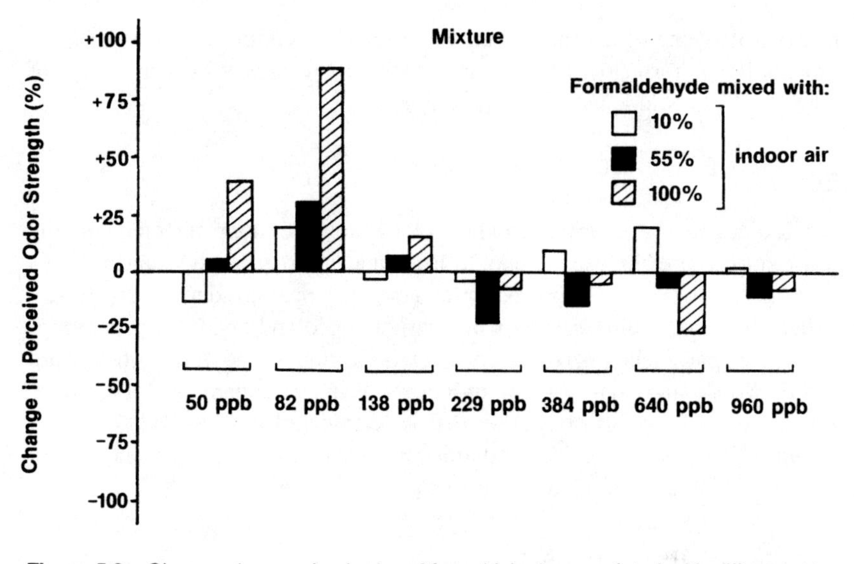

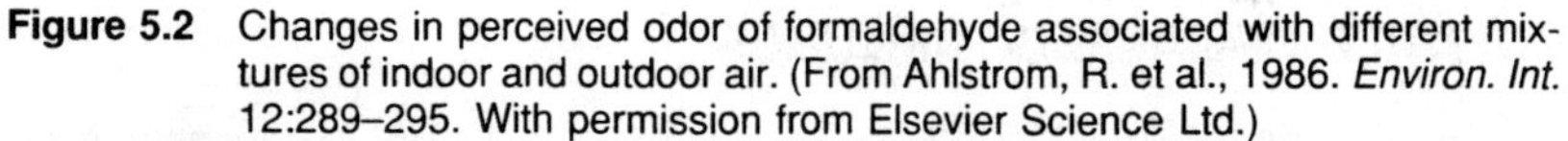

Figure 5.2 Changes in perceived odor of formaldehyde associated with different mixtures of indoor and outdoor air. (From Ahlstrom, R. et al., 1986. *Environ. Int.* 12:289–295. With permission from Elsevier Science Ltd.)

Potentiation

As a consequence of the relatively low formaldehyde levels reported in most sick building investigations, a working group of the Commission of European Communities[68] has suggested that it is unlikely that formaldehyde by itself is responsible for SBS symptoms but that it may be a contributor by potentiation of other factors. Significant potentiation of the odor of formaldehyde in the air of a sick building has been reported by Alhstrom et al.[69,70] at a concentration of 100 μg/m³ (82 ppb), which is the European threshold guideline for formaldehyde.[59] A fourfold increase in perceived odor strength for formaldehyde was observed when mixed with 100% indoor air from a sick building as compared to formaldehyde mixed with 10% sick building air. A significant change occurred also at a formaldehyde concentration of 67 μg/m³ (55 ppb).[70] At significantly higher concentrations, the odor strength of formaldehyde was perceived as being only marginally stronger. This effect can be seen in Figure 5.2. Such facilitation is known to occur near the sensory threshold.[71] The olfactory response to low concentrations of formaldehyde in the air of a sick building may be significant since the olfactory system interacts with the trigeminal nerve response or common chemical sense.

Though formaldehyde levels in office and other institutional buildings are considered to be low, they are often close to and in some cases, exceed the California IAQ[60] and on the order of one third to one half of the European threshold guideline.[59] In light of the VOC mixture studies of Molhave[73] and the fact that it is one of the most abundant VOCs in building environments, the probability of formaldehyde potentiation by even modest concentrations of other VOCs would appear to be probable.

Sources

Major sources of formaldehyde in office and other nonresidential building environments are likely to include furniture made from pressed-wood materials such as hardwood plywood, particleboard, and medium-density fiberboard (MDF), storage cabinets made from particleboard and/or MDF, countertops made from particleboard, reception area workstations made from particleboard and/or MDF, and room office dividers with particleboard or MDF cores. Formaldehyde may also be released from acid-cured finishes used on wood furniture. There is also evidence that formaldehyde can be generated by office equipment[74] and from the decomposition of VOCs.[58]

VOLATILE ORGANIC COMPOUNDS

In addition to formaldehyde, a large variety of organic compounds have been reported to be present in office and other nonresidential nonindustrial buildings. Organic compounds, particularly volatile organic compounds (VOCs), have been suspected to be causal or risk factors for SBS symptoms. Reasons for suspecting VOCs in causing SBS symptoms include the fact that (1) many VOCs have the potential to cause both sensory irritation and central nervous system symptoms characteristic of SBS, (2) they are found in significantly higher concentrations (2–100 × higher) in indoor compared to outdoor environments, and (3) because of the large number of VOCs present in indoor air, they may cause symptoms as a result of additive and/or multiplicative effects.

Organic compounds found in indoor air include very volatile organic compounds (VVOCs) with boiling points which range from less than 0°C to 50–100°C, volatile organic compounds (VOCs) with boiling points in the range of 50–100 to 240–260°C, semivolatile organic compounds (SVOCs) with boiling points in the range of 240–260 to 380–400°C, and organic compounds which are solids (POM) and have boiling points in excess of 380°C. Because of their great variety, relative abundance, and gas-phase presence in indoor air, VOCs have received major attention as potential causal agents of SBS symptoms.

Volatile organic compounds in indoor air include aliphatic hydrocarbons which may be straight, branched chain, or cyclic; aromatic hydrocarbons; halogenated hydrocarbons (primarily chlorine or fluorine); and oxygenated hydrocarbons such as aldehydes, alcohols, ketones, esters, ethers, and acids. These are emitted by a wide variety of sources including building materials and furnishings, consumer products, building maintenance materials, humans, office equipment, and tobacco smoking.[75]

TVOC Theory

A potential causal relationship between exposures to a large number of VOCs at low individual concentrations and SBS symptoms has been suggested by Molhave et al.[72,73] and supported by human exposure studies conducted by Molhave and his co-workers[72,73,76,78] and USEPA scientists.[79-86] The major tenet of the TVOC theory is that sensory irritation reported by individuals in problem buildings are due to the combined effects of the many VOCs found in indoor air. The biological mechanism suggested for such sensory irritation is stimulation of the trigeminal nerve system.

Common Chemical Sense

The common chemical sense is one of two olfactory mechanisms by which humans perceive and respond to odors. The chemical sense organ consists of free endings of trigeminal nerves which are distributed throughout the nasal cavity. It also includes trigeminal nerves in facial skin and in other skin areas.[87] Trigeminal nerves can be stimulated by both chemical and physical agents including, in the latter case, touch and temperature. Stimulation of the chemical sense can result in irritation described as a burning, stinging, or smarting feeling. It can also result in changes in heart rate, cough, sneezing, respiratory frequency, and tearing.[87]

The primary human olfaction mechanism which provides for normal sensations of smell and taste is based in the olfactory epithelium located at the highest part of the nasal cavity. It consists of millions of yellow-pigmented bipolar receptor cells which connect directly to the olfactory bulbs of the brain. The two olfaction organs stimulate different parts of the brain. However, the sensory responses of the two organs are difficult to separate since some chemical exposures stimulate both systems. The most important outcome of trigeminal nerve stimulation is that similar qualitative responses are elicited irrespective of the chemical exposure. As a consequence, the same "symptom" may be produced by many different chemical exposures.

Exposure Studies

Molhave and his colleagues[72] have conducted a number of human studies to determine potential effects of exposures to VOC mixtures. In their first study, subjects who had previously complained of SBS symptoms were exposed to a mixture of 22 VOCs. These VOCs and their relative proportions are reported in Table 5.1. Sixty-two subjects were exposed to total VOC (TVOC) concentrations of 0, 5, and 25 mg/m^3 (toluene equivalent) in groups of four in a double-blind experiment for durations of 2.75 hours. Exposure concentrations were selected to represent clean air, the average, and the highest concentration measured in new Danish houses. Perceptions of air quality and odor intensity were observed to increase with VOC concentration. The perception

Table 5.1 VOC mixture used in Danish human exposure studies.

Compound	Weight ratio
n-Hexane	1.0
n-Nonane	1.0
n-Decane	1.0
n-Undecane	0.1
1-Octene	0.01
1-Decene	1.0
Cyclohexane	0.1
3-Xylene	10.0
Ethylbenzene	1.0
1,2,4 Trimethylbenzene	0.1
n-Propybenzene	0.1
α-Pinene	1.0
n-Pentanal	0.1
n-Hexanal	1.0
Iso-propanol	0.1
2-Butanol	1.0
2-Butanone	0.1
3-Methyl-3-Butanone	0.1
4-Methyl-2-Petanone	0.1
n-Butylacetate	10.0
Ethoxyetlylacetate	1.0
1,2 Dicloroethane	1.0

From Molhave, L. et al., 1991. *Atmos. Environ.* 25A:1283–1293. With permission.

of mucous membrane irritation was equal at both exposure concentrations and significantly higher than with clean air. Decreased scores on the digit span test, a measure of short-term memory impairment, were observed for test subjects at both exposure concentrations.

This study was the first of its kind to demonstrate that a mixture of VOCs in the concentration range reported for new or renovated buildings can cause changes in perceived air quality, mucous membrane irritation, and neurobehavior. These initial results have been for the most part confirmed in other exposures studies conducted by the Danish group led by Molhave[73,76-78] and in exposures studies conducted by the USEPA.[79-83] Exposed subjects (25 mg/m^3 for 2.75 hours) in the USEPA study found the odor of VOCs unpleasantly strong and reported increased headache and general discomfort as well as decreased perception of air quality. Unlike Molhave and his colleagues, the USEPA group was not able to show any effects of TVOC exposures on performance as measured by neurobehavioral tests.[79,82]

Because Otto et al.[79,80] were not able to replicate the neurobehavioral results of impaired memory reported by Molhave et al.,[72] the USEPA group[81-83] conducted additional studies in which 26 male and 15 female subjects were exposed to a 25 mg/m^3 TVOC mixture and to putatively clean air. Though they observed increased eye, nose, and throat irritation, greater odor intensity and unpleasantness, and reduced perception of air quality during exposure, there were no

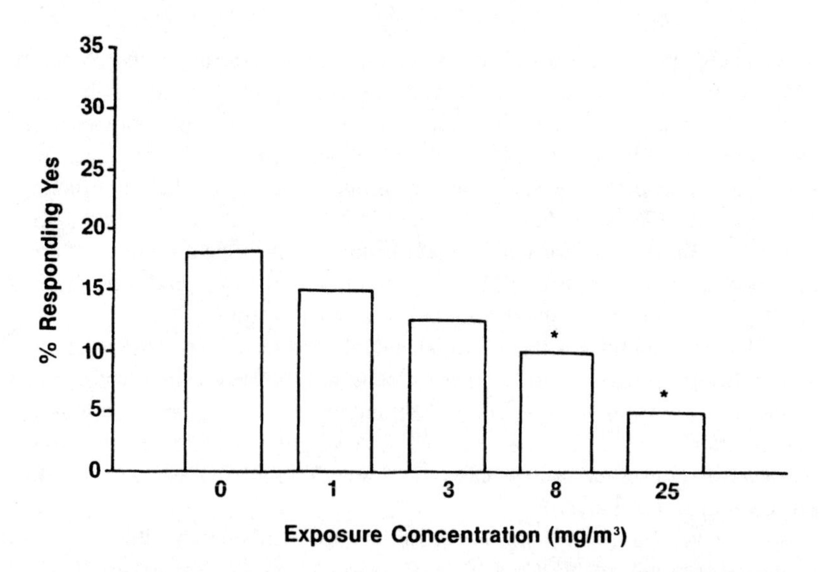

Figure 5.3 Dose-response relationship between TVOC exposure and air quality acceptability. (From Molhave, L., 1990. *Proceedings Indoor Air Quality International Symposium: The Practitioner's Approach to Indoor Air Quality Investigations.* With permission.)

differences in sensory effects observed between males and females.[81] Again they observed no effect of the exposure on neurobehavioral performance.[82]

A major criticism of the Danish study has been the choice of test subjects. Subjects were selected to simulate the worst case. As such, they were not representative of a normal population of building occupants. In the USEPA study, young adult male subjects were selected who had no known sensitivity to chemicals. An exclusively young adult male population would, of course, not be representative of a normal building population either. Despite the differences in subject populations, the USEPA study substantially confirmed the findings of Molhave[72] at a mixed VOC concentration of 25 mg/m^3.

Molhave et al.[73] followed up their initial exposure studies with the intent of confirming their findings and establishing a dose-response relationship. In a second series of studies, the exposure mixture remained the same; however, a broader range of TVOC exposure concentrations were employed (0, 1, 3, 8, and 25 mg/m^3), for an exposure duration of 50 minutes. Significant differences in ratings for acceptability of odor intensity from clean air conditions were reported for TVOC concentrations of 3, 8, and 25 mg/m^3; acceptability of air quality at 8 and 25 mg/m^3; nose irritation at 8 and 25 mg/m^3, and general well-being at 25 mg/m^3. These results show different apparent thresholds for the variables evaluated. Responses to the TVOC mixtures appear to be dose-related (Figure 5.3).

Molhave et al.[76,77] conducted a third series of exposures to compare responses of individuals claiming to be indoor air (SBS sensitive) to those of normal subjects. Exposures were conducted with both clean air and 25 mg/m^3 TVOCs for periods of 2.75 hours. Both groups reported significantly worse odor, reduced air quality, and more irritation of the mucous membranes of the eyes, nose, and throat on exposure. There was, however, a tendency to a stronger response for the SBS exposure group. Significant objective findings were observed. These included a decrease in lung function (FEF_{25}) in the SBS-sensitive subjects, and for both subject groups an increased number of polymorphonuclear neutrophils (PMN) in tear fluid of the eyes and a diminished ability to learn as determined from psychological testing.

Objective symptoms have also been evaluated in several USEPA studies. A significant increase in inflammatory cells were observed in 14 of subjects exposed to 25 mg/m^3 for 4 hours.[83] PMN increases were observed both immediately after exposure and 18 hours later. A significant increase in eye blink rate was also observed at exposures of 12 and 24 mg/m^3. There was no effect on respiratory frequency.[84]

Molhave et al.,[78] in more recent studies, exposed human subjects to four different mixtures of VOCs at 0, 1.7, 5 and 15 mg/m^3 for a period of 60 minutes. Compounds comprising the mixtures included the 22 VOCs used previously and compounds selected from these in a 1:1 proportion. The three submixtures are indicated in Table 5.2. One mixture M_x contained compounds with vapor pressures below three torr, while the other two generally contained compounds of higher vapor pressure. Air quality, odor intensity, and sensory irritation were evaluated by means of a comfort questionnaire and potentiometer ratings. The threshold for acute effects was observed to be below 1.7 mg/m^3 for all three mixtures. Mixtures containing compounds with higher vapor pressures were more potent in causing sensory effects, whereas those with higher thermodynamic activity (based on partial pressure of substance divided by its vapor pressure) indicated higher potency for objective symptoms such as skin-humidity and foam formation in the eyes.

Irritation and Odor Effects

In the initial study of Molhave and his co-workers,[72] eye, nose, and throat irritation were observed to increase during exposure with no observable decline or adaptation by the end of the exposure. This suggested that both the olfactory and trigeminal systems were contributing to perceptions reported by test subjects.

The USEPA group[82] evaluated sensory irritation and olfactory effects of exposures to 25 mg/m^3 TVOC mixture. Eye and nose irritation, headache and drowsiness were observed to increase during exposure. Perceived irritation increased sharply during the initial part of the exposure (75 minutes), peaked at 150 minutes, and remained constant for the rest of the exposure period (total of 340 minutes). As such, no adaption was evident. In contrast, odor intensity ratings rose sharply from the beginning to the midpoint of the exposure and then declined about 30% by the end of the exposure. A smaller but significant

Table 5.2 Compounds included in submixtures.

M_x	M_y	M_z
2-Xylene	*n*-Nonane	*n*-Hexane
Ethyl benzene	*n*-Decane	Cyclohexane
n-Pentanal	*n*-Undecane	1,2-Dichoroethane
n-Hexanal	1,2,4 Trimethylbenzene	2-Propanol
n-Butanol	*n*-Propylbenzene	2-Butanone
3-Methyl-2-butanone	1-Octene	4-Methyl-2-pentanone
	1-Decene	
	α-Pinene	
	n-Butylacetate	

Note: Relative proportions 1:1 (mg/mg).

From Hudnell, K. et al., 1993. *Proceedings Sixth International Conference on Indoor Air Quality and Climate.* Vol. 1. Helsinki.

decrease in dissatisfaction with air quality was also observed. Significant decreases in dissatisfaction with air quality have also been reported by the Danish group.[76-78] The observed differences in the time course of responses indicated that irritation was not a simple function of odor intensity. However, declines in ratings of odor intensity and dissatisfaction with air quality suggested that both irritation level and odor intensity contribute to perceived air quality.

The USEPA group[84] conducted additional studies of the time course of odor and irritation effects. In this case, healthy male subjects were exposed to concentrations of 6, 12, and 24 mg/m^3 TVOC mixture for 4 hours. Perceived odor intensity peaked relatively rapidly and then declined. The magnitude of adaptation was 60, 48, and 38% at TVOC levels of 6, 12, and 24 mg/m^3. Perceived odor intensity not only showed a greater degree of adaptation at lower TVOC concentrations but also took longer to reach a constant level. Sensory irritation magnitude increased as a function of dose with ratings of 68, 114, and 359% at exposure levels of 6, 12, and 24 mg/m^3, respectively. The time course of eye, nose, and throat irritation did not parallel that of odor intensity. The magnitude of perceived air quality was also observed to vary significantly among exposure levels. Air quality was perceived to improve during exposure at lower exposure levels where eye irritation effects were relatively small and adaption to odor was large. However, at 24 mg/m^3 where odor adaptation was relatively small and eye irritation, severe, air quality was perceived to decrease, indicating that both odor and irritation intensity affect perceptions of air quality.

The USEPA group[84] concluded that TVOC exposure stimulated both the olfactory and trigeminal nerve systems and suggested that their interaction produced the observed effects.[88] They also suggested that reported effects of the addition of an odorant to an irritant which reduced perceived irritation[89] and the addition of an irritant to an odorant which reduced perceived odor intensity[90] accounted for the observation that odor magnitude decreased by about 30% whereas such odor adaptation typically reaches a level of 50–70%.[91]

Neurotoxic Effects

The Danish and USEPA exposure studies provide strong evidence of a relationship between exposures to VOC mixtures, sensory irritation, and feelings of discomfort. They are less convincing as to the cause of the general-type or neurotoxic-like symptoms of headache and fatigue. Headache has been reported in but one of the TVOC exposure studies and only at 25 mg/m^3. Molhave[92] suggests that there is a significant difference in the frequency of headaches in problem buildings and those described as controls when concentrations were less than 3 mg/m^3 and that this lower threshold, as compared to exposure studies, is due to the interaction of other exposures or the effect of longer exposure durations typical of office environments.

Headache may not occur as a direct result of trigeminal nerve stimulation. Molhave[92] suggests that headaches may occur secondarily as a result of stress associated with the body's attempts to override unwanted sensory information and to maintain protective reflexes. This is similar to a theory proposed by Berglund et al.[93] that SBS symptoms may arise from multisensory adaptation to indoor air which is achieved by either exhaustive stimulation or by a sensory deprivation of signals important to optimal levels of sensory variation. As a result of extreme homogenization, the indoor environment may lose all recognizable stimulus patterns resulting in sensory confusion and strain on the system attempting to interpret sensory signals.

TVOC Sensitivity

Molhave and his co-workers[76,77] reported that test subjects who complained of experiencing SBS symptoms prior to controlled TVOC exposures were apparently more sensitive than normal subjects. This is consistent with the findings of Kjaergaard et al.[94] who observed a lower threshold for sensory irritation (associated with high level CO_2 exposures of the eyes) in SBS subjects. This chemical sense response was not observed to be dependent on sex or whether the subject was a smoker or not. This is in contrast to the studies of Cain[90] who observed that CO_2-induced chemical sense responses on breathing rate were significantly related to gender (females had lower thresholds than males) and smoking (nonsmokers had lower thresholds than smokers).

Application of TVOC Concept

Molhave[92] proposed a tentative dose-response relationship for discomfort and health effects for TVOC exposures (Table 5.3). In his dose-response model, the no effect level would be about 0.2 mg/m^3, and above 3 mg/m^3 TVOCs discomfort or health effects would always be expected. Molhave[73] suggests that complaints will occur in almost all cases when building TVOCs are above 3 mg/m^3. Above 25 mg/m^3, neurotoxic effects other than headache would be expected. In the range of 0.20–3.0 mg/m^3, both irritation and discom-

Table 5.3 Proposed dose-response relationship between discomfort/health effects and exposures to TVOC mixtures.

TVOC concentration (mg/m³)	Response	Exposure range
<0.20	No effects	Comfort range
0.20–3.0	Irritation/ discomfort possible	Multifactorial exposure range
3.0–25.0	Irritation and discomfort; headache possible	Discomfort range
>25.0	Neurotoxic effects in addition to headaches	Toxic range

From Molhave, L., 1990. *Proceedings Fifth International Conference on Indoor Air Quality and Climate.* Vol. 5. Toronto.

fort would be expected to occur if other factors interact with TVOC exposures (e.g., temperature, concentration of contaminants). This is described as the multifactorial exposure range. The multifactorial range would include TVOC concentrations reported in the Danish Town Hall Study,[14] the sick building investigations of Norback et al.,[16,17] and the sick building investigation of Turiel et al.[6]

Molhave and Nielsen[96] have recently cautioned against the use of the TVOC concept as a generic indicator of potential health risks in buildings. It is at best an indicator of the risk of nonspecific sensory irritation to nonreactive VOCs within a limited range of vapor pressures (excluding VVOCs and SVOCs).

Systematic Building Studies

A number of investigators have attempted to determine whether TVOC levels are associated with SBS symptoms. Both the Danish Town Hall Studies[14,56] and California Healthy Building Study[11] failed to show relationships between area VOC levels and retrospectively reported symptoms. Significant relationships were, however, observed between VOC levels and SBS symptoms in the studies of Hodgson et al.[66,97] and Norback et al.[16,17] In the former, significant associations were observed between VOC levels determined at each building occupant's workstation and simultaneously reported symptom intensity. Norback et al.[17] reported significant positive correlations between SBS symptom prevalence rates and the logarithmic value of the TVOC concentration in 11 buildings investigated. In a study of problem schools, Norback et al.[16] observed that symptom prevalence was not only associated with log TVOC levels (Figure 5.4) but also to the concentration of terpenes, *n*-alkanes (C_8-C_{11}), and butanols. TVOCs varied from 0.05–1.38 mg/m³ in the 11-building investigation and 0.07–0.18 mg/m³ in problem school buildings. In the former case, TVOC levels were in the multifactorial range proposed by Molhave;[91] in the latter case, they would have been in the no-effects range.

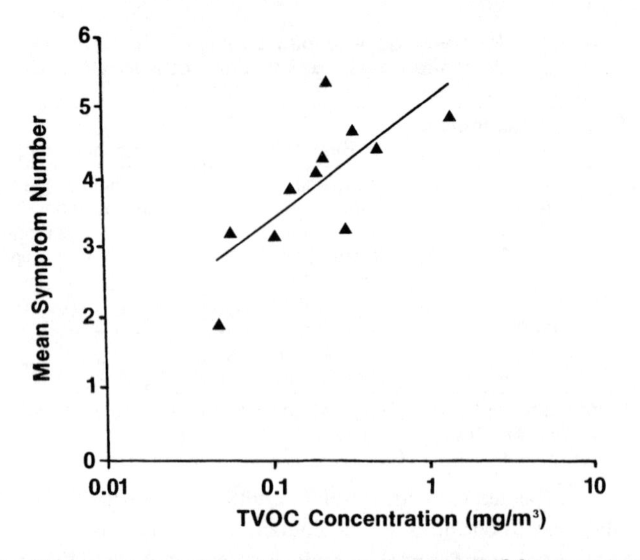

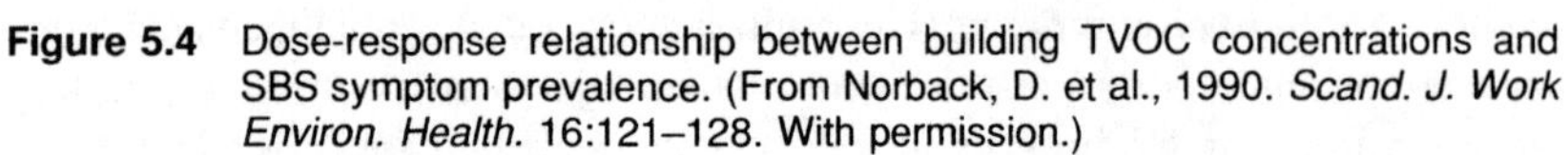

Figure 5.4 Dose-response relationship between building TVOC concentrations and SBS symptom prevalence. (From Norback, D. et al., 1990. *Scand. J. Work Environ. Health.* 16:121–128. With permission.)

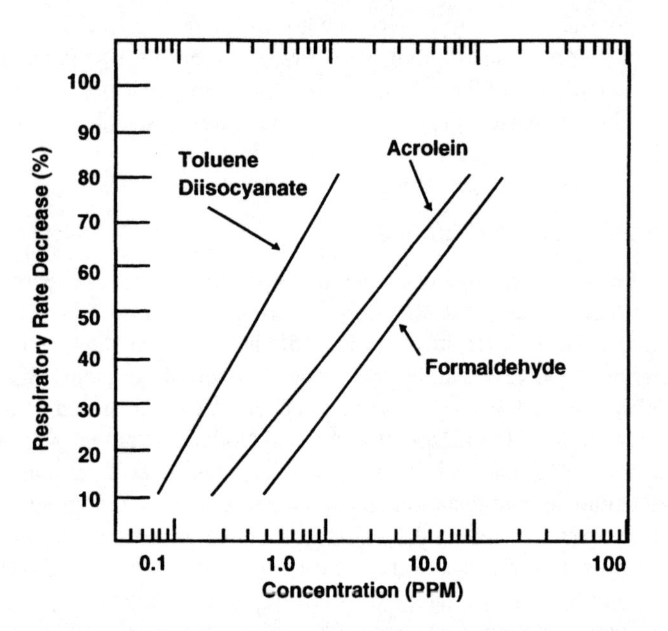

Figure 5.5 Log-linear dose-response relationships between irritant chemical exposures and mouse respiratory responses. (From Alarie, Y., 1981. *Environ. Health Perspect.* 42:9–13.)

The log-linear relationship between VOC levels and SBS symptoms observed by Norback et al.[16,17] are consistent with toxicological studies using mouse bioassays[40,98] which show that sensory irritation is a log-linear phenomenon (Figure 5.5).

Additional evidence to support a causal relationship between the prevalence of SBS symptoms and TVOC exposures has been reported by Berglund et al.[99] They observed a highly significant (r = 0.96) correlation between the prevalence change over the course of a day of eight SBS symptoms with the mean concentration of 34 VOCs measured in the building exhaust air of a sick library building. They also observed a significant but less strong (r = 0.67) correlation between percent dissatisfied with air quality and average TVOC concentration. Baird et al.[100] were able to distinguish between "sick" and "healthy" school buildings by means of chemical classification. Using cluster analyses which evaluated the pattern of presence or absence of VOC chemicals in different locations, they were able to determine the presence of ten critical chemicals which distinguished the two buildings. They identified six of these including tetrachloroethene, α-pinene, 1,1,1 trichloroethane, butylacetate, *n*-dodecane, and *n*-tridecane.

Sundell et al.,[58] as indicated in the discussion on formaldehyde observed some rather interesting and unexpected results. Elevated TVOC levels in office building spaces appeared to be associated with reduced SBS symptom prevalence. However, a significant dose-response relationship was observed between SBS symptoms and the difference between TVOC levels in office rooms and intake air (Figure 5.6). Symptoms of fatigue, feeling heavy-headed, hoarse or dry throat, and facial dry skin were most frequently associated with "lost" TVOCs which were associated with increases in room formaldehyde levels.

Several investigators have attempted to determine the potential relationship between TVOC concentrations and ratings of perceived air quality as determined by a trained panel of judges. The studies of Fanger et al.[8] showed a relatively poor correlation between perceived air quality and TVOC concentration. In the human exposure studies of Wolkoff et al.,[101] some correlation was observed between subjective evaluations of air quality and TVOC concentrations associated with building materials. The association was diminished in the presence of strong irritants such as formaldehyde and odorous compounds.

VOC Concentrations

Volatile organic compounds found in the air of offices and other nonresidential, nonindustrial building spaces vary considerably relative to types of compounds and concentrations of individual VOCs assessed as TVOCs. Individual VOCs in building air samples vary from as few as 20 to several hundred. Concentrations are typically in the low parts per billion range whereas TVOCs are in the low parts per million or mg/m^3 range. Most individual VOCs in

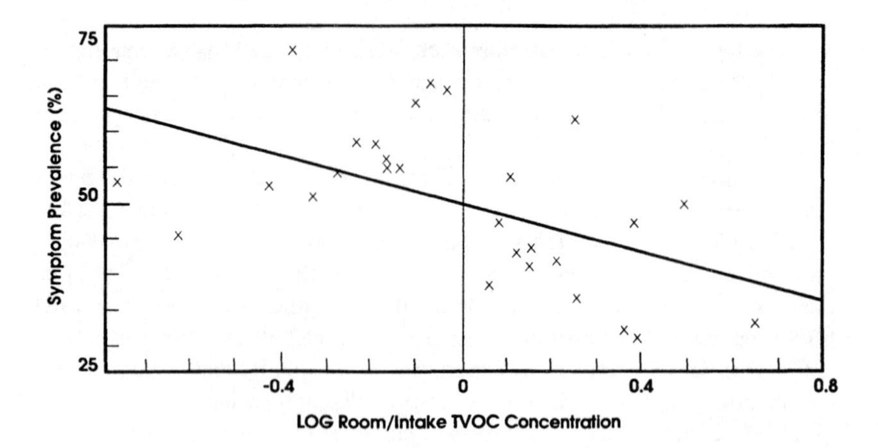

Figure 5.6 Dose-response relationships between "lost" TVOCs and SBS symptom prevalence. (From Sundell, J. et al., 1993. *Indoor Air.* 3:82–92. With permission.)

indoor air are several orders of magnitude lower than occupational guideline values.

Reported TVOC concentrations in the Danish Town Hall Study[14] averaged 1.56 mg/m^3 (range 0.43–2.63 mg/m^3) when measured by charcoal sampling and 0.5 mg/m^3 (range 0.1–1.2 mg/m^3) by Tenax sampling. Norback et al.[16] reported concentrations of VOCs which averaged 0.130 mg/m^3 (range 0.70–0.180 mg/m^3) in primary school buildings and 0.380 mg/m^3 (range 0.05–1.38 mg/m^3) in their 11-building investigation.[17] De Bortoli et al.[57] conducted measurements of VOCs in 10 buildings of the European Parliament. Reported TVOCs as determined by a flame ionization detector had a median value of 0.22 mg/m^3 with a maximum concentration of 3.93 mg/m^3. Compounds most frequently observed were *n*-hexane, *n*-heptane, benzene, toluene, 1,3-xylene + 1,4-xylene, methyl chloroform, and tetrachloroethene.

"DUST"

Though concentrations of particulate-phase matter (typically referred to as "dust") are rarely assessed in problem building investigations, there is, nevertheless, increasing evidence from epidemiological studies that airborne dust or dust on building surfaces is a risk factor for SBS symptoms. The term "dust" here includes particulate-phase materials in the broadest sense including a large range of sizes, types, sources, chemical compositions, etc.

Expression of "Dust" Concentrations

In the various studies conducted, airborne "dust" concentrations were quantified by a number of different techniques. For airborne dusts, concentrations may be reported as total suspended particles (TSP), respirable suspended particles (RSP), ultraviolet particulate matter (UVPM), particle numbers, or specific particle fractions such as man-made mineral fibers (MMMF). TSP concentrations represent the largest range of particles including those which are respirable (<3.0 μm), inhalable (<10.0 μm), and those that are non-inhalable (>10.0 μm). Particles less than 3.0 μm are the most biologically significant, whereas larger particles contribute disproportionately to particle concentrations expressed on a mass basis. RSP values include particles in the size range which can enter and be deposited in lung tissue. Though UVPM concentrations also reflect particle concentrations in the respirable size range, RSP and UVPM values are not comparable because the latter is dependent on the light scattering and absorbing characteristics of particulate matter being measured.

Systematic Studies

Airborne Dusts

Armstrong et al.[102] conducted a cross-sectional epidemiological study of a problem high-rise office building. They observed from multivariate statistical analyses that SBS symptoms were significantly associated with TSP levels. On the other hand, Skov et al.[14] failed to observe any relationship between TSP levels and SBS symptom prevalence.

Hodgson et al.[66,103] conducted an intensive cross-sectional epidemiological study of a problem building. RSP levels measured in the breathing zone of those surveyed were the strongest predictor of SBS symptoms in multivariate analyses used to evaluate relationships between SBS symptoms and environmental parameters. Nelson et al.,[104] in a study of the USEPA headquarters building, observed that log 9-hour RSP concentrations were associated with nervous system symptoms in males in one of several regression models evaluated. In the school study of Norback et al.,[16] RSP concentrations were associated with the incidence of new SBS symptoms. In a school study in Norway, Hanssen[105] observed no association between RSP levels and mucous membrane symptoms despite reportedly high RSP values. Hedge et al.[106] also reported no apparent association between SBS symptom prevalence and RSP and UVPM values in nine buildings. Skov et al.[14] inexplicably observed decreased symptom prevalence rates with increasing dust concentrations as determined from size-defined particle counts.

Indirect evidence to support a potential causal relationship between airborne dust levels in an individual's workstation and SBS symptoms is suggested from the studies of Hedge et al.[107,108] Using filtration systems installed in specially designed workstations, Hedge et al.[107] observed significant reductions in particle levels which appeared to be associated with significant improvements in occupant ratings of lethargy and marginal improvements in breathing difficulty, dizziness, nasal congestion, and dry cough. In a follow-up study,[108] the breathing zone filtration system decreased submicronic (<0.25

μm) particle counts by 80%. This was observed to be associated with significant reductions in perceived workplace dustiness and complaints of poor IAQ. Significant reductions in complaints of lethargy, dry skin, and headache were also reported. Work-related symptoms decreased from 3.84 to 2.05 symptoms per person. Certified sickness absence decreased by 61%.

Surface Dusts

A number of investigators have attempted to evaluate health risks associated with settled or deposited dust on interior building surfaces and SBS symptoms. Skov et al.[14] observed significant associations between floor dust levels and mucous membrane irritation symptoms and the macromolecular organic fraction (MOD) of floor dust with both mucous membrane and general symptoms. They were not, however, able to show an association between floor dust levels and airborne dust concentrations. The concentration of MOD in airborne dust could not be determined.

Kjaergaard and Brandt[109] conducted studies in which they attempted to evaluate dust sedimentation rates at workplace desks and objective eye symptoms. They observed significant correlations between dust sedimentation determined by an optical method and tear film stability ($r = 0.72$), foam formation ($r = 0.66$), corneal epithelial damage ($r = 0.58$), and PMN ($r = 0.53$). In microscopic assessments of dust sedimentation rates, only corneal eye damage and PMN were observed to be significantly correlated. These correlations were very strong ($r = 0.83$ and 0.93, respectively).

Hansen[110] and Hedge et al.[106] reported that SBS symptom prevalence could be related to levels of MMMF in settled dust but not to airborne MMMF. Hedge et al.[106] observed a relatively high correlation ($r = 0.75$) between aggregate symptom scores and the mass of settled MMMF. Because there was no apparent association between airborne and settled MMMF, Hedge et al.[106] suggested that random disturbance and resuspension of MMMF may be an important exposure factor that could be reduced by office cleaning. The perceived quality of cleaning was a significant risk factor for mucous membrane irritation in the 1-year follow-up of the Danish Town Hall Study.[56]

The potential effect of cleaning of surface dust on the prevalence of SBS symptoms has been investigated in double-blind studies conducted by scientists at the Building Research Establishment in the United Kingdom. In initial studies,[111,112] a rigorous program of surface cleaning which included steam cleaning of carpets and soft furnishings and wet wiping of hard surfaces was observed to significantly reduce symptom prevalence (from 4.37 to 2.62 symptoms/person). In another study,[113] a full cleaning regime which also included high efficiency vacuuming of files was observed to significantly reduce symptom prevalence rates when compared to a before-cleaning period and to a control group.

Based on the results of such studies and others, Raw[114] has proposed that indoor surface pollution by particulate dusts be considered a major risk factor for SBS symptoms. Such dust could cause symptoms as a result of toxic effects, irritation, or immunological mechanisms. The potential toxic or irritative effects may be increased as a result of the adsorption of gases and vapors.

A causal relationship between indoor surface dust and SBS symptoms is further supported by the fact that personal exposures of office workers have been shown to be 3–5 times greater than dust levels measured from area sampling.[112] Raw[114] suggests that individuals create their own "dust cloud" as settled dust is stirred up during work activities.

Further evidence of a potential relationship between surface dust and SBS symptoms is suggested from USEPA headquarter buildings studies. Significant associations were reported for perceived dustiness and the following symptoms: headache; nasal, chest, eye, and throat symptoms; fatigue; chills and fever; difficulty concentrating; dizziness; dry skin and four comfort concerns.[115]

Dust Concentrations

As reported previously, airborne particle concentrations are determined by a variety of procedures and, as such, may reflect different aspects of indoor particle contamination. In the studies of Armstrong et al.,[102] TSP levels in 17 of 29 areas were reported to exceed 60 μg/m³ with the highest reported concentration of 1.07 mg/m³. In the studies of Skov et al.,[14] TSP levels were relatively high with an average of 201 μg/m³ with a range of 86–382 μg/m³.

Respirable suspended particles represent a smaller fraction of TSP particle mass. Hodgson and Collopy[103] reported median RSP levels of 30 μg/m³ with a range of nondetectable (<10 μg/m³) to 90 μg/m³ in their problem building investigation. Norback et al.[16] reported relatively low RSP levels in problem schools with a median value of 15 μg/m³ and a range of 8–24 μg/m³. In the surface cleaning studies of Raw et al.,[112] personal dust levels were 250 μg/m³ and were 3–5 times those measured in area samples. Hedge et al.[116] studied RSP and UVPM concentrations in buildings with different smoking policies. They observed RSP and UVPM concentrations of 14.8 ± 14.0 μg/m³ and 0.2 ± 22.9 μg/m³ in smoking-prohibited buildings with average RSP concentrations in the range of 12–30 μg/m³ in nonsmoking areas of buildings with smoking-restricted policies. When smoking was allowed at the workstation, the RSP concentration reportedly averaged 2.6 ± 14.0 μg/m³ and the UVPM, 10.2 ± 22.9 μg/m³. Turner et al.[117] reported results of RSP measurements in 585 office buildings. Concentrations averaged 20 ± 17.6 μg/m³ in no-smoking-observed areas and 46 ± 56.9 μg/m³ in smoking-observed areas. Hanssen[105] in his school study reported relatively high RSP levels which appeared to be affected by the presence of carpeting and building occupancy (Table 5.4). In the office simulation study of Wolkoff et al.,[74] total particle concentrations averaged 50 ± 43 μg/m³ and 44 ± 27 μg/m³ (in two trials) in the chamber where office equipment was being actively used. These concentrations were approximately double those in a control chamber.

Table 5.4 Respirable particle levels (RSP) in school buildings associated with the type of floor covering and building occupancy.

	RSP (μg/m³) during school hours		RSP (μg/m³) after school hours	
School	With carpet	Without carpet	With carpet	Without carpet
A	108–230	130–200	50	50
B	230	120	50–60	70–80
C	120–140	100	60	60
D	230	—	100	—
E	320	—	20–30	—
F	280	160	60–100	—
G	380	—	—	—

From Hanssen, S.O., 1993. *Proceedings Sixth International Conference on Indoor Air Quality and Climate.* Vol. 5. Helsinki.

RECAPITULATION

Indoor spaces are contaminated by both gas/vapor- and particulate-phase substances. Contaminants that have received major attention in epidemiological studies as potential risk factors for SBS symptoms include bioeffluents such as CO_2, formaldehyde, VOCs, and "dust".

Though bioeffluents are widely believed to contribute to comfort problems in buildings, there is little direct evidence to implicate bioeffluents as determined from CO_2 levels in buildings. Potential effects of bioeffluents on human health and comfort based on human pheromones and aversion responses has been suggested.

With a few exceptions, formaldehyde levels in office and public-access buildings have been considered to be too low to cause reported SBS symptoms. Results reported from the Office Illness Project of Northern Sweden indicate that formaldehyde levels as low as 31 μg/m³ (25 ppb) may be a risk factor for SBS symptoms. However, formaldehyde levels were observed to covary with "lost" VOCs (from air intakes to room air) which were also significantly associated with increased symptom prevalence.

Because of the wide variety of VOCs present in indoor air, their sensory irritancy and neurotoxicity and potential for additive and/or multiplicative effects, VOC mixtures have received considerable attention in both human exposure and in systematic building studies. Danish investigators have proposed that exposures to mixtures of VOCs in the range of those found in buildings can cause sensory irritation and even neurotoxic effects in humans. Sensory effects at relatively low levels of exposure to VOC mixtures in controlled chamber studies have been demonstrated by Danish investigators and confirmed by USEPA scientists. As a consequence, the total volatile organic compound (TVOC) concentration in building environments has been proposed to be a risk factor for SBS symptoms.

Several systematic building studies have attempted to evaluate the potential association between TVOC concentrations and SBS symptoms. Mixed results have been reported. Those that evaluated SBS symptom prevalence with the log values of VOC concentration observed significant associations. These are consistent with toxicological studies which demonstrate that sensory irritancy of chemicals is a log-linear function of contaminant concentration. The results of the Office Illness Project of Northern Sweden are notable because significant associations were observed with lost or depleted VOCs and SBS symptoms rather than VOC levels in occupied spaces.

Though "dust" levels in buildings are rarely measured in problem building investigations, a variety of epidemiological studies have implicated either airborne or deposited dust as a risk factor for SBS symptoms. Significant associations have been reported for both TSP and RSP levels, the mass of floor dust, the macromolecular or organic fraction of floor dust, general surface dust, and MMMF in settled dust.

REFERENCES

1. National Research Council. 1981. *Indoor Air Pollutants.* National Academy Press. Washington, D.C.
2. Bell, S. and B. Khati. 1983. "Indoor Air Quality in Office Buildings." *Occup. Health in Ontario.* 4:103–118.
3. Rajhans, G. 1983. "Indoor Air Quality and CO_2 Levels." *Occup. Health in Ontario.* 4:160–167.
4. ASHRAE Standard 62–1989. 1989. "Ventilation for Acceptable Indoor Air Quality." American Society of Heating, Refrigerating and Air-Conditioning Engineers. Atlanta.
5. Godish, T. 1993. Unpublished data.
6. Turiel, I. et al. 1983. "The Effects of Reduced Ventilation on Indoor Air Quality in an Office Building." *Atmos. Environ.* 17:51–64.
7. Wang, T.C. 1975. "A Study of Bioeffluents in a College Classroom." *ASHRAE Trans.* 81:32–44.
8. Fanger, P.O. et al. 1988. "Air Pollution Sources in Offices and Assembly Halls Quantified by the Olf Unit." *Energy and Buildings.* 12:7–19.
9. Fanger, P.O. 1989. "The New Comfort Equation for Indoor Air Quality." *ASHRAE J.* October, 33–38.
10. Hawkins, L.H. and T. Wang. 1991. "The Office Environment and the Sick Building Syndrome." 365–371. In: *Proceedings of IAQ '91: Healthy Buildings.* American Society of Heating, Refrigerating and Air-Conditioning Engineers. Atlanta.
11. Fisk, W.J. et al. 1993. "The California Healthy Building Study. Phase 1: A Summary." 279–284. In: *Proceedings of the Sixth International Conference on Indoor Air Quality and Climate.* Vol. 1. Helsinki.
12. Valbjorn, O. and N. Korsgaard. 1984. "Headache and Mucous Membrane Irritation — An Epidemiological Study." 249–254. In: *Proceedings of the Third International Conference on Indoor Air Quality and Climate.* Vol. 3. Stockholm.
13. Taylor, P.R. 1984. "Illness in an Office Building with Limited Fresh Air Access." *J. Environ. Health.* 47:24–27.

14. Skov, P. et al. 1990. "Influence of Indoor Air Climate on the Sick Building Syndrome in an Office Environment." *Scand. J. Work Environ. Health.* 16:367–371.
15. Skov, P. et al. 1989. "Influence of Personal Characteristics, Job-Related Factors and Psychosocial Factors on the Sick Building Syndrome." *Scand. J. Work Environ. Health.* 15:286–295.
16. Norback, D. et al. 1990. "Volatile Organic Compounds, Respirable Dust, and Personal Factors Related to Prevalence and Incidence of Sick Building Syndrome in Primary Schools." *Br. J. Ind. Med.* 47:733–741.
17. Norback, D. et al. 1990. "Indoor Air Quality and Personal Factors Related to the Sick Building Syndrome." *Scand. J. Work Environ. Health.* 16:121–128.
18. Burge, P.S. et al. 1990. "Sick Building Syndrome. Environmental Comparisons of Sick and Healthy Buildings." 479–484. In: *Proceedings of the Fifth International Conference on Indoor Air Quality and Climate.* Vol. 6. Toronto.
19. Norback, D. et al. 1989. "The Prevalence of Symptoms Associated with Sick Buildings and Polluted Industrial Environments as Compared to Unexposed Reference Groups Without Expressed Dissatisfaction." *Environ. Int.* 15:85–94.
20. Broder, I. et al. 1990. "Building-Related Discomfort as Associated with Perceived Rather than Measured Levels of Indoor Environmental Variables." 221–226. In: *Proceedings of the Fifth International Conference on Indoor Air Quality and Climate.* Vol.1. Toronto.
21. Tamblyn, R.M. et al. 1993. "Big Air Quality Complainers — Are Their Office Environments Different from Workers with No Complaints?" 133–138. In: *Proceedings of the Sixth International Conference on Indoor Air Quality and Climate.* Vol. 1. Helsinki.
22. Thorstenson, E. et al. 1990. "Air Pollution Sources and Indoor Air Quality in Schools." 531–536. In: *Proceedings of the Fifth International Conference on Indoor Air Quality and Climate.* Vol. 1. Toronto.
23. Whitten, W.K. et al. 1968. "Estrus-Inducing Pheromone of Male Mice: Transport by Movement of Air." *Science.* 161:584–585.
24. Comfort, A. 1971. "Likelihood of Human Pheromones." *Nature.* 230:432–433.
25. Brooksbank, B.W.L. et al. 1974. "The Detection of 5 alpha-androst-16-en-3-alpha-ol in Human Axilliary Sweat." *Experientia.* 30:864.
26. Brooksbank, B.W.L. et al. 1961. "The Estimation of Androst-16-en-3-alpha-ol in Human Urine. Partial Synthesis of Androstenol and Its Beta-glucose-duronic Acid." *Biochem. J.* 80:488–496.
27. Kwan, T.K. et al. 1989. "Odorous Androst-16-enes and Other C_{19} Steroids in Human Semen." *Biochem. Soc. Trans.* 17:749–750.
28. Gower, D.B. et al. 1988. "The Significance of Odorous Steroids in Axillary Odour." 47–75. In: Van Toller, S. and G.H. Dogg (Eds.). *Perfumery: The Psychology and Biology of Fragrance.* Chapman and Hall. London.
29. Preti, G. et al. 1986. "Human Axillary Secretions Influence Women's Menstrual Cycles: The Role of Donor Extract on Females." *Hormones and Behavior.* 20:474–482.
30. Cutler, W.B. et al. 1986. "Human Axillary Secretions Influence Women's Menstrual Cycles: The Role of Donor Extract from Men." *Hormones and Behavior.* 20:463–473.
31. Wilson, H.C. 1987. "Female Axillary Secretions Influence Women's Menstrual Cycles: A Critique." *Hormones and Behavior.* 21:536–546.

32. Preti, G. 1987. "Reply to Wilson." *Hormones and Behavior*. 21:547–550.
33. Benton, D. 1982. "The Influence of Androstenol—a Putative Human Pheromone—on Mood Throughout the Menstrual Cycle." *Biol. Psychol.* 15:249–256.
34. Gustavson, A.R. et al. 1987. "Androstenol, a Putative Human Pheromone, Affects Human (Homo sapiens) Male Choice Performance." *J. Comp. Psychol.* 101:210–212.
35. Kirk-Smith, M.D. and D.A. Booth. 1980. "Effect of Androstenone on Choice of Location in Other's Presence." 397–400. In: Van der Starre, H. (Ed.). *Olfaction and Taste*. Information Retrieval. London.
36. Clark, T. 1978. "Whose Pheromone Are You?" *World Med.* July 26, 21–3.
37. Kirk-Smith, M.D. et al. 1978. "Human Social Attitudes Affected by Androstenol." *Res. Comm. Psychol., Psych. and Behav.* 3:379–383.
38. Cowley, J.J. et al. 1977. "The Effect of Two Odorous Compounds on Performance an Assessment of People Test." *Psychoneuroendocrinology* 2:159–172.
39. McClintock, M.K. 1983. "Synchronizing Ovarian and Birth Cycles by Female Pheromones." 159–178. In: Muller-Schwarze, D. and R. Silverstein (Eds.). *Chemical Signals in Vertebrates*. Plenum Press. New York.
40. Doty, R.L. et al. 1975. "Changes in the Intensity and Pleasantness of Human Vaginal Odors During the Menstrual Cycle." *Science*. 190:1316–1317.
41. Kane, L.E. and Y. Alarie. 1977. "Sensory Irritation to Formaldehyde and Acrolein During Single and Repeated Exposures in Mice." *Am. Ind. Hyg. Assoc. J.* 38:509–522.
42. Kulle, T.J. and G.P. Cooper. 1975. "Effects of Formaldehyde and Ozone on the Trigeminal Nasal Sensory System." *Arch. Environ. Health*. 30:237–243.
43. Bach, B. et al. 1987. "Human Reactions During Controlled Exposures to Low Concentrations of Formaldehyde—Performance Tests." 620–624. In: *Proceedings of the Fourth International Conference on Indoor Air Quality and Climate*. Vol. 2. West Berlin.
44. Broder, I. et al. 1988. "Comparison of Health and Home Characteristics of Houses Among Control Homes and Homes Insulated with Urea-formaldehyde Foam. II. Initial Health and House Variables and Exposure Response Relationships." *Environ. Res.* 45:156–178.
45. Ritchie, I.M. and R.H. Lehnen. 1987. "Formaldehyde-Related Health Complaints of Residents Living in Mobile and Conventional Houses." *Am. J. Pub. Health*. 77:323–328.
46. Dally, K.A. et al. 1981. "Formaldehyde Exposure in Nonoccupational Environments." *Arch. Environ. Health*. 33:277–284.
47. Olsen, J.H. and M. Dossing. 1982. "Formaldehyde-Induced Symptoms in Day Care Centers." *Am. Ind. Hyg. Assoc. J.* 43:366–370.
48. Godish, T. et al. 1990. "Residential Formaldehyde—Increased Exposure Levels Aggravate Adverse Health Effects." *J. Environ. Health*. 53:34–37.
49. Ritchie, I.M. and R.H. Lehnen. 1985. "An Analysis of Formaldehyde Concentrations in Mobile and Conventional Homes." *J. Environ. Health*. 47:300–305.
50. Hanrahan, L.P. et al. 1985. "Formaldehyde Concentrations in Wisconsin Mobile Homes." *JAPCA*. 35:1164–1167.
51. Anderson, I. et al. 1975. "Indoor Air Pollution Due to Chipboard Used as a Construction Material." *Atmos. Environ.* 9:1121–1127.

52. Syrotynski, S. 1985. "Prevalence of Formaldehyde Concentrations in Residential Settings." 127–136. In: Walkinshaw, D.S. (Ed.). *Transactions: Indoor Air Quality in Cold Climates*. Air Pollution Control Association. Pittsburgh.
53. Godish, T. 1988. "Formaldehyde Sources and Levels." *Comments Toxicol.* 2:115–134.
54. Crandall, M.S. et al. 1990. "Library of Congress and USEPA Indoor Air Quality and Work Environment Study: Environmental Survey Results." 597–602. In: *Proceedings of the Fifth International Conference on Indoor Air Quality and Climate*. Vol. 4. Toronto.
55. Turk, B. H. et al. 1989. *Indoor Air Quality Measurements in Commercial Buildings. Vol. 1: Measurement Results and Interpretation*. Bonneville Power Administration, Portland, OR.
56. Skov, P. and O. Valbjorn. 1990. "The Danish Town Hall Study — A One-year Follow Up." 787–791. In: *Proceedings of the Fifth International Conference on Indoor Air Quality and Climate*. Vol. 1. Toronto.
57. De Bortoli, M. et al. 1990. "Investigation of Volatile Organic Compounds to Air Quality Complaints in Office Buildings of the European Parliament." 695–670. In: *Proceedings of the Fifth International Conference on Indoor Air Quality and Climate*. Vol. 2. Toronto.
58. Sundell, J. et al. 1993. "Volatile Organic Compounds in Ventilating Air in Buildings at Different Sampling Points in the Building and their Relationship with the Prevalence of Occupant Symptoms." *Indoor Air*. 3:82–92.
59. World Health Organization. 1987. "Air Quality Guidelines for Europe." World Health Organization Regional Publications. European Series No. 23. Copenhagen.
60. California Air Resources Board. 1991. "Indoor Air Quality Guideline No. 1. Formaldehyde in the House."
61. Kalinic, N. et al. 1984. "Formaldehyde Levels in Selected Indoor Microenvironments." 145–148. In: *Proceedings of the Third International Conference on Indoor Air Quality and Climate*. Vol. 3. Stockholm.
62. Kalinic, N. and K. Sega. 1993. "Formaldehyde Exposure in Kindergartens." 597–600. In: *Proceedings of the Sixth International Conference on Indoor Air Quality and Climate*. Vol. 1. Helsinki.
63. Vargova, M. et al. 1993. "The Evaluation of Possible Adverse Effects of Nonoccupational Exposure to Formaldehyde." 613–618. In: *Proceedings of the Sixth International Conference on Indoor Air Quality and Climate*. Vol. 1. Helsinki.
64. Dement, J.M. et al. 1984. "An Evaluation of Formaldehyde Sources, Exposures and Possible Remedial Actions in Two Office Environments." 99–104. In: *Proceedings of the Third International Conference on Indoor Air Quality and Climate*. Vol. 3. Stockholm.
65. Wanner, H.U. and M. Kuhn. 1986. "Indoor Air Pollution by Building Materials." *Environ. Int.* 12:311–315.
66. Hodgson, M.J. et al. 1992. "Sick Building Symptoms, Work Stress, and Environmental Measures." 47–56. In: *Proceedings of IAQ '92: Environments for People*. American Society of Heating, Refrigerating and Air-Conditioning Engineers. Atlanta.

67. Falk, J. et al. 1993. "Dose-Response Study of Formaldehyde on Nasal Mucous Swelling. A Study on Residents with Nasal Distress at Home." 585–589. In: *Proceedings of the Sixth International Conference on Indoor Air Quality and Climate*. Vol. 1. Helsinki.
68. Molina, C. et al. 1989. "Sick Building Syndrome, A Practical Guide." Report No. 4. Commission of the European Communities. Brussels-Luxembourg.
69. Ahlstrom, R. and B. Berglund. 1984. "Odor Interaction Between Formaldehyde and the Indoor Air of a Sick Building." 461–466. In: *Proceedings of the Third International Conference on Indoor Air Quality and Climate*. Vol. 3. Stockholm.
70. Ahlstrom, R. et al. 1986. "Formaldehyde Odor and Its Interaction with the Air of a Sick Building." *Environ. Int.* 12:289–295.
71. Engen, T. 1982. *The Perception of Odors*. Academic Press. New York.
72. Molhave, L. et al. 1986. "Human Reactions to Low Concentrations of Volatile Organic Compounds." *Environ. Int.* 12:157–176.
73. Molhave, L. 1990. "The Sick Building Syndrome (SBS) Caused by Exposures to Volatile Organic Compounds (VOCs)." 1–18. In: Weekes, D.M. and R.B. Gammage (Eds.). *Proceedings of the Indoor Air Quality International Symposium: The Practitioner's Approach to Indoor Air Quality Investigations*. American Industrial Hygiene Association. Akron, OH.
74. Wolkoff, P. et al. 1992. "A Study of Human Reactions to Office Machines in a Climate Chamber." *J. Exp. Anal. Environ. Epidemiol.* Suppl. 1:71–96.
75. Girman, J.R. 1989. "Volatile Organic Compounds and Building Bakeout." 695–712. In: Cone, J.E. and M.J. Hodgson (Eds.). *Problem Buildings: Building-Associated Illness and the Sick Building Syndrome*. Hanley and Belfus, Inc. Philadelphia.
76. Molhave, L. et al. 1989. "Subjective Reactions to Volatile Organic Compounds as Air Pollutants." *Atmos. Environ.* 25A:1283–1293.
77. Kjaergaard, S.K. et al. 1991. "Human Reactions to a Mixture of Indoor Volatile Organic Compounds." *Atmos. Environ.* 25A:1417–1426.
78. Molhave, L. et al. 1993. "Human Response to Different Mixtures of Volatile Organic Compounds." 555–560. In: *Proceedings of the Sixth International Conference on Indoor Air Quality and Climate*. Vol. 1. Helsinki.
79. Otto, D.A. et al. 1990. "Neurotoxic Effects of Controlled Exposure to a Complex Mixture of Volatile Organic Compounds." U.S. Environmental Protection Agency—Health Effects Research Laboratory. EPA/600/1–90/00/.
80. Otto, D.A. et al. 1990. "Neurobehavioral and Sensory Irritant Effects of Controlled Exposure to a Complex Mixture of Volatile Organic Compounds." *Neurotox. Teratol.* 12:1–4.
81. Hudnell, H.K. et al. 1992. "Exposure of Humans to a Volatile Organic Mixture. II. Sensory." *Arch. Environ. Health.* 47:31–38.
82. Otto, D.A. et al. 1992. "Exposure of Humans to a Volatile Organic Mixture I. Behavioral Assessment." *Arch. Environ. Health.* 47:23–30.
83. Koren, H.S. et al. 1992. "Exposure of Humans to a Volatile Organic Mixture. III. Inflammatory Response." *Arch. Environ. Health.* 47:39–43.
84. Hudnell, K. et al. 1993. "Time Course of Odor and Irritation Effects in Humans Exposed to a Mixture of 22 Volatile Organic Compounds." 567–572. In: *Proceedings of the Sixth International Conference on Indoor Air Quality and Climate*. Vol. 1. Helsinki.

85. Prah, J.D. et al. 1993. "Pulmonary, Respiratory, and Irritant Effects of Exposure to a Mixture of VOCs at Three Concentrations in Young Men." 607–612. In: *Proceedings of the Sixth International Conference on Indoor Air Quality and Climate*. Vol. 1. Helsinki.
86. Otto, D. et al. 1993. "Neurobehavioral and Subjective Reactions of Young Men and Women to a Complex Mixture of Volatile Organic Compounds." 59–64. In: *Proceedings of the Sixth International Conference on Indoor Air Quality and Climate*. Vol. 1. Helsinki.
87. Berglund, B. et al. 1986. "Assessment of Discomfort and Irritation from the Indoor Air". 138–149. In: *Proceedings of IAQ '86: Managing Indoor Air for Health and Energy Conservation*. American Society of Heating, Refrigerating and Air-Conditioning Engineers. Atlanta.
88. Hudnell, H.K. et al. 1990. "Odour and Irritation Effects of a Volatile Organic Compound Mixture." 263–268. In: *Proceedings of the Fifth International Conference on Indoor Air Quality and Climate*. Vol. 1. Toronto.
89. Cain, W.S. et al. 1986. "Irritation and Odor from Formaldehyde: Chamber Studies." 126–137. In: *Proceedings of IAQ '86: Managing Indoor Air for Health and Energy Conservation*. American Society of Heating, Refrigerating and Air-Conditioning Engineers. Atlanta.
90. Cain, W.S. and C.L. Murphy. 1980. "Interaction Between Chemoreceptive Modalities of Odor and Irritation." *Nature*. 284:254–257.
91. Cain, W.S. 1974. "Perception of Odor Intensity and the Time-Course of Olfactory Adaptation." *ASHRAE Trans*. 80:53–75.
92. Molhave, L. 1990. "Volatile Organic Compounds, Indoor Air Quality and Health." 15–33. In: *Proceedings of the Fifth International Conference on Indoor Air Quality and Climate*. Vol. 5. Toronto.
93. Berglund, B. et al. 1984. "Characterization of Indoor Air Quality and Sick Buildings." *ASHRAE Trans*. 90, Pt. 1:1045–1055.
94. Kjaergaard, S. et al. 1990. "Common Chemical Sense of the Eyes—Influence of Smoking, Age, and Sex." 257–262. In: *Proceedings of the Fifth International Conference on Indoor Air Quality and Climate*. Vol. 1. Toronto.
95. Cain, W.S. 1987. "A Functional Index of Human Sensory Irritation." 661–665. In: *Proceedings of the Fourth International Conference on Indoor Air Quality and Climate*. Vol. 2. West Berlin.
96. Molhave, L. and G.D. Nielsen. 1992. "Interpretations and Limitations of the Concept 'Total Volatile Organic Compounds' (TVOC) as an Indicator of Human Responses to Exposures of Volatile Organic Compounds (VOC) in Indoor Air." *Indoor Air*. 2:65–77.
97. Hodgson, M.J. et al. 1991. "Symptoms and Microenvironmental Measures in Nonproblem Buildings." *J. Occup. Med*. 33:527–533.
98. Alarie, Y. 1981. "Dose-Response Analysis in Animal Studies: Prediction of Human Responses." *Environ. Health Perspect*. 42:9–13.
99. Berglund, B. et al. 1990. "A Longitudinal Study of Perceived Air Quality and Comfort in a Library Building." 489–494. In: *Proceedings of the Fifth International Conference on Indoor Air Quality and Climate*. Vol. 1. Toronto.
100. Baird, J. C. et al. 1987. "Distinguishing Between Healthy and Sick Preschools by Chemical Classification." *Environ. Int*. 13:167–174.

101. Wolkoff, P. et al. 1991. "A Study of Human Reactions to Emissions from Building Materials in Climate Chambers. Part II. VOC Measurements, Mouse Bioassay, and Decipol Evaluation in the 1–2 mg/m^3 TVOC Range." *Indoor Air*. 389–403.
102. Armstrong, C.W. et al. 1989. "Sick Building Syndrome Traced to Excessive Total Suspended Particulates (TSP)." 3–7. In: *Proceedings of IAQ '89: The Human Equation: Health and Comfort*. American Society of Heating, Refrigerating and Air-Conditioning Engineers. Atlanta.
103. Hodgson, M.J. and P. Collopy. 1991. "Symptoms and the Micro-Environment in the Sick Building Syndrome: A Pilot Study." 8–16. In: *Proceedings of IAQ '91: Healthy Buildings*. American Society of Heating, Refrigerating and Air-Conditioning Engineers. Atlanta.
104. Nelson, C.J. et al. 1991. "EPA's Indoor Air Quality and Work Environment Survey: Relationships of Employees' Self- Reported Health Symptoms with Direct Indoor Air Quality Measurements." 22–32. In: *Proceedings of IAQ '91: Healthy Buildings*. American Society of Heating, Refrigerating and Air-Conditioning Engineers. Atlanta.
105. Hanssen, S.O. 1993. "Increased Ventilation Reduces General Symptoms but Not Sensory Reactions." 33–38. In: *Proceedings of the Sixth International Conference on Indoor Air Quality and Climate*. Vol. 5. Helsinki.
106. Hedge, A.R. et al. 1993. "Effects of Man-Made Mineral Fibers in Settled Dust on Sick Building Syndrome in Air- Conditioned Offices." 291–296. In: *Proceedings of the Sixth International Conference on Indoor Air Quality and Climate*. Vol. 1. Helsinki.
107. Hedge, A.R. et al. 1991. "Breathing-Zone Filtration Effects on Indoor Air Quality and Sick Building Syndrome Complaints." 351–355. In: *Proceedings of IAQ '91: Healthy Buildings*. American Society of Heating, Refrigerating and Air-Conditioning Engineers. Atlanta.
108. Hedge, A.R. et al. 1993. "Effects of Furniture Integrated Breathing Zone Filtration System on Indoor Air Quality, Sick Building Syndrome, Productivity, and Absenteeism." 383–388. In: *Proceedings of the Sixth International Conference on Indoor Air Quality and Climate*. Vol. 5. Helsinki.
109. Kjaergaard, S. and J. Brandt. 1993. "Objective Human Conjunctival Reactions to Dust Exposure, VDT-Work, and Temperature in Sick Buildings." 41–46. In: *Proceedings of the Sixth International Conference on Indoor Air Quality and Climate*. Vol. 1. Helsinki.
110. Hansen, L. 1989. "Monitoring of Symptoms in Estimating the Effect of Intervention in the Sick Building Syndrome: A Field Study." *Environ. Int.* 15:159–162.
111. Leinster, P. et al. 1990. "A Modular Longitudinal Approach to the Investigation of Sick Building Syndrome." 287–292. In: *Proceedings of the Fifth International Conference on Indoor Air Quality and Climate*. Vol. 1. Toronto.
112. Raw, G.J. et al. 1991. "A New Approach to the Investigation of Sick Building Syndrome." 339–343. In: *Proceedings of the CIBSE National Conference*. London.
113. Raw, G.J. et al. 1993. "Sick Building Syndrome: Cleanliness is Next to Healthiness." *Indoor Air*. 3:237–245.
114. Raw, G.J. 1993. "Indoor Surface Pollution: A Cause of Sick Building Syndrome." British Research Establishment, United Kingdom.

115. Wallace, L.A. et al. 1991. "Workplace Characteristics Associated with Health and Comfort Concerns in Three Office Buildings in Washington, D.C." 56–60. In: *Proceedings of IAQ '91: Healthy Buildings*. American Society of Heating, Refrigerating and Air-Conditioning Engineers. Atlanta.
116. Hedge, A.R. et al. 1993. "Effects of Restrictive Smoking Policies on Indoor Air Quality and Sick Building Syndrome: A Study of 27 Air-Conditioned Offices." 517–522. In: *Proceedings of the Sixth International Conference on Indoor Air Quality and Climate*. Vol. 1. Helsinki.
117. Turner, S. et al. 1992. "The Measurement of Environmental Tobacco Smoke in 585 Office Environments." *Environ. Int.* 18:19–28.

6 CONTAMINANTS OF BIOLOGICAL ORIGIN

Both viable and non-viable contaminants of biological origin (biogenic) have been reported to cause building-related illness (BRI) such as hypersensitivity pneumonitis, humidifier fever, Legionnaires' disease, and asthma. In addition, it is probable that exposures to a variety of biological contaminants in building environments contribute to symptoms of allergic rhinitis (common allergy) in individuals who are atopic, that is, those who have a genetic predisposition to allergy. Biogenic contaminants responsible for causing BRI and/or allergic rhinitis include bacteria and their endotoxins, amoeba, fungal reproductive and vegetative structures, and fecal wastes and parts of dust mites and insects. In addition, a variety of biogenic contaminants have been suggested to be risk factors for sick building symptoms. These include both viable and non-viable organisms themselves or their antigens, microbial products such as glucans, endotoxins, mycotoxins, volatile organic compounds, and organic matter, macromolecular organic dust (MOD), found in particulate matter deposited on floor and indoor surfaces.

HYPERSENSITIVITY DISEASES

Hypersensitivity pneumonitis (HP) and its European relative humidifier fever (HF) represent building-related health problems which have a characteristic symptomatology and an etiology which can be confirmed by clinical laboratory testing. Outbreaks of HP in office and other nonresidential environments appear to be relatively uncommon, representing less than 5% of the health hazard evaluations of office, school, and hospital buildings conducted by NIOSH investigators.[1-3] NIOSH investigations, however, are usually initiated after major outbreaks involving a number of individuals and are unlikely to include buildings where only a few cases are present. Hypersensitivity pneumonitis may, as a consequence, be more common in buildings than NIOSH investigations would indicate.

Hypersensitivity Pneumonitis

Hypersensitivity pneumonitis is a pulmonary disease caused by an immunological and possibly cytotoxic response to the inhalation of a variety of biogenic aerosols. Acute responses are characterized by respiratory and systemic symptoms with an onset 4–6 hours after exposure. Typical symptoms include fever, chills, myalgia, malaise, shortness of breath, and cough. Recovery is spontaneous, usually occurring within 18 hours after exposure. Symptom severity depends on dose, individual sensitivity, and the nature of previous exposures. Symptoms may diminish in severity on continued exposure with reccurrence of severe symptoms after a period of nonexposure (e.g., over a weekend). Pathological effects occur in lung tissue with the infiltration of lung interstia and the formation of fibroblasts. In its chronic form, irreversible lung damage may occur with impairment of gas exchange.[4] It is also described as extrinsic allergic alveolitis, indicating that the alveoli or air sacs of the lungs become inflamed during the course of the disease.

Hypersensitivity pneumonitis represents a disease syndrome in which similar symptoms are observed on exposure to a variety of etiological agents. Ailments which can be described as types of HP include farmer's lung, brown lung (cotton workers' disease), pigeon fancier's disease, swine handler's disease, Pontiac fever, and the various hypersensitivity outbreaks reported for office and institutional environments. The manifestations of HP have typically been associated with high-level exposures to organic dusts and biological aerosols.

Epidemic-like outbreaks or individual cases of HP in building environments have been reported to occur as a result of exposures and sensitization to thermophilic actinomycetes,[5,6] mold,[7,8] thermotolerant bacteria,[9] and serotypes of bacteria that cause Legionnaires' disease.[10,11] In a number of cases where HP has been diagnosed, investigators have failed to identify causal agents.[12-14] Various components of building heating, ventilating, and air-conditioning (HVAC) systems have been associated with HP outbreaks. These include condensate drip pans,[8,15] ductwork,[5,6] humidifiers,[6,9] air washers,[15] and open spray cooling units.[15] Outbreaks have also been reported to have been caused by exposure to moldy file folder materials and other materials damaged by water intrusions.[12,15]

Cases of HP have been reported for exposures to the organism (*Legionella pneumophila*) that causes Legionnaires' disease.[10,11] Identification of this causal relationship was based on an outbreak of an acute febrile myalgia affecting 144 employees in a Pontiac, Michigan, health department building in 1968. The disease syndrome is now known as Pontiac fever. The etiological agent was identified from tissue samples almost a decade later, after the cause of Legionnaires' disease was identified. Pontiac fever resembles HP in that symptoms are similar including chills, fever, headache, and myalgia. It differs in that it has an unusually high attack rate (95%) and symptoms last 2–5 days. After the initial illness, affected individuals do not become symptomatic again. A contaminated evaporative condensor was believed to be the source of the disease organism which was spread through the building by the air-conditioning system. Other outbreaks of Pontiac fever have been reported for workers in an office building,[16] an automobile assembly plant,[17] and workers during routine maintenance inside a steam turbine condensor.[18]

Humidifier Fever

Humidifier fever is a respiratory disease that is closely related to HP. It has been exclusively reported in Europe in association with contaminant exposures produced by water spray humidifiers. Symptoms are influenza-like occurring at the end of the work shift or the evening after work and decreasing in severity as the work week progresses. Unlike HP, it is a relatively benign disease with no evidence of any long-term lung tissue damage. In general, investigations of HF outbreaks fail to identify a specific causal agent,[19,20] but exposures to nonpathogenic amoeba[19] or endotoxins from Gram negative bacteria[21] have been suggested as potential causal factors.

LEGIONNAIRES' DISEASE

Legionnaires' disease (LD) was first identified (and therefore so named) after an extensive investigation of illness and deaths among attendees of an American Legion convention in a hotel in Philadelphia, Pennsylvania, in 1976.[22] The disease syndrome is characterized by high fever, headache, and malaise and may also include cough and gastrointestinal problems. If untreated, it may progress to lung consolidation, respiratory failure, and death. Approximately 15% of 221 infected individuals at the Philadelphia American Legion convention died of the disease. Though the mortality rate was high, the attack rate was relatively low. The incubation period was 2 to 10 days. A variety of factors appeared to predispose individuals to the disease. These included middle age, cigarette smoking, excessive alcohol use, existing respiratory disease, malignancy, and immunosuppressive drugs.[23]

Since its initial identification as the cause of the epidemic outbreak of acute febrile illness in Philadelphia, cases of LD have been widely reported. It has been found to be a relatively common cause of human pneumonia.[24] A number of cases have reportedly occurred in hotels[25,26] and hospitals,[27,28] and sporadic cases have been observed in the general population. Studies of sporadic community-acquired LD implicate tobacco smoking, alcohol consumption, residing near excavation sites, and travel prior to disease onset as risk factors (29).

Legionnaires' disease is caused by *Legionella pneumophila*, a bacterium widely distributed in streams and lakes. Because *L. pneumophila* is somewhat chlorine tolerant, it is commonly found in potable water systems. As a consequence, it may not be removed in normal water treatment processes. It may grow in water environments which provide a favorable temperature, physical protection, and sufficient nutrients. Such environments include hot water storage tanks and recirculation lines, and hydraulic heat rejection systems such as cooling towers, evaporative condensors, and closed-loop recirculating cooling systems.[24]

Cooling Waters

In the first identified case of LD in Philadelphia, exposures appeared to have resulted from contaminated aerosols from a cooling tower. Since that time a number of outbreaks have been associated with aerosol drift from cooling towers and evaporative condensors.[30-34] The warm water in such systems provides excellent growing conditions for the causal organism.

Mechanical draft cooling towers (Figure 6.1) are used to dissipate heat from air-conditioning systems of hospitals, schools, and large office buildings.[35] In operation, warm water is pumped to the top of the cooling tower where it is atomized. As water gives off heat, small droplets coalesce into larger droplets and collect in a sump at the base from which it is recirculated. Large volumes of air are moved through the tower by mechanical fans to maximize cooling. Some of the aerosolized water is lost as drift. This drift has the potential for carrying infective *L. pneumophila* into buildings through outside air intakes.

L. pneumophila is commonly found in cooling tower waters. Witherell et al.,[36] for example, sampled waters from 184 cooling towers in Vermont. Twenty-two of these were positive for the LD organism. Of 54 continuously operating units, 15% were positive compared to 4% of 76 seasonally operated units. Tukki et al.[37] in Finland studied the occurrence of *L. pneumophila* in cooling tower and humidifier systems. It was observed to be present in waters of 14 of 30 cooling towers but only in 1 of 25 humidifiers sampled. Differences were apparently due to differences in water temperatures of the two systems (25.5 vs. 14°C). Seidel et al.[38] in Germany investigated both the presence of *L. pneumophila* in the waters of various environments and in aerosols and splashwaters generated by them (Table 6.1). *L. pneumophila* was observed to be present in 22 of 178 water samples taken from cooling towers of large-scale power plants, one out of 138 aerosol samples, and 11 of 547 aerosol and splashwater samples; in air-conditioning system cooling towers, *L. pneumophila* was observed in 6 of 30 water samples, 5 of 72 aerosol samples, and 4 of 60 aerosol and splashwater samples. Interestingly, highest recoveries occurred in whirlpool waters (9/14), hand basins (136/141), pools used for hydrotherapy (8/8), and shower water (69/80).

Bentham and Broadbent[39] observed that small cooling towers (<100 tons, 300 KW capacity) were predominantly implicated in LD outbreaks in Australia. These occurred primarily in autumn after a period of system shut-down. *L. pneumophila* concentrations in tower waters were observed to increase signifi-

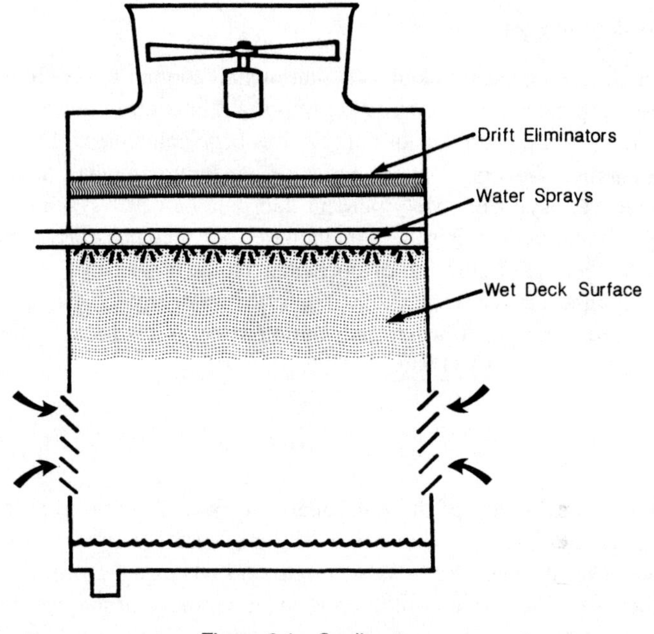

Figure 6.1 Cooling tower.

Table 6.1 Presence of *Legionella* spp. in water, aerosols and splashwaters of different environments.

	Positive *Legionella.* spp. per sample collected		
Environment	**Water**	**Aerosols**	**Aerosols and splashwaters**
Large-scale power plant	22/178	1/138	11/547
Activated sludge sewage treatment	2/34	1/74	0/92
Agricultural irrigation plants	0/35	0/41	0/10
Air-conditioning systems with cooling towers	6/30	5/72	4/60
Showers	69/80	27/137	5/42
Hand basins	136/141	0/32	59/350
Whirlpool	9/14	0/20	—
Hydrotherapy	8/8	0/7	0/4

From Seidel, K. et al., 1987. *Proceedings Fourth International Conference on Indoor Air Quality and Climate.* Vol. 1. West Berlin.

cantly when cooling towers were put into service after a short period of weather-related inactivity. This was probably due to biofilm and sediment disturbance on system start-up. A general relationship between increasing surface area/volume ratios of cooling tower systems and increasing summertime counts of *L. pneumophila* in basin cooling waters was observed.

Potable Water Systems

Though LD has been linked to contaminated cooling towers in a number of disease outbreaks, other sources are important in contributing to exposures as well. In hospitals, the focus of concern has been contaminated water distribution systems. Vickers et al.[40] conducted an environmental survey of 15 hospitals in Pennsylvania. Sixty percent had potable water systems contaminated by *L. pneumophila*. Risk factors included elevated hot water temperature (<60°C), vertical configuration of hot water tanks, older tanks, elevated calcium and magnesium concentrations, and river water supply. The presence of *L. pneumophila* in potable water distribution systems has been epidemiologically linked to cases of LD.[41-43]

ASTHMA

Asthma is a disease of the pulmonary or lower respiratory system. It is characterized by episodal constriction of the respiratory airways resulting in chest tightness, shortness of breath, cough, and wheezing. Symptoms, which vary in severity, may occur within an hour of exposure or may be delayed in onset for 4 to 12 hours. The incidence of asthma has been estimated to be 4–5% of the population.[44,45] In the United States 10–12 million individuals would presumably have asthma. A number of factors have been shown to cause or contribute to the disease. These include chemical irritants, exercise, cold air, respiratory infections, aspirin, and allergens.[46]

Asthma has been strongly associated with exposures to common allergens. A large percentage of asthmatic patients are atopic for allergens such as house dust, mold, and pollen.[46] Asthma, in these cases, results from an immunological response to the inhalation of airborne allergens which are primarily of biological origin. Such allergens must be in sufficient concentrations to induce immunological sensitization in those who are genetically prone (atopic). Prolonged exposure to allergens increases the risk of sensitization, and continued exposure, the probability that the asthmatic condition will develop once sensitization has occurred. Continued exposure also increases the severity of asthma.[46] Allergens produced by dust mites and mold appear to be the most important causes of asthma and initiators of asthmatic attacks.

Only a relatively few cases of inhalant allergen-induced asthma have been reported for nonresidential building environments. These include an outbreak of asthmatic-type symptoms in a printing plant in Great Britain.[47] Thirty-five workers complained of chest tightness, wheezing, and shortness of breath. Fifteen were clinically diagnosed as having asthma. This outbreak was related to contaminated humidifiers and humidifier antigens, though no single microorganism or antigen was determined to be responsible. Finnegan and Pickering[48] reported several cases of asthma in an air-conditioned office building. In one instance, asthma was associated with exposures to contaminated humidifier water which also produced four cases of HF including the asthmatic subject.

In a second case, no causal exposure could be determined. In both instances, affected workers had no known history of asthma prior to building-related exposures.

A relationship between the severity of asthmatic symptoms in children in Copenhagen, Denmark, and their school building environment has been reported by Hansen et al.[49] Asthmatic children in schools with carpeting had more severe asthmatic symptoms than asthmatic children in school buildings with no carpeting. In another study, Ibsen et al.,[50] comparing registrations with school health services and school design, observed that school buildings with textile floor coverings had significantly more cases of asthma.

CHRONIC ALLERGIC RHINITIS

Chronic allergic rhinitis is a relatively nonserious ailment affecting the upper respiratory system with characteristic symptoms of sinus inflammation, runny nose, sneezing, and eye irritation. It is caused by immunological sensitization to antigens found in house dust; mold spores and hyphal fragments; animal danders, saliva, and urine; insect excreta and parts; pollen; and allergens found in food. The primary type of sensitization is described as a type I reaction involving IgE-mediated responses.[51] The prevalence rate for chronic allergic rhinitis, or common allergy, has been estimated by Anderson and Korsgaard[44] to be about 8% in Denmark and worldwide. However, Shata et al.[52] have estimated that 21% of the population of West Germany is afflicted by common allergy. The latter is consistent with the observations of Norback et al.[53] who reported that signs of atopy are found in about 18–20% of individuals in sick and control buildings. Pickering[54] has recently reported that the prevalence of positive skin tests to common environmental allergens doubled from 23 to 46% in 30–60-year-olds over an 8-year period, indicating a very high prevalence of atopy at least in the general population of the United Kingdom.

ALLERGY AND ALLERGENS AS RISK FACTORS FOR SBS SYMPTOMS

Mucous membrane symptoms reported in both problem and systematic building investigations are similar to those associated with common allergic rhinitis. Given these similarities, (1) are reported SBS symptoms in part due to responses to allergen exposures of atopic individuals and/or (2) are atopic individuals as a group more sensitive to irritant gas/vapor- and particulate-phase contaminants which cause SBS symptoms?

The question of atopy has been addressed in Chapter 2. Almost all systematic studies have observed that atopy is a significant risk factor for SBS. As an example, Levy et al.[55] compared individuals with self-reported atopy to those with no reported indicators of atopy in a building population of approximately

2200 individuals. They observed a significant increased risk for general symptoms (RR = 1.8) and mucous membrane symptoms (RR = 3.0) for atopic individuals. Molina et al.[56] clinically evaluated two groups of employees working in naturally ventilated and fully air-conditioned buildings. The former included 38 and 48 putatively asymptomatic and symptomatic individuals, respectively. On clinical examination, significantly higher prevalence rates of rhinitis and/or sinusitis were observed in the population of study individuals from the air-conditioned building (67.3% vs. 7.9%). A significantly higher number of individuals in the air-conditioned building group tested positive to one or more common allergens with house dust being the most significant, (Table 6.2).

Though atopy appears to be a major risk factor for SBS symptoms, it is not clear whether this is due to exposures to allergens directly, that atopic individuals are more sensitive to gas/vapor- or particulate-phase substances/materials which may be the primary cause or causes of SBS symptoms, or both.

Limited evidence is available that atopic individuals are more sensitive to irritants such as formaldehyde. Higher prevalence rates of putatively formaldehyde-induced symptoms have been reported among atopic individuals exposed to formaldehyde in residential environments.[57-59]

It is probable that some portion of building populations reporting allergy-type symptoms are both immunologically sensitive and exposed to allergens present in their work environment. Such exposures may include allergens associated with dust mites, mold, and components of floor dust described as MOD.

Dust Mites

Exposure to allergens produced in the fecal wastes of dust mites appears to be the single most important cause of asthma and chronic allergic rhinitis in residential environments.[44] The two most important species contributing to human exposures are *Dermatophagoides pteronyssinus*, the European dust mite, and *D. farinae*, commonly found in North America. Dust mites are members of the spider family, which, because of their small size (250–300 µm) cannot be seen with the unaided eye. They eat sloughed off human skin and, not surprisingly, are common in environments where humans are present. Dust mite populations thrive in warm, humid places and do particularly well in residential environments with wall-to-wall carpeting.[60,61]

Systematic Building Studies

Neither dust mite populations nor mite antigen levels in floor dust have been assessed in the many cross-sectional epidemiological studies conducted to evaluate associations between SBS symptom prevalence rates and a large number of potential risk factors.

Table 6.2 Atopic status of a reference asymptomatic and symptomatic populations of two buildings determined from skin prick tests.

Allergen	Reference group	Air-conditioned building group	Statistical significance
Grass	2.6	14.3	NS
House dust	5.3	30.6	**
Dust mites	7.9	24.5	*
Other	13.1	8.2	NS
At least one positive	21.1	40.8	*

Note: * P = <0.05; ** P = <0.01.

From Molina, C. et al., 1993. *Proceedings Sixth International Conference on Indoor Air Quality and Climate.* Vol. 1. Helsinki.

Evidence to implicate dust mites and their allergens as a potential causal factor for SBS symptoms is limited to the experimental longitudinal studies conducted under the auspices of the British Research Establishment[62-64] and the case study of Lundblad.[65] In the former, an intensive cleaning program was implemented in a 16-story office building to determine whether such a mitigation effort would reduce both mite populations and SBS symptom prevalence. The dust cleaning regime which included carpet shampooing was observed to significantly reduce dust mite populations on seats and hard surfaces but not in carpeting. More notable was the significant reduction in symptom prevalence associated with the implementation of cleaning measures. The building symptom index declined from a pre-cleaning value of 3.40 to 2.14 in the post-cleaning period. Significant reductions in the prevalence of individual symptoms such as dry eyes, stuffy nose, dry throat, and dry skin were also observed. Lundblad[65] has reported results of a case investigation of upper respiratory symptoms in 2 occupants of an 11-story office building. After one of these individuals was diagnosed as having dust mite allergy, carpeting in both areas was treated with a specially formulated biocide which binds mite excrement and other particles containing allergens for easier vacuuming after application of the granular material to carpeting. Following miticidal application, symptoms in both employees were reported to have ceased with no significant reoccurrence.

Though the above studies suggest that dust mites and their allergens may play a role in SBS symptoms, results should be interpreted cautiously. In the BRE studies,[62-64] mite antigen levels were not assessed, and there were no significant reductions in mite levels in carpeting, a major nurturing environment for dust mites. Dust mite allergen levels are more directly related to allergic rhinitis and asthma than dust mite populations.[66] In the case study of Lundblad, semi-quantitative measurements of mite antigen levels in carpet dust indicated that mite infestation was light. Mite populations in the BRE study were also relatively low compared to those common in residential environments which appear to be the cause of allergy and asthmatic symptoms. In both studies, the effect of mite population reduction could not be distinguished from reductions in surface dust levels, which appear to be a significant risk factor for SBS symptoms.[64,67]

Dust Mite Populations

Only a few studies have been conducted to assess mite populations in nonresidential, nonindustrial indoor environments. In addition to the British office building study,[62-64] mite populations have been assessed in wards of hospitals in the United Kingdom,[68] in Southwest Ohio nursing homes,[69] and in University of Hawaii dormitories.[70] Mite levels in these studies were, for the most part, in the same range (from a few to about 100 mites/gram floor dust). In three of these studies, residential samples were also collected. Mite populations were as much as an order of magnitude higher in residential environments. Factors responsible for the relatively low dust mite populations in nonresidential buildings assessed have been suggested to include the presence of vinyl floors[69] and more rigorous cleaning practices.[68,69] Differences may have also been related to relative humidity which is a major limiting factor in the growth of mite populations.[60,61,71]

Mold

Mold is a term used commonly to describe a number of biologically similar organisms known as fungi. The largest percentage of fungal species extant on the earth's surface today are organisms of decay, particularly of materials of plant origin. These decay organisms are described as being saprophytic. Other fungi may be obligate or opportunistic pathogens of plants, animals, or humans. The reproductive propagules (spores) and, to a lesser extent, vegetative fragments of fungi (hyphal) are a common component of aerosol fractions of both ambient and indoor air. As such, they represent a major form of air contamination and potential human inhalation exposure.

Atopic individuals may develop an immunological sensitization as a result of exposure to one or more fungal allergens. Once sensitized to allergens of a particular fungal species, an individual may develop typical allergy-type symptoms whenever re-exposures are of sufficient magnitude. Symptoms can only develop from exposures to allergens to which an individual has been sensitized. It is not, however, uncommon for atopic individuals to have become sensitized to allergens from several or more fungal species as well as other allergens.

Presumptive Evidence of a Causal Link

Evidence from a number of investigations is suggestive of a potential association between mold exposures in office or institutional buildings and allergy-type respiratory symptoms. These include studies in which outbreaks of HP,[7,8] or potential exposures to toxicogenic fungi[72] have been reported.

Because of the levels of mold contamination reported, it is probable that some percentage of the building population may have experienced allergy symptoms as well.

Several studies have assessed the prevalence of respiratory symptoms among staff and/or children in day-care centers in Finland in which significant water damage, moisture problems, and mold contamination were experienced. In a study of 268 female staff of 30 day-care centers, Ruotsalainen et al.[73] observed that respiratory symptoms were more common among individuals in buildings with water damage and mold odor compared to those where only water damage was reported. However, only eye irritation was significantly higher in the former.

In another study, Koskinen et al.[74] reported that half of the children attending a moldy day-care center had chronic respiratory symptoms (rhinitis, nasal congestion, cough) and recurring sinusitis, otitis (ear infection), tonsillitis, and bronchitis. The health status of 16 children was evaluated the year after they left the moldy day-care center. Though the prevalence of rhinitis and nasal congestion remained high, cough decreased. A significant decline of sinusitis, otitis, and tonsillitis was also observed. Cough, nasal congestion, rhinitis, sore throat, hoarseness, eye irritation, and headache were significantly more prevalent in a second less moldy day-care center than in a reference day-care center.

Ahlen and Skange[75] conducted an investigation of work-related illness symptoms including throat irritation, eye irritation, headache, and lethargy among museum personnel. Most of the affected individuals tested positive to the antigens of *Phoma* spp., one of the mold taxa isolated from museum air. None of the spouses of affected individuals tested positive to *Phoma*, indicating that the exposure was work-related.

Cross-sectional studies[76-78] which have implicated fully air-conditioned buildings as a risk factor for SBS symptoms provide indirect evidence of a potential association between SBS symptoms and biological agents, which may include mold and its antigens. Such systems produce condensate waters which provide a favorable growth environment on organic dusts deposited in condensate drip pans,[8,15] HVAC system filters,[52,79-81] and porous acoustical/thermal system insulation.[82]

Schata et al.[52] and Elixmann et al.[79] have reported a strong association between SBS symptoms and exposure to fungal allergens in a hospital environment. Patients who developed allergy-type symptoms in air-conditioned rooms were tested by skin prick and inhalation challenge. Of 150 individuals tested, 135 were positive for allergens of fungi isolated from HVAC system filters. Filters were observed to serve as a medium upon which fungi grew and produced allergens which passed into supply air downstream.[74] This last case represents the broader concern of mold exposures and associated symptoms in buildings where gross mold contamination of building materials and/or systems is not evident.

Systematic Health Studies

Only a relatively few epidemiological studies have attempted to evaluate potential associations between SBS symptom prevalence rates and exposure to mold in buildings under study. All studies,[83-85] save the study by Harrison et al.,[86] observed no relationship between viable mold levels and SBS symptom prevalence.

Harrison et al.[86] conducted a study of SBS symptoms in 15 office buildings. When the relationship between viable/culturable mold levels and SBS symptom prevalence was evaluated across the spectrum of buildings investigated, it was found to be negative; that is, symptom prevalence appeared to increase with decreasing mold levels. However, when buildings were segregated into different ventilation types (naturally ventilated, mechanically ventilated, and fully air-conditioned), they observed significant positive relationships between mold levels and SBS symptom prevalence including blocked nose, dry throat, dry skin, and mean symptom frequency.

As indicated by the studies of Harrison et al.,[86] potential causal relationships between mold levels and SBS symptoms may be confounded by the ventilation status of a building. As a consequence, the lack of any apparent association between mold levels and SBS symptoms in other reported studies should be viewed with some caution. Further confounding the problem of establishing a significant relationship is that viable/culturable mold values are, at best, a relatively crude measure of human exposure to mold and its antigens. A major limitation to using total viable/culturable mold counts is that one ignores qualitative and quantitative differences in mold species which comprise samples. The importance of this is evident in the studies of Su et al.[87] who observed associations between reported hay fever-type symptoms in children and two specific fungal types, *Cladosporium* and yeast, whereas no significant association was observed for total viable/culturable mold levels.

In addition, viable/culturable mold values typically only represent a fraction of the total mold particle level present in a building space at a given time. Total mold particle levels vary from a few to tenfold higher than viable/culturable values.[88] Since allergenicity is independent of viability/culturability, these differences can be quite significant relative to the potential health risk associated with mold and mold allergen exposures.

Even if total mold particle counts were conducted for purposes of exposure assessment, they themselves may only be crude indicators of exposures to mold allergens. This is evident in the studies of Elixmann et al.[79] who observed that despite the fact that HVAC system filters were effective in reducing airborne mold levels, they nevertheless became a significant source of allergens which contaminated supply air downstream and were apparently the cause of respiratory symptoms in patients in an asthma hospital.

Exposure Assessments

A number of investigators have conducted mold assessment studies in office and institutional buildings. These assessments have, for the most part, attempted to quantify airborne concentrations of viable/culturable mold spore/particle levels and, in some cases, determine the various mold taxa present. Assessments have included mold concentrations in surface dusts as well.

Assessments of mold levels in office buildings have been conducted in the California Healthy Building Study,[85] the United Kingdom study of Harrison et al.,[86] the Finnish studies of Nevalainen et al.,[89] and in the Danish Town Hall Study.[83] In general, airborne mold levels reported as colony-forming units per cubic meter (CFU/m^3) were observed to be relatively low in all four studies. In the California study,[84] mold levels averaged 72 ± 10, 59 ± 20, and 12 ± 4.9 CFU/m^3 in naturally ventilated, mechanically ventilated, and air-conditioned buildings, respectively. Significantly lower levels were observed in fully air-conditioned buildings. Similar results were observed in the 15 office building U.K. study which reported median values of 277, 36 and 26 CFU/m^3 in naturally ventilated, mechanically ventilated, and fully air-conditioned buildings. Nevalainen et al.[89] reported average mold levels of 40 CFU/m^3 (range = 10–4700) in 70 office buildings. Mold levels in 14 Danish town hall buildings[83] averaged 32 CFU/m^3 (range = 0–111).

Mold assessments have also been conducted in a variety of institutional buildings. Nevalainen et al.[89] reported average levels of 70 CFU/m^3 (range = 10–340) in 11 Finnish day-care centers. Strom et al.[90] conducted both viable and total mold particle counts in a population of healthy and sick school buildings in Sweden. Viable/culturable mold concentrations were similar for the two building categories with concentrations ranging from 150–2700 CFU/m^3. Total mold particles were considerably higher, ranging from $10–80 \times 10^3$ mold particles/m^3. Mouilleseux et al.[91] reported average viable mold levels of 100 CFU/m^3 with a few samples exceeding 1000 CFU/m^3 in ten Paris schools.

In most cases when mold assessments are conducted in buildings, outdoor levels are observed to be considerably higher than indoor levels. This is evident in the California study[85] where indoor/outdoor ratios in naturally ventilated, mechanically ventilated, and air-conditioned buildings were 0.72, 0.62, and 0.12, respectively. The lower the value, the higher the outdoor concentration. Holt,[92] in a study of three multi-storied office buildings, reported indoor/outdoor ratios which averaged 31%. Morey and Jenkins,[93] in a study of seven office buildings, observed that indoor levels were only 10–25% of those measured outdoors at the same time. Maroni et al.[94] observed no differences in indoor/outdoor concentrations in ten naturally and seven mechanically ventilated buildings under wintertime conditions in Italy.

Few office and institutional buildings studies report the types and quantities of mold taxa present. Yang et al.[95] collected several thousand indoor and outdoor samples in unidentified workspace buildings in 14 states and the

District of Columbia. Approximately 75% of indoor samples had concentrations less than 200 CFU/m^3 with a range from the limit of detection (35 CFU/m^3) to 14,000 CFU/m^3. Dominant genera included *Cladosporium*, *Aspergillus*, *Penicillium*, *Basidiomycetes*, and *Alternaria*. In studies conducted by Morey and Jenkins,[93] *Aspergillus*, *Penicillium*, *Cladosporium*, *Aureobasidium*, and *Chaetomium* were the dominant fungal genera found indoors. In general, the composition of fungi in indoor samples was considerably different from those in outdoor samples.

Mold levels in buildings described above are in the main relatively low. Significantly higher concentrations can occur when a building is subject to major mold infestations. Such infestations occur as a consequence of water intrusions, high relative humidities, etc.[15] Higher concentrations have also been reported when aggressive sampling is conducted.[93] For example, when HVAC system induction units, internal insulation, or drip pans are disturbed, dramatic increases in indoor mold levels have been observed with indoor/outdoor ratios of 0.5 to 100.

Several investigators have determined mold levels in HVAC system filters and dusts. Martikainen et al.[80] sampled mold from intake filters in 11 buildings. Fungal spore counts varied from 7×10^2–2.5×10^5 CFU/g filter. Most commonly isolated genera were *Cladosporium* (47%) and *Penicillium* (24%). Microbial growth was not observed to be associated with the length of filter use. Significant fungal growth occurred on intake filters despite the fact that filters themselves had low nutrient and water contents. This suggests that the medium for growth was the matter trapped on filters.

Mold types isolated from air-conditioning filters in the studies of Shata et al.[79] are identified in Table 6.3. These species were observed to grow in the filters and sporulate on the downstream side, suggesting that filters could serve as a source of supply air contamination. Sverdrup and Nyman[96] reported that filters in outdoor air intakes of 11 buildings were regularly wet and supported fungal growth. Mold levels were, however, relatively low indoors (<50 CFU/m^3) with higher concentrations in outdoor air. Surface samples taken from the supply air duct had very low mold concentrations, indicating that the duct was relatively uncontaminated.

Morey and Williams[82] conducted studies of porous thermal insulation as potential amplification sites for mold and other microorganisms. This fiberglass insulation is commonly used to line the inside of plenums and ductwork immediately downstream of cooling deck coils. In one building, *Penicillium* and thermophilic actinomycete levels were 2–3 orders of magnitude higher than outdoors. Porous thermal insulation was determined to be the amplification site. In a second building, high mold levels were observed only after mechanical disturbance of porous insulation lining the cooling coil deck plenum. In a third building, elevated *Cladosporium* and *Epicoccum* levels were observed when the metal housing on an induction unit was gently disturbed. Typically, if the porous insulation was wet downstream, mold levels were high; if it was dry, elevated mold levels occurred only on disturbance.

Table. 6.3 Fungal species isolated from HVAC system filters.

Penicillium chrysogenum
P. brevicompactum
P. frequentans
Cladosporium hebarum
Aspergillus versicolor
A. fumigatus
A. repens
A. amstelodami
A. flavus
A. ruber
Alternaria tenius
Aureobasidium pullulans
Wallemia sebi

From Shata, M. et al., 1989. *Environ. Int.* 15:177–179. With permission.

Valbjorn et al.[81] observed a wide range of mold spore concentrations in dust taken from supply air ducts (70–6200 CFU/g dust) and from filter dust (70–3400 CFU/g dust). Dominant fungi isolated included *Penicillium*, *Chaetomium*, *Aspergillus*, *Alternaria*, *Mucor*, and *Rhizopous*. Dust was found to be "cleaner" in supply air as compared to exhaust air.

Macromolecular Organic Dust

The organic and presumably immunogenic fraction of floor and building surface dust has been implicated as a potential causal or risk factor for SBS symptoms. Nexo et al.[97] in Denmark reported cases of extreme fatigue, slight fever, and malaise in two individuals who associated symptoms with their work environment. After surveying an additional ten individuals, five were determined to have had work-related symptoms and four of five had precipitating antibodies against dust collected from poorly cleaned wall-to-wall carpeting. Symptoms resolved after carpet removal. Gravesen,[98] in investigations of office and school buildings, reported that complaints typically came from occupants of carpeted rooms. Gravesen[99] proposed that SBS symptoms were more prevalent in offices and schools which had high occupant density, used needle-felt carpeting, were subject to dust accumulation because of aspects of building design and construction, and had a reduced cleaning budget. Carpeting was suggested to be a reservoir for organic dust.

A significant relationship between the MOD fraction of floor dust with both mucous membrane and general symptoms has been reported in the Danish Town Hall Study.[83] Gravesen and Skov[100] have proposed that inhalation exposure of MOD may result in antigen-induced effects on the immune system.

Gravesen and Skov[100] studied MOD levels in dust from office buildings and schools. The MOD contents of carpet dust were significantly higher than dust collected from bare floors. Samples of floor dust contained proteins, rigid carbohydrates, and DNA molecules.

In more recent studies, Gravesen et al.[101] attempted to characterize the MOD fraction of floor dust. Human serum albumin (HSA) appeared to be the dominant component, showing a significant correlation (r = 0.65) with the amount of MOD. The HSA present appeared to be changed or modified (partially denatured or complexed with other molecules, either proteins or low molecular weight substances). HSA was suggested to act as a carrier for mycotoxins, endotoxins, or other molecules which may cause inflammatory or toxic reactions. It is interesting to note that HSA can bind with formaldehyde to produce a sensitizing antigen and antibodies in humans.[102]

Observed high MOD levels in textile floor coverings are notable. Studies have shown increased symptom prevalence rates in asthmatic children in carpeted schools as compared to those with bare floors[49] and a higher prevalence rate of asthma and allergy in schools with textile floor coverings.[50] In the school/office building study of Gravesen,[98] mold levels in carpet dust were also significantly higher than in the dust of bare floors. Predominant fungi included *Penicillium*, *Alternaria*, *Aspergillus*, and *Cladosporium*.

The nature of the potential causal relationship between MOD and SBS symptoms is not known. MOD is discussed here with other allergenic materials, such as dust mite wastes and mold spores, and vegetative fragments because of its immunogenicity.

OTHER BIOLOGICAL CONTAMINANT-RELATED HEALTH CONCERNS

Of the clinically diagnosable disease/health problems discussed previously, only allergic rhinitis appears to be directly and/or indirectly associated with sick building syndrome. Allergy symptoms result from exposures which provoke immunological responses. Here we discuss other exposures whose effects are not likely to be immunological in the classical antigen-antibody sense.

Bacteria

Though a number of bacteria are pathogenic and immune responses are initiated as a result of infections, bacteria apparently do not cause allergy-type responses characteristic of foreign materials associated with dust mites, mold, pollen, animal danders, etc. As a consequence, bacterial exposures in indoor air would not be expected to cause symptom responses similar to those associated with mold exposures. Nevertheless, potential associations between bacterial levels and SBS symptoms have been assessed in several systematic building studies. Additionally, a number of studies have attempted to assess bacteria levels in office and institutional buildings in HVAC system filter materials.

Systematic Building Studies

Potential associations between viable airborne bacteria levels and SBS symptoms have been evaluated in several systematic building investigations. No significant relationships between viable bacteria levels and SBS symptoms were observed in the Danish Town Hall Study[83] and California Healthy Building Study.[85] Harrison et al.[86] did observe significant positive correlations between total viable/culturable bacterial levels and SBS symptoms within different building ventilation groups. Inconsistent results were observed for bacteria grown at 36°C.

Assessment of Bacteria in Indoor Spaces

Holt[92] conducted air sampling for viable/culturable bacteria in three noncomplaint multi-storied office buildings. Levels varied from 0.3 CFU/m³ to 274 CFU/m³ with a mean value of 65 CFU/m³. Feeley et al.[103] reported that bacteria levels varied from 30–138 CFU/m³ in three complaint office buildings. In the U.K. Office Building Study of Harrison et al.,[86] bacteria levels were observed to be related to ventilation system type with average concentrations of 678, 233, and 200 CFU/m³ in naturally ventilated, mechanically ventilated, and fully air-conditioned buildings, respectively. Results were similar to those reported for mold. Highest levels were found in the naturally ventilated buildings which had the lowest prevalence rate of SBS symptoms, and lowest bacteria levels were found in the fully air-conditioned buildings with the highest prevalence rate of SBS symptoms. Relatively little differences in indoor viable bacterial levels (average values of 120–180 CFU/m³) in different building ventilation types were observed in the California Healthy Building Study.[85]

Strom et al.[90] in Sweden compared both viable and total bacterial levels in the air of healthy and sick school and office buildings. There were no differences in levels between healthy and sick buildings in terms of either viable or total bacterial cell counts. Schools appeared to have had higher viable bacterial cell counts than office buildings, while office buildings had slightly higher total bacterial levels. Dominant bacteria (including actinomycetes) identified included *Bacillus cereus, B. mucoides, B.* spp., *Micrococcus* spp., *Pseudomonas aeruginosa, P. alcaligenes, P. mesophilica, Streptomyces griseus*, and *S.* spp.

Nevalainen et al.[104] in Finland conducted sampling for airborne bacteria and mesophilic actinomycetes in 50 building environments including homes, schools, offices, and day-care centers. Actinomycetes are fungal-like organisms which are found in large numbers in soil. Though fungal-like in their growth habit, their primitive cell structure is closer to that of bacteria which they are classified as. Actinomycete levels ranged from 2–240 CFU/m³. In problem buildings, actinomycete colonies were observed in over 70% of air samples collected. They were only occasionally observed in samples of an apparently healthy building. As a consequence, a potential causal role for mesophilic actinomycetes for SBS symptoms in subarctic environments (where they are not a normal wintertime microflora of indoor urban air) has been suggested. Higher average bacteria levels were also reported in problem environments (GM = 600 CFU/m^3) compared to controls (GM = 380 CFU/m^3).

Nevalainen et al.[89] conducted sampling of bacteria in 11 day-care centers and compared their results to those in 31 office buildings. The geometric mean (GM) of 11 day-care centers was 1690 CFU/m^3 with a range of 180–8500 CFU/m^3; for the 31 offices, the GM value was 60 CFU/m^3 with a range of 10–340 CFU/m^3. Day-care centers appear to have significantly higher bacteria levels than office environments. This was suggested to be due to actively playing children producing high levels of airborne bacteria in inadequately ventilated buildings.

Martikainen et al.[80] conducted studies of microbial growth on HVAC system filter materials. Counts of bacteria varied from 3×10^3–1.9×10^5 CFU/g dry filter material. The most common bacterial genera recovered from filters were *Bacillus* (42%) and *Pseudomonas* (20%). Valbjorn et al.[81] conducted studies of viable bacteria in filter materials and dust in supply air ducts. The range of viable bacteria reported for filter materials was 100–6700 CFU/g filter and 50–5000 CFU/g dust. Sverdrup and Nyman[96] conducted bacterial sampling of filters in 12 buildings. They reported average bacterial levels of 35,000 CFU/g filter with a range of 2×10^3–7×10^4 CFU/g. They also conducted air sampling for viable bacteria at various locations within HVAC system components, the outdoor air and the room supplied. The mean bacteria values (CFU/m^3) were air intake channel, 13; supply duct, 33; exhaust duct, 50; in the room, 190; and outdoor air.[44] These values indicated that despite high bacteria levels found on filters, the occupied room itself was the primary source of viable bacteria contamination.

Microbial Products

Microorganisms produce a variety of products during their growth and development. Some of these are part of the organism itself or excreted by it to the environment. There is increasing evidence to suggest that exposure to some of these products (or by-products) may have significant health effects. The association between bacterial endotoxin exposure and HP is an example of one such health effect. In addition to bacterial endotoxins, other microbial products have been either reported to cause building-related symptoms or have been suggested as potential causal factors. These include fungal glucans, mycotoxins, and VOCs produced in microbial metabolism.

Endotoxins and Glucans

Endotoxins and glucans are polysaccharides produced in the cell walls of Gram negative bacteria and fungi, respectively.[105] Sverdrup and Nyman[96] conducted monitoring studies of airborne endotoxin and glucan levels in two relatively dirty office buildings in Sweden. Air sampling in one building indicated both the presence of Gram negative bacteria and endotoxin, with endotoxin concentrations of 75 ng/m^3. In a second building, relatively high concentrations of the fungal genus *Alternaria* were observed, with 0.8 ng/m^3 glucan found in occupied zones of the building.

Studies with organic dusts in farm environments[88] have shown that bacterial endotoxins can cause acute pulmonary function changes and inflammation of mucous membranes resulting in irritation of the eyes, nose, and throat, hoarseness, and dry cough.[106] Glucans have been reported to cause inflammation, stimulation of the reticulo-endothelial system, and activation of macrophages which are scavenging cells in the lungs.[107]

Rylander et al.[105] conducted two case studies to evaluate the potential role of endotoxins and glucans in causing SBS symptoms. Endotoxin and glucan levels were measured in 18 apartments where occupants had complained of subjective symptoms and four apartments in which no symptoms were reported. Levels of endotoxin in the 22 apartments ranged from 0.0 to 18.1 ng/m^3; the values of glucan ranged from 0.01 to 11.0 ng/m^3. Concentrations of endotoxin were significantly correlated ($r = 0.73$) with glucan levels. Data presented in Table 6.4 indicate a dose-response relationship between endotoxin levels and cough, breathing difficulty, and tiredness; glucan levels showed a relationship with irritation of the nose, hoarseness, and tiredness. Skin problems seemed to be more prevalent in high endotoxin and glucan exposure categories.

In another case study, Rylander[105] investigated a day-care center where personnel complained of irritation of the eyes, nose, and throat as well as cough and tiredness. Levels of airborne endotoxin and glucan were significantly higher in those areas of the building where symptoms were reported. Significantly lower endotoxin and glucan levels were observed when no human activity occurred. This observation indicated that endotoxin and glucan present in particulate matter became airborne as a result of human activity and air movement of sufficient velocity.

Rylander et al.[105] proposed that since exposure to endotoxin was related to cough and itchy eyes and glucan to irritation of the nose and hoarseness, it was likely that such exposures caused an irritation/inflammation of the respiratory airways. They further suggested that such responses are similar to those in non-office work environments described by them as "mucous membrane irritation syndrome".[108]

Mycotoxins

Mycotoxins are relatively large (200+ M.W.) and complex organic molecules which are associated with fungal hyphae or spores.[109] They are produced as secondary metabolites when fungal colonies begin to exhaust available substrate nutrients.[110] Human or animal exposure to mycotoxins may result in

Table 6.4 Symptom prevalence (%) of Swedish apartment dwellers related to exposures to endotoxin and glucan levels.

Concentration	Endotoxin			Glucan		
(ng/m³)	<0.1	0.1–0.2	>0.2	<0.6	0.6–2.0	>2.0
N =	10	17	8	10	14	11
Symptoms (%)						
Itchy eyes	50	47	50	20	57	64
Irritated nose	20	29	38	10	36	36
Hoarseness	20	29	50	30	9	36
Cough	20	12	38	10	14	36
Difficulty breathing	20	18	38	10	21	36
Tiredness	30	41	88	10	43	91
Headache	50	12	25	20	1	36
Nausea	20	6	38	0	1	27
Frequent colds	10	0	0	0	2	0
Sleep disturbance	30	18	63	20	21	55
Dry skin	30	6	38	20	21	18
Face itching	40	12	38	10	36	27
Itchy scalp	30	35	50	30	43	36
Backache	30	12	25	10	21	27
White fingers	20	6	25	10	14	18

From Rylander, R. et al., 1989. *Proceedings Present and Future of Indoor Air Quality.* Brussels. With permission.

poisonings described as mycotoxicoses. Historically, mycotoxicoses have been associated with mold contamination of cereals and grains. Many, if not most, species of fungi produce toxins which allow them to compete in substrate colonization. The antibiotic properties of penicillin and actinomycete-produced streptomycin are prime examples of this phenomenon. The most common fungi involved in mycotoxicoses are *Fusarium*, *Aspergillus*, and *Penicillium*. Mycotoxicoses produced by species in these genera have been historically associated with the consumption of agricultural products.

The best-known mycotoxins are those produced by species of the form genus *Fusarium.* A number of *Fusarium* species produce substances called trichothecenes.[111] Because of their high human toxicity, they allegedly have been used as chemical warfare agents.

Because mycotoxins are high molecular weight substances, they are not volatile. As such, they are not likely to be airborne except in the particulate phase. Human exposures would therefore most likely result from the inhalation of contaminated dusts, mold spores, and fungal vegetative fragments. Spores of some species have been reported to contain very high concentrations of mycotoxins. For example, spores of *F. gramineaum* and *F. sporotrichoides* have been found to contain 30 ppm deoxynivalenol and 50 ppm T-2 toxin,[110] spores of *Stachybotrys atra* 15 ppm of 3 toxins of that species,[112] and spores of *Aspergillus flavus* >600 ppm aflatoxins.[113] Tobin et al.[110] have suggested that high concentrations of mycotoxins on particles or in/on spores could reach lung tissue on inhalation where they would readily be absorbed through the mucous membranes of the respiratory tract. Mycotoxins in the lung could interfere with cell-mediated immunity and contribute to diffuse alveolitis.

Several investigators have reportedly associated illness with indoor exposure to toxigenic fungi including *Trichoderma viride*, *P. veridicatum*, and *S. atra*.[114-116] Croft et al.[116] reported a case of apparent mycotoxin-related illness among occupants of a home near Chicago, Illinois. After the house had suffered water damage from a leaking roof, occupants over several years became increasingly ill with complaints of headache, sore throat, hair loss, flu symptoms, diarrhea, fatigue, dermatitis, and general malaise. *S. atra* was observed to have infested a cold air duct and a second floor ceiling. Extracts of the ceiling material contained a number of highly toxic trichothecene mycotoxins which Croft et al.[116] concluded were the cause of the illness. On removal of contaminated materials, workmen experienced typical signs of trichothecene toxicosis including irritated eyes and throat and severe skin rash around the eyes and nose.

Johanning et al.[72] reported results of an investigation of a problem office building heavily contaminated with *S. atra*. The building had been subject to flooding and drain water backup. Affected individuals experienced a variety of central nervous system symptoms, eye and skin irritation, symptoms of the upper or lower respiratory tract, and unusual fatigue. The prevalence rate was highest in offices with significant tested mold contamination including both *S. atra* and *Aspergillus* species. Of 48 employees immunologically tested, 12 had elevated IgE antibodies with 7 positive to IgE-specific antibody tests to fungi. Four individuals were positive to *S. atra* and were observed to be clustered in an office area where toxigenic fungi were found. Significant contamination of sheet rock wallboard, air handling system insulation, and paper products was observed. Samples contained the potent cytotoxin saratoxin H, a member of the trichothecene mycotoxin group. Such toxins inhibit protein synthesis and cause immunosuppression.

In another investigation, Morey[117] reported a case of a 12,000 m^2 building in the southeastern United States which had a history of elevated relative humidity, visual presence of fungi on interior furnishings, and reports by occupants of building-related allergy-type illness. A significant infestation of vinyl-covered perimeter walls (dominated by *A. versicolor*) and ceiling tiles (dominated by *S. atra*) was observed. *A. versicolor* has been suspected to be toxigenic as well. The building was subsequently vacated and subject to intensive remediation.

Mainville et al.[118] reported results of an investigation of what they described as "exhaustion syndrome" in Quebec City hospital personnel. Over 300 individuals were diagnosed with another 300 cases suspected. Wipe sampling of building interior surfaces and interior HVAC system components was conducted. Seventeen different species of fungi including *S. atra* and *Trichoderma viride* were isolated. Both *S. atra* and *T. viride* are known to produce trichothecenes. They concluded that the syndrome which was similar to "chronic fatigue" was due to exposure to fungal toxins. Levels of contamination by these two toxigenic fungi either in numbers or trichothecene levels were not reported.

The investigations of Croft et al.,[116] Johanning et al.,[72] and Morey[117] involved cases where mold infestation by toxigenic fungi was heavy and, as a consequence, exposure to mycotoxins could be expected to occur as well. In most instances, exposures to mold and any mycotoxins would be expected to be significantly lower, particularly in so-called sick buildings. The potential effects of chronic, low-level exposures to mycotoxins is not known.

In assessments of mold contamination of office and institutional buildings previously described, mold levels were usually relatively low. In those cases where fungal taxa were reported, toxigenic species such as *S. atra* were apparently not observed. *S. atra* requires special culture media for isolation and would not normally be observed using ordinary sampling and culturing methods.

Microbial VOCs

Fungi, bacteria, and actinomycetes are known to produce a variety of VOCs. Fungal VOCs include a complex mixture of alcohols, esters, aldehydes, and paraffinic and aromatic hydrocarbons.[119] Kaminsky et al.[120] characterized VOC emissions from *A. flavus*. Reported fungal VOCs included 3-methylbutanol, 3-octanone, 3-octanol, 1-octen-3-ol, 1-octanol, and *cis*-2-octen-1-ol. Odor commonly associated with mold growth was suggested to be primarily due to 1-octen-3-ol. Kaminsky et al.[121] studied the emissions of VOCs from a number of imperfect fungi including *Aspergillus*, *Penicillium*, *Alternaria*, *Fusarium*, and *Cephalosporium*. The first five compounds previously reported by Kaminsky et al.[120] were observed to comprise 67–97% of all fungal VOCs observed. Other fungal VOCs were tentatively identified. These included butyl alcohol, isobutyl alcohol, butyl acetate, amyl acetate, octyl acetate, pyridine, hexanol, nonanone, and benzaldehyde. Miller et al.[119] identified VOCs emitted by seven common filamentous fungi. Compounds detected included heptane, benzene, octane, toluene, 2-hexanone, 3-methyl-1-butanol, hexanol, nonane, ethyl benzene, 3-pentanol, ethenyl benzene, 2-heptanone, 2-butoxy ethanol, benzaldehyde, naphthalene, decanol, and cyclohexane-1-methyl-4-1-methyl-ethenyl. Approximately a dozen of these compounds were detected in air samples collected in each of 52 houses. Eighty-nine percent contained 2-hexanol and 2-heptanone and 44%, 3-methyl-1-butanol. Only 2-hexanol, 2-heptanol, and 2-methyl-1-butanol were considered to be fungal in origin as most of the other compounds were suggested to be common in indoor air.

Bjurman[122] studied emissions of fungal VOCs emitted from a sample of thermal-insulating material taken from a house infested with mold. Significant emissions of ethylhexanol were reported. Other compounds detected were phenylacetaldehye and acetophonone. In additional studies, mineral wool insulation was inoculated with *P. brevicompactum* and *Chaetomium globosum*. Emissions included acetophenone, phenylacetaldehyde, octan-3-one, and octanoic acid. Octanoic acid was suggested to be the cause of the unpleasant odor associated with the infested sample. Geosmin was not detected despite the fact that *C. globosum* has the ability to produce it.

Table 6.5 Results of partial least squares analysis of VOC emissions from the fiber fraction of floor dust and "average mucous membrane irritation" and difficulty concentrating.

Average mucous membrane irritation		Difficulty concentrating	
VOC	**Modeling power**	**VOC**	**Modeling power**
2-Methylpropanal	0.640	Pentanoic acid	0.562
Hexanoic acid	0.586	Hexanoic acid	0.558
2-Alkanone $M^+ = 142$	0.586	Hexanal	0.526
3-Methylbutanal	0.498	35.8	0.492
35.8	0.420	Heptanoic acid	0.422
Octane	0.412	2-Methylbutanal	0.418
Pentanoic acid	0.375	Butryic acid	0.380
Heptanoic acid	0.362	Benzaldehyde	0.366
2-Undecanon	0.332		
5-Methyl-3-methylene-5-Hexene-2-on $M^+ = 124$	0.305		

From Wilkins, C.K. et al., 1993. *Indoor Air.* 3:283–290. With permission.

Wilkins et al.[123] characterized VOC and SVOC emissions from office dust samples collected in nine Danish town halls. Dominant compounds emitted included aldehydes ($C_{2\text{-}11}$) and carboxylic acids ($C_{2\text{-}14}$). Lipid decomposition by microorganisms or de novo synthesis was suggested as a potential origin of these substances. The potential association between VOC emissions from the fiber fraction of floor dust and symptoms of mucous membrane irritation and difficulty concentrating was evaluated by partial least squares (PLS) analysis. Results are summarized in Table 6.5. PLS analysis of VOC data combined with percentage complaints for six mucous membrane symptoms explained 80% of the variation in complaint data. Most of the compounds of high modeling power could have been produced as a result of the biological degradation of fatty acids or by de nova microbial synthesis.

A potential causal role for fungal VOCs in causing health complaints has been suggested by Samson[124] and Miller.[119] Fungal VOCs are suspected to cause acute respiratory response varying from stuffiness to wheezing. Strom et al.[90] have proposed that exposures to geosmin, a volatile compound with an earthy odor produced by the actinomycete *Streptomyces griseus* and several fungi including *Chaetomium*, may be responsible for sick building complaints in Swedish houses. Geosmin was suggested to cause mucous membrane irritation.

Norback et al.[125,126] evaluated the potential association between 2-ethyl-hexanol (a compound commonly produced by microorganisms) and SBS symptoms. No relationship was observed in the concentration range of 1–10 $\mu g/m_3$ for SBS symptoms or dampness. Emissions of 2-ethyl-hexanol were suggested to have been associated with polyvinyl floor covering. In a third study, Norback et al.[127] observed no apparent relationship between SBS symptoms and 2-ethyl-hexanol levels. Surprisingly, a positive association was observed between levels of 2-ethyl-hexanol and perceived air quality. Additionally, a significant relationship between exposures to 1-octen-3-ol, a microbial metabolite, and facial dermal symptoms and feelings of dustiness and high humidity were observed.

RECAPITULATION

Contaminants of biological origin have been shown to cause outbreaks of building-related illness which have a specific etiology and can be diagnosed clinically. These include hypersensitivity pneumonitis, humidifier fever, Legionnaires' disease, and asthma.

Both HP and HF are flu-like illnesses that have been associated with heavily contaminated components of HVAC systems or infested materials associated with water incursions. Biological contaminants which have reportedly caused HP include a variety of mold species, thermophilic actinomycetes, and strains of *L. pneumophila*. Humidifier fever reported in Europe has been reportedly caused by exposures to nonpathogenic amoebae or bacterial endotoxins from cold mist humidifiers. The two ailments differ — HP can result in progressive lung injury, while HF causes no permanent effects on lung tissue. A unique form of HP has been reported to be caused by exposures to strains of *L. pneumophila* and is known as Pontiac fever.

Legionnaires' disease is a relatively severe lung disease caused by exposures to infectious bacteria, *L. pneumophila*. The disease syndrome is pneumonia-like and has a fatality rate of approximately 20%. It is a building-related ailment in that exposures may result from the entrainment of contaminated cooling tower and evaporative condensor aerosols into building supply air. Exposures may also occur from potable water systems.

Asthma is a severe respiratory disease which is induced as a result of exposure to a variety of causal agents, most importantly antigenic materials from organisms such as dust mites and mold. Asthma has only rarely been associated with exposures in office and institutional workspaces. Cases of asthma can be expected in building environments where heavy infestations of mold have occurred.

Chronic allergic rhinitis has a specific etiology and is subject to clinical diagnosis. It is caused by exposures to a variety of allergenic materials and its association with residential environments has been well documented. Little information is available relative to exposures to allergenic materials and the elicitation of symptoms in office, institutional, and other public access buildings. It is notable that mucous membrane symptoms associated with SBS are indistinguishable from those associated with common allergy. Given the relatively high prevalence of atopy in the general population, it is probable that some portion of SBS symptoms are due to allergen exposures. Additionally, atopy in and of itself appears to be a risk factor for SBS symptoms.

Potential allergens which may be found in workplace environments are dust mite fecal wastes, mold spores, and fungal hyphal fragments and MOD.

Both dust mite and mold concentrations in office and institutional buildings are relatively low compared to those reported in residential environments. This would suggest that exposures are not sufficient in magnitude to cause symptoms. Some studies are suggestive of a causal link between dust mite-related exposures and SBS symptoms. There is some evidence to indicate that mold allergens can be produced on and pass through HVAC system filters and cause allergy symptoms even though airborne mold levels remain relatively low.

Exposure to MOD in floor dust has been suggested to be a risk factor for SBS symptoms. MOD has been shown to be significantly associated with both mucous membrane and general symptoms in the Danish Town Hall Study. MOD appears to be immunogenic in nature and consists of protein, DNA molecules, etc. Modified human serum albumin appears to comprise the dominant portion of the MOD fraction.

Other biological contaminants have been evaluated as potential risk factors or contributors to SBS symptoms. Exposures to bacteria and microbial metabolites have received limited study as potential causal factors of SBS symptoms. Most cross-sectional epidemiological studies have failed to show any relationship between viable bacteria levels and SBS symptoms. However, several complaint investigations have shown significant associations between endotoxins and glucan exposures and several SBS symptoms. Endotoxins and glucans are polysaccharide substances produced in the cell walls of Gram negative bacteria and fungi, respectively.

Other microbial metabolites of concern include mycotoxins and VOCs. Several cases of mycotoxin exposures and associated health effects have been reported in residential and office buildings. In each case, mycotoxin exposure resulted from heavy infestation of building materials with toxigenic fungi. The potential relationship between mycotoxins and SBS symptoms in building environments with no frank evidence of mold infestation has yet to be evaluated.

Microorganisms such as bacteria, actinomycetes, and fungi produce a variety of VOCs or semi-volatile VOCs during the course of metabolism. Exposure to these metabolites has been suggested as a potential contributing factor to SBS symptoms. Only a relatively few studies have been conducted to both characterize microbial VOCs (or semi-VOCs) and to evaluate their potential health effects. The most notable findings to date are those that report a relationship between putatively microbially derived metabolites in floor dust and SBS symptoms. These compounds consist primarily of aldehydes and carboxylic acids and differ considerably in their volatility.

REFERENCES

1. Melius, J. et al. 1984. "Indoor Air Quality — The NIOSH Experience." *Ann. ACGIH: Evaluating Office Environmental Problems.* 10:3–8.

2. Wallingford, K.M. and J. Carpenter. 1986. "Field Experience Overview: Investigating Sources of Indoor Air Quality Problems in Office Buildings." 448–453. In: *Proceedings IAQ '86: Managing Indoor Air for Health and Energy Conservation*. American Society of Heating, Refrigerating and Air-Conditioning Engineers. Atlanta.
3. Seitz, T.A. 1990. "NIOSH Indoor Air Quality Investigations 1971–1988." 163–171. In: Weekes, D.M. and R.B. Gammage (Eds.). *Proceedings Indoor Air Quality International Symposium: Practitioner's Approach to Indoor Air Quality Investigations*. American Industrial Hygiene Association. Akron, OH.
4. Fink, J. 1983. "Hypersensitivity Pneumonitis." 1085–1099. In: Middleton, E. et al. (Eds.). *Allergy: Principles and Practice*. 2nd ed. C.V. Mosby, St. Louis, MO.
5. Banaszak, E.F. et al. 1970. "Hypersensitivity Pneumonitis Due to Contamination of an Air-Conditioner." *New Engl. J. Med.* 283:271–276.
6. Patterson, R. et al. 1981. "Hypersensitivity Lung Disease Presumptively Due to *Cephalosporium* in Homes Contaminated by Sewage Flooding or by Humidifier Water." *J. Aller. Clin. Immunol.* 68:128–132.
7. Solley, G.O. and R.E. Hyatt. 1980. "Hypersensitivity Pneumonitis Induced by *Penicillium* Species." *J. Aller. and Clin. Immunol.* 65:65–70.
8. Bernstein, R.S. et al. 1983. "Exposures to Respirable Airborne *Penicillium* from a Contaminated Ventilation System: Clinical, Environmental, and Epidemiological Aspects." *Am. Ind. Hyg. Assoc. J.* 44:161–169.
9. Kohler, P.F. et al. 1976. "Humidifier Lung: Hypersensitivity Pneumonitis Related to Thermotolerant Bacterial Aerosols." *Chest*. Suppl. 69:294–296.
10. Glick, T.H. et al. 1978. "Pontiac Fever. An Epidemic of Unknown Etiology in a Health Department. I. Clinical and Epidemiological Aspects." *Am. J. Epidemiol.* 107:149–160.
11. Kaufmann, A.F. et al. 1981. "Pontiac Fever: Isolation of the Etiologic Agent (*Legionella pneumophila*) and Demonstration of its Mode of Transmission." *Am. J. Epidemiol.* 114:337–347.
12. Hodgson, M.J. et al. 1985. "Pulmonary Disease Associated with Cafeteria Flooding." *Arch. Environ. Health.* 40:96–101.
13. Arnow, P.M. et al. 1978. "Early Detection of Hypersensitivity Pneumonitis in Office Workers." *Am. J. Med.* 64:236–241.
14. Hodgson, M.J. et al. 1987. "An Outbreak of Recurrent Acute and Chronic Hypersensitivity Pneumonitis in Office Workers." *Am. J. Epidemiol.* 125:631–638.
15. Morey, P.R. et al. 1984. "Environmental Studies in Moldy Office Buildings: Biological Agents, Sources, Preventative Measures." *Ann. ACGIH: Evaluating Office Environmental Problems.* 10:21–36.
16. Friedman, S. et al. 1987. "Pontiac Fever Outbreak Associated with a Cooling Tower." *Am. J. Pub. Health.* 77:568–572.
17. Herwaldt, L.A. et al. 1984. "A New *Legionella* species, *Legionella feelei* Species Nova, Causes Pontiac Fever in an Automobile Plant." *Ann. Intern. Med.* 100:333–338.
18. Fraser, D.W. et al. 1979. "Nonpneumonic, Short Incubation-Period Legionellosis (Pontiac Fever) in Men Who Cleaned a Steam Turbine Condenser." *Science.* 205:690–691.

19. Edwards, J.H. 1980. "Microbial and Immunological Investigations and Remedial Action After an Outbreak of Humidifier Fever." *Br. J. Ind. Med.* 37:55–62.
20. Ganier, M. et al. 1980. "Humidifier Lung: Another Outbreak in Office Workers." *Chest.* 77:183–187.
21. Rylander, R. et al. 1980. "Humidifier Fever and Endotoxin Exposure." *Clin. Aller.* 8:511–516.
22. Imperato, P.J. 1981. "Legionellosis and the Indoor Environment." In: *Proceedings Symposium Health Aspects of Indoor Air Pollution. Bulletin New York Acad. Med.* 57:922–935.
23. Ager, B.P. and J.A. Tickner. 1983. "The Control of Microbiological Hazards Associated with Air-Conditioning and Ventilation Systems." *Ann. Occup. Hyg.* 27:341–358.
24. Muraca, P.W. et al. 1988. "Legionnaires' Disease in the Work Environment: Implications for Environmental Health." *Am. Ind. Hyg. Assoc. J.* 49:584–590.
25. Rosmini, F. et al. 1984. "Febrile Illness in Successive Cohorts of Tourists at a Hotel on the Italian Adriatic Coasts: Evidence for a Persistent Focus of *Legionella* Infection." *Am. J. Epidemiol.* 119:124–134.
26. Grist, N.R. et al. 1979. "Legionnaires' Disease and the Traveller." *Ann. Int. Med.* 90:563–564.
27. Best, M. et al. 1983. "Legionellaceae in the Hospital Water Supply: Epidemiological Link with Disease and Evaluation of a Method for Control of Nosocomial Legionnaires' Disease and Pittsburgh Pneumonia Agent." *Lancet.* 2:307–310.
28. Fisher-Hoch, S.P. et al. 1981. "Investigation and Control of an Outbreak of Legionnaires' Disease in a District General Hospital." *Lancet.* 1:932–936.
29. Storch, G. et al. 1979. "Sporadic Community-Acquired Legionnaires' Disease in the United States: A Case Control Study." *Ann. Intern. Med.* 90:596–600.
30. Klaucke, D.N. et al. 1984. "Legionnaires' Disease: The Epidemiology of Two Outbreaks in Burlington, VT, 1980." *Am. J. Epidemiol.* 119:382–341.
31. Buehler, J.W. et al. 1985. "Prevalence of Antibodies to Legionella Pneumophila Among Workers Exposed to a Contaminated Cooling Tower." *Arch. Environ. Health.* 40:207–210.
32. Politi, B.D. et al. 1979. "A Major Focus of Legionnaires' Disease in Bloomington, Indiana." *Ann. Intern. Med.* 90:587–591.
33. Cordes, L.G. et al. 1980. "Legionnaires' Disease Outbreak at an Atlanta, Georgia Country Club: Evidence for Spread From an Evaporative Condenser." *Am. J. Epidemiol.* 111:425–431.
34. Dondero, T.J. et al. 1980. "An Outbreak of Legionnaires' Disease Associated with a Contaminated Air-Conditioning Cooling Tower." *New Engl. J. Med.* 302:365–370.
35. Miller, R.P. 1979. "Cooling Towers and Evaporative Condensers." *Ann. Int. Med.* 90:667–670.
36. Witherell, L.E. et al. 1986. "*Legionella* in Cooling Towers." *J. Environ. Health.* 49:134–139.
37. Tukki, A. et al. 1990. "*Legionella* in Cooling Tower and Humidifier Systems: The Effects of Temperature and Nutrient Concentrations." 547–549. In: *Proceedings of the Fifth International Conference on Indoor Air Quality and Climate.* Vol. 4. Toronto.

38. Seidel, K. et al. 1987. "*Legionellae* in Aerosols and Splashwaters in Different Habitats." 690–693. In: *Proceedings of the Fourth International Conference on Indoor Air Quality and Climate*. Vol. 1. West Berlin.
39. Bentham, R.H. and C.R. Broadbent. 1993. "An Explanation for Autumn Outbreaks of Legionnaires' Disease Associated with Small Cooling Towers." 329–332. In: *Proceedings of the Sixth International Conference on Indoor Air Quality and Climate*. Vol. 4. Helsinki.
40. Vickers, R.M. et al. 1987. "Determinants of *L. pneumophila* Contamination of Water Distribution Systems: 15-Hospital Prospective Study." *Infect. Control*. 8:357–363.
41. Best, M. et al. 1983. "Legionellaceae in the Hospital Water Supply — Epidemiological Link with Disease and Evaluation of a Method of Control of Nosocomial Legionnaires' Disease and Pittsburgh Pneumonia." *Lancet*. ii:307–310.
42. Johnson, J.T. et al. 1985. "Nosocomial Legionellosis Uncovered in Surgical Patients with Head and Neck Cancer; Implications for Epidemiologic Reservoir and Mode of Transmission." *Lancet*. ii:298–300.
43. Plouffe, J.F. et al. 1983. "Subtypes of *Legionella Pneumophila* Subgroup 1 Associated with Different Attack Rates." *Lancet*. ii:649–650.
44. Anderson, I. and J. Korsgaard. 1984. "Asthma and the Indoor Environment — Assessment of the Health Implications of High Indoor Humidity." 79–86. In: *Proceedings of the Third International Conference on Indoor Air Quality and Climate*. Vol. 1. Stockholm.
45. American Thoracic Society. 1990. "Environmental Controls and Lung Disease." Report to the ATS Workshop on Environmental Controls and Lung Disease. Santa Fe, NM. March 24–26, 1988. *Am. Rev. Respir. Dis*. 142:915–939.
46. Reed, C.E. and R.G. Toconley. 1983. "Asthma — Classification and Pathogenesis." 811–831. In: Middleton, E. et al. (Eds.) *Allergy: Principles and Practice*. 2nd ed. C.V. Mosby. St. Louis, MO.
47. Burge, P.S. et al. 1985. "Occupational Asthma in a Factory with a Contaminated Humidifier." *Thorax*. 40:248–254.
48. Finnegan, M.J. and C.A.C. Pickering. 1984. "Occupational Asthma and Humidifier Fever in Air-Conditioned Buildings." 257–262. In: *Proceedings of the Third International Conference on Indoor Air Quality and Climate*. Vol. 3. Stockholm.
49. Hansen, L. et al. 1987. "Carpeting in Schools as an Indoor Pollutant." 727–731. In: *Proceedings of the Fourth International Conference on Indoor Air Quality and Climate*. Vol. 2. West Berlin.
50. Ibsen, K.K. et al. 1981. "The Indoor Environment and Symptoms of Disease. Assessment of the Conditions of Schools in Copenhagen." *Ugeskr. Lalger*. 143:1919–1923. (In Danish-English Summary).
51. Lowenstein, H. et al. 1979. "Airborne Allergens — Identification of Problems and the Influence of Temperature, Humidity and Ventilation." 111–125. In: *Proceedings of the First International Indoor Climate Symposium*. Danish Research Institute. Copenhagen.
52. Schata, M. et al. 1989. "Allergies to Molds Caused by Fungal Spores in Air-Conditioning Equipment." *Environ. Int*. 15:177–179.

53. Norback, D. et al. 1989. "The Prevalence of Symptoms Associated with Sick Buildings and Polluted Industrial Environments as Compared to Unexposed Reference Groups Without Expressed Dissatisfaction." *Environ. Int.* 15:85–94.
54. Pickering, C.A.C. 1993. "Allergy and Environmental Hypersensitivity Related to the Indoor Environment." 13–19. In: *Proceedings of the Sixth International Conference on Indoor Air Quality and Climate*. Vol. 1. Helsinki.
55. Levy, F. et al. 1993. "Gender and Hypersensitivity as Indicators of Indoor-Related Health Complaints in a National Reference Population." 357–362. In: *Proceedings of the Sixth International Conference on Indoor Air Quality and Climate*. Vol. 1. Helsinki.
56. Molina, C. et al. 1993. "Sick Building Syndrome and Atopy." 369–374. In: *Proceedings of the Sixth International Conference on Indoor Air Quality and Climate*. Vol. 1. Helsinki.
57. Garry, V.F. 1980. "Formaldehyde in the Home. Environmental Disease Perspectives." *Minn. Med.* 63:107–111.
58. Liu, K.S. et al. 1987. "Irritant Effects of Formaldehyde in Mobile Homes." 610–614. In: *Proceedings of the Fourth International Conference on Indoor Air Quality and Climate*. Vol. 2. West Berlin.
59. Krzyzanowski, M. et al. 1990. "Chronic Respiratory Effects of Indoor Formaldehyde Exposure." *Environ. Res.* 52:117–125.
60. Arlian, L.G. et al. 1982. "The Prevalence of House Dust Mites *Dermatophagoides* spp. and Associated Environmental Conditions in Ohio." *J. Aller. Clin. Immunol.* 69:527–533.
61. Woodford, P.J. et al. 1979. "Population Dynamics of *Dermatophagoides* spp. in Southwest Ohio Homes." 197–204. In: Rodriguez, J.G. (Ed.). *Recent Advances in Acarology*. Vol. II. Academic Press. New York.
62. Leinster, P. et al. 1990. "A Modular Longitudinal Approach to the Investigation of Sick Building Syndrome." 287–292. In: *Proceedings of the Fifth International Conference on Indoor Air Quality and Climate*. Vol. 1. Toronto.
63. Raw, G.J. et al. 1991. "A New Approach to the Investigation of Sick Building Syndrome." 339–343. In: *Proceedings of CIBSE National Conference*. London.
64. Raw, G.J. et al. 1993. "Sick Building Syndrome: Cleanliness is Next to Healthiness." *Indoor Air*. 3:237–245.
65. Lundblad, F.P. 1991. "House Dust Mite Allergy in an Office Building." *Appl. Occup. Environ. Hyg*. 6:94–96.
66. Report of a Second International Workshop. 1989. "Dust Mite Allergens and Asthma." *J. Aller. Clin. Immunol.* 83:416–427.
67. Raw, G.J. 1993. "Indoor Surface Pollution: A Cause of Sick Building Syndrome." Building Research Establishment, United Kingdom.
68. Blythe, M.E. et al. 1975. "Study of Dust Mites in Three Birmingham Hospitals." *Br. Med. J.* 1:62–64.
69. Vyszenski-Moher, D.L. et al. 1986. "Prevalence of House Dust Mites in Nursing Homes in Southwest Ohio." *J. Aller. Clin. Immunol.* 77:745–748.
70. Massey, D.G. et al. 1988. "House Dust Mites in University Dormitories." *Ann. Allergy*. 61:229–231.
71. Korsgaard, J. 1982. "Preventative Measures in House Dust Allergy." *Am. Rev. Respir. Dis.* 125:80–84.

72. Johanning, E. et al. 1993. "Clinical Epidemiological Investigation of Health Effects Caused by *Stachybotrys atra* Building Contamination." 225–230. In: *Proceedings of the Sixth International Conference on Indoor Air Quality and Climate*. Vol. 4. Helsinki.
73. Ruotsalainen, R. et al. 1993. "Water Damage and Moisture Problems as Determinants of Respiratory Symptoms Among Workers in Day-Care Centers." 317–322. In: *Proceedings of the Sixth International Conference on Indoor Air Quality and Climate*. Vol. 4. Helsinki.
74. Koskinen, O. et al. 1993. "Respiratory Symptoms and Infections Among Children in a Day–Care Center with Mold Problem." 231–236. In: *Proceedings of the Sixth International Conference on Indoor Air Quality and Climate*. Vol. 1. Helsinki.
75. Ahlen, A. and F. Skange. 1993. "Exposure Relevant Mould Antigens/Allergens — A Necessary Tool for Reliable Demonstration of Immunological Response." 181–184. In: *Proceedings of the Sixth International Conference on Indoor Air Quality and Climate*. Vol. 1. Helsinki.
76. Burge, P.S. et al. 1987. "Sick Building Syndrome: A Study of 4373 Office Workers." *Ann. Occup. Hyg*. 31:493–504.
77. Mendell, M.J. and A.H. Smith. 1990 "Consistent Pattern of Elevated Symptoms in Air-Conditioned Office Buildings: A Re-analysis of Epidemiological Studies." *Am. J. Pub. Health*. 80:1193–1199.
78. Zweers, T. et al. 1992. "Health and Indoor Climate Complaints of 7043 Office Workers in 61 Buildings in the Netherlands." *Indoor Air*. 2:127–136.
79. Schata, M. et al. 1989. "Allergies to Molds caused by Fungal Spores in Air Conditioning Equipment." *Environ. Int*. 15:177–179.
80. Martikainen, P.J. et al. 1990. "Microbial Growth on Ventilation Filter Materials." 203–206. In: *Proceedings of the Fifth International Conference on Indoor Air Quality and Climate*. Vol. 3. Toronto.
81. Valbjorn, O. et al. 1990. "Dust in Ventilation Ducts." 361–364. In: *Proceedings of the Fifth International Conference on Indoor Air Quality and Climate*. Vol. 3. Toronto.
82. Morey, P.R. and C. Williams. 1990. "Porous Insulation in Buildings: A Potential Source of Microorganisms." 529–533. In: *Proceedings of the Fifth International Conference on Indoor Air Quality and Climate*. Vol. 4. Toronto.
83. Skov, P. et al. 1990. "Influence of Indoor Climate on the Sick Building Syndrome in an Office Environment." *Scand. J. Work Environ. Health*. 16:367–371.
84. Menzies, R. et al. 1993. "Impact of Exposure to Multiple Contaminants on Symptoms of Sick Building Syndrome." 363–368. In: *Proceedings of the Sixth International Conference on Indoor Air Quality and Climate*. Vol. 1. Helsinki.
85. Fisk, W.J. et al. 1993. "The California Healthy Building Study, Phase 1: A Summary." 279–284. In: *Proceedings of the Sixth International Conference on Indoor Air Quality and Climate*. Vol. 1. Helsinki.
86. Harrison, J. et al. 1992. "An Investigation of the Relationship Between Microbial and Particulate Indoor Air Pollution and the Six Building Syndrome." *Respir. Med*. 225–235.
87. Su, H.J. et al. 1990. "Examination of Microbiological Concentrations and Association with Childhood Respiratory Health." 21–26. In: *Proceedings of the Fifth International Conference on Indoor Air Quality and Climate*. Vol. 2. Toronto.

88. Burge, H.A. et al. 1977. "Comparative Recoveries of Airborne Fungal Spores by Viable and Nonviable Modes of Volumetric Collection." *Mycopathologia.* 61:27–33.
89. Nevalainen, A. et al. 1987. "Airborne Bacteria, Fungal Spores, and Ventilation in Finnish Day-Care Centers." 678–680. In: *Proceedings of the Fourth International Conference on Indoor Air Quality and Climate.* Vol. 1. West Berlin.
90. Strom, G. et al. 1990. "The Sick Building Syndrome. An Effect of Microbial Growth in Building Constructions?" 173–178. In: *Proceedings of the Fifth International Conference on Indoor Air Quality and Climate.* Vol. 1. Toronto.
91. Mouillesaux, A. et al. 1993. "Microbial Characterization of Air Quality in Classrooms." 195–200. In: *Proceedings of the Sixth International Conference on Indoor Air Quality and Climate.* Vol. 4. Helsinki.
92. Holt, G.L. 1990. "Seasonal Indoor/Outdoor Fungi Ratios and Indoor Bacteria Levels in Non–complaint Office Buildings." 33–38. In: *Proceedings of the Fifth International Conference on Indoor Air Quality and Climate.* Vol. 2. Toronto.
93. Morey, P.R. and B.A. Jenkins. 1989. "What Are Typical Concentrations of Fungi, Total Volatile Organic Compounds, and Nitrogen Dioxide in an Office Environment?" 67–71. In: *Proceedings of IAQ '89: The Human Equation: Health and Comfort.* American Society of Heating, Refrigerating and Air-Conditioning Engineers. Atlanta.
94. Maroni, M. et al. 1993. "Microbial Contamination in Buildings: Comparison Between Seasons and Ventilation Systems." 137–142. In: *Proceedings of the Sixth International Conference on Indoor Air Quality and Climate.* Vol. 4. Helsinki.
95. Yang, C.S. et al. 1993. "Airborne Fungal Populations in Non-Residential Buildings in the United States." 219–224. In: *Proceedings of the Sixth International Conference on Indoor Air Quality and Climate.* Vol. 4. Helsinki.
96. Sverdrup, C.F. and E. Nyman. 1990. "A Study of Micro-organisms in the Air of 12 Swedish Buildings." 583–588. In: *Proceedings of the Fifth International Conference on Indoor Air Quality and Climate.* Vol. 4. Toronto.
97. Nexo, E. et al. 1983. "Extreme Fatigue and Malaise Syndrome Caused by Badly-Cleaned Wall-to-Wall Carpets?" *Ecol. Dis.* 2:415–418.
98. Gravesen, S. 1987. "Microbiological Studies on Carpets Versus Hard Floors in Non-industrial Occupations." 668–672. In: *Proceedings of the Fourth International Conference on Indoor Air Quality and Climate.* Vol. 1. West Berlin.
99. Gravesen, S. 1987. "Microbial and Dust Pollution in Non-industrial Work Places." *Adv. Aerobiology.* 51:279–282.
100. Gravesen, S. and P. Skov. 1988. "Indications for Organic Dust as an Etiological Factor in the Sick Building Syndrome." *Allergy.* Suppl. 43:60.
101. Gravesen, S. et al. 1993. "Particle Characterization of the Components in the Macromolecular Organic Dust (MOD) Fraction and Their Possible Role in the Sick Building Syndrome (SBS)." 33–36. In: *Proceedings of the Sixth International Conference on Indoor Air Quality and Climate.* Vol. 4. Helsinki.
102. Patterson, R. et al. 1986. "Human Antibodies Against Formaldehyde — Human Serum Albumin Conjugates." *Int. Arch. Appl. Immunol.* 79:53–59.

103. Feeley, J.C. et al. 1988. "Bioaerosol Study Results from IAQ Evaluations of Three Office Buildings." 354–359. In: *Proceedings of IAQ '88: Engineering Solutions to Indoor Air Problems.* American Society of Heating, Refrigerating and Air-Conditioning Engineers. Atlanta.
104. Nevalainen, A. et al. 1990. "Mesophilic Actinomycetes—The Real Indoor Air Problem?" 203–206. In: *Proceedings of the Fifth International Conference on Indoor Air Quality and Climate.* Vol. 1. Toronto.
105. Rylander, R. et al. 1989. "The Importance of Endotoxin and Glucan for Symptoms in Sick Buildings." 219–226. In: Bieva, C. J. et al. (Eds.). *Proceedings: Present and Future of Indoor Air Quality.* Brussels.
106. Rylander, R. 1986. "Lung Disease Caused by Organic Dusts in the Farm Environment." *Am. J. Ind. Med.* 10:221–227.
107. Di Luzio, N.R. 1979. "Lysozyme, Glucan-Activated Macrophages and Neoplasia." *J. Reticuloendoth. Soc.* 26:67–81.
108. Rylander, R. and Y. Peterson. 1989. "Organic Dusts and Disease." *Am. J. Ind. Med.* 17:106–120.
109. Miller, J.D. 1990. "Fungi as Contaminants in Indoor Air." 51–64. In: *Proceedings of the Fifth International Conference on Indoor Air Quality and Climate.* Vol. 5. Toronto.
110. Tobin, R.S. et al. 1987. "Significance of Fungi in Indoor Air." 718–722. In: *Proceedings of the Fourth International Conference on Indoor Air Quality and Climate.* Vol. 1. West Berlin.
111. Jarvis, B. B. 1990. "Mycotoxins and Indoor Air Quality." 201–214. In: Morey, P.R. et al. (Eds.). *Biological Contaminants in Indoor Environments.* ASTM 1071. American Society for Testing and Materials. Philadelphia.
112. Sorenson, W.W. et al. 1987. "Trichothecene Mycotoxins in Aerosolized Conidia of *Stachybotrys atra.*" *Appl. Environ. Microbiol.* 53:1370–1375.
113. Wicklow, D.T. and O.L. Shotwell. 1983. "Intrafungal Distribution of Aflatoxins Among Conidia and Sclerotia of *Aspergillus flavus* and *Aspergillus parasiticus.*" *Can. J. Microbiol.* 29:1–5.
114. Tobin, R.S. et al. 1987. "Significance of Fungi in Indoor Air." *Can. J. Pub. Health.* 78:S1–S32.
115. Emmanuel, J.A. et al. 1975. "Pulmonary Mycotoxicosis." *Chest.* 67:293–297.
116. Croft, W.A. et al. 1986. "Airborne Outbreak of Trichothecene Toxicosis." *Atmos. Environ.* 20:548–552.
117. Morey, P.R. 1993. "Use of Hazard Communication Standard and General Duty Clause During Remediation of Fungal Contamination." 391–395. In: *Proceedings of the Sixth International Conference on Indoor Air Quality and Climate.* Vol. 4. Helsinki.
118. Mainville, C. et al. 1990. "Mycotoxins and Exhaustion Syndrome in a Hospital." 3–13. *Proceedings: International CIB W67 Symposium. Energy, Moisture and Climate in Buildings.* Montreal.
119. Miller, J.D. et al. 1988. "Fungi and Fungal Products in Some Canadian Houses." *Int. Biodeteration.* 24:103–120.
120. Kaminsky, E. et al. 1974. "Identification of the Predominant Volatile Compounds Produced by *Aspergillus flavus.*" *Appl. Microbiol.* 24:721–726.
121. Kaminsky, E. et al. 1974. "Volatile Flavor Compounds Produced by Molds of *Aspergillus, Penicillium* and Fungi Imperfecti." *Appl. Microbiol.* 27:1001–1004.

122. Bjurman, J. 1993. "Thermal Insulation Materials, Microorganisms and the Sick Building Syndrome." 339–344. In: *Proceedings of the Sixth International Conference on Indoor Air Quality and Climate*. Vol. 4. Helsinki.
123. Wilkins, C.K. et al. 1993. "Characterization of Office Dust by VOC and TVOC Release — Identification of Potential Irritant VOCs by Principal Least Squares Analysis." *Indoor Air*. 3:283–290.
124. Samson, R.A. 1985. "Occurrence of Moulds in Modern Living and Working Environments." *Eur. J. Epidemiol.* 1:54–61.
125. Norback, D. et al. 1990. "Indoor Air Quality and Personal Factors Related to the Sick Building Syndrome." *Scand. J. Work Environ. Health.* 16:121–128.
126. Norback, D. et al. 1993. "Exposure to Volatile Organic Compounds (VOC) in the General Swedish Population and its Relation to Perceived Air Quality and Sick Building Syndrome (SBS)." 573–578. In: *Proceedings of the Sixth International Conference on Indoor Air Quality and Climate*. Vol. 1. Helsinki.
127. Norback, D. et al. 1990. "Volatile Organic Compounds, Respirable Dust, and Personal Factors Related to Prevalence and Incidence of Sick Building Syndrome in Primary Schools." *Br. J. Ind. Med.* 47:733–741.

7 DIAGNOSING PROBLEM BUILDINGS

Based on results of problem-building investigations, systematic building studies, and a variety of anecdotal reports, health and comfort problems associated with work/building environments appear to be relatively common. In many cases, prevalence rates of building/work-related symptoms are apparently relatively low (<20%) and occupants do not overtly express dissatisfaction with building air quality. In other instances, symptom prevalence rates may be relatively high (>50%) and complaints to building management may be numerous and intense. Illness complaints may occur after some incident (e.g., in the USEPA headquarters building, the installation of new carpeting) or manifestation of unusual workspace odors or illness. Generally speaking, the need to conduct building investigations for the purpose of diagnosing the nature of a suspected problem only occurs when occupant complaints are sufficient to convince building management that alleged health problems and their cause or causes need investigation. As a general rule, building management is skeptical that health problems are real.

ROLE OF PUBLIC AND PRIVATE GROUPS

The task of investigating building-related health and sometimes odor complaints may fall on a variety of local, state, and federal governmental agencies or to private consulting firms which provide industrial hygiene (IH) or specialized indoor air quality (IAQ) diagnostic services. In the United States, building investigations are conducted by both governmental agencies and private consultants. In Canada, in the Nordic countries, and countries in the European Economic Community, building investigations are more commonly conducted by public agencies.

Requests for governmental assistance in the United States are initially directed to local or state health departments. The local health agency may conduct the investigation itself if it has the resources. When it does not (which is often the case), assistance requests are referred to the state health department or in some instances to the state Department of Labor or similar entity which provides IH investigation of workplace health complaints. If the local or state agency fails to diagnose and resolve the problem, it may be referred to the Health Hazard Evaluation Division of the National Institute of Occupational Safety and Health (NIOSH). NIOSH has conducted multiple hundreds of investigations of workplace complaints in a variety of office and institutional work environments since the early 1970s.

Building management may in some instances be wary of calling in a governmental agency perceived to have potential regulatory authority in the matter. As a consequence, it may choose to engage private consultants. Building management may also find the use of private consultants desirable because they may be more responsive to their needs.

ROLE OF INDUSTRIAL HYGIENE

Historically, the task of conducting investigations in response to IAQ complaints in the United States has fallen on the profession of industrial hygiene. Industrial hygienists have played a leadership role in providing diagnostic services both within state and federal agencies and in private consulting practice. This has been the case because industrial hygienists have both training and experience in conducting measurements of contaminant levels in workspace environments and investigating sources of worker exposure problems.

The application of IH practice to building investigations has been relatively unsuccessful. This has been in good measure due to the fact that IH typically focuses on measuring contaminant levels to determine compliance with federal permissible exposure limits (PELs) or threshold limit values (TLVs) of the American Conference of Governmental Industrial Hygienists. In the early history of problem-building investigations, many industrial hygienists operated "by the book". If levels of measured substances were below PELs or TLVs, then there was no problem despite what the health facts may have been.

The application of traditional IH practices in conducting problem-building investigations has been criticized by Woods et al.[2] who claim that they are not appropriate for identifying and mitigating IAQ problems. They propose that this is the case because (1) sources of indoor contaminants are usually numerous and diffuse (e.g., bioeffluents and tobacco smoke from occupants and volatile organic compounds from building materials and furnishings), (2) levels of potential causal contaminants may be several orders of magnitude lower than PELs or TLVs, and (3) contaminant control in mechanically ventilated buildings is achieved by using general dilution rather than local exhaust ventilation commonly applied in industrial workspaces.

Despite these criticisms and the widely held view in the IAQ research community that traditional IH investigative practices are not only inappropriate but detrimental to successful diagnoses and mitigation of sick building problems, industrial hygienists nevertheless have played and will continue to play a major role in conducting problem building investigations. It is evident that IH has to adapt to the new realities and challenges involved in investigating health concerns in nonindustrial workplaces. Investigative protocols developed by the NIOSH and the American Industrial Hygiene Association (AIHA) represent major efforts by industrial hygienists to develop systematic investigative procedures which are less constrained than traditional practices.[3,4]

PROBLEM-BUILDING INVESTIGATION

Techniques and protocols employed in problem building investigations typically reflect the unique backgrounds of those providing such services. They also reflect the resources of the governmental agencies and private consultants providing the services and, of course, the resources of those requesting services on a private fee basis.

In a theoretical sense, it would be desirable for all investigators to use a diagnostic protocol that maximizes the probability that the cause or causes will be identified and the problem resolved. This indicates a need for a single universal protocol or a consensus on diagnostic elements common to all investigative protocols.

Early building investigations were, for the most part, conducted on an *ad hoc* basis. They were, in many cases, limited to conducting measurements of some common contaminants or inspection of HVAC systems and their operation. Such *ad hoc* investigations were in most instances not up to the task. This was particularly true when investigations were conducted by different individuals in the same public agency or private consulting group. A need for standardizing investigative procedures was both obvious and compelling.

A number of investigative protocols have been developed and used in the United States, Canada, and Western Europe. As would be expected, they share many common elements. There are, however, considerable differences among them which reflect different philosophies, the training of those conducting investigations, specific details of investigation conduct, and emphasis on potential causal factors.

American Investigative Protocols

The NIOSH Protocol

The National Institute of Occupational Safety and Health (NIOSH) has developed an investigative protocol to serve the needs of its health hazard evaluation teams.[3] It has been described as being solution-oriented with an emphasis on systematically excluding a narrowing range of possibilities. This exclusion hierarchy (in order) includes the evaluation of physical, chemical, and microbiological factors.

Requests for assistance are typically received by telephone. Based on information obtained from these requests, NIOSH staff may respond in one of three ways. Potential responses include (1) providing self-help evaluation materials and being available for telephone consultation, (2) providing an initial evaluation and problem-solving recommendations or further study on a self-help basis, and (3) conducting a full-scale investigation.

Because NIOSH employs a multi-disciplinary approach, investigation teams include an industrial hygienist, an epidemiologist, and a professional familiar with the operation and maintenance of heating, ventilating, and air-conditioning (HVAC) systems. A typical NIOSH investigation includes an initial background assessment, followed by an on-site investigation. A follow-up site investigation is conducted if NIOSH staff determine that it is needed.

In the background assessment, investigators obtain (by telephone) information on the building involved, the nature of symptoms and complaints, and a chronology of the problem. Such information may be collected by the use of a standardized questionnaire or checklists. Reports from previous investigations are requested and reviewed. Information obtained in the background assessment is used to structure the on-site investigation.

The NIOSH site investigation consists of an opening conference, a walk-through survey, personal interviews, phase I environmental monitoring, and a closing conference. In the opening conference, the nature of the problem is initially ascertained from discussions with employers, building management, employees, and an individual knowledgeable with the operation and maintenance of the building's HVAC systems. The walk-through survey focuses on the identification of potential sources of emissions and inspection of HVAC systems with a special emphasis on provision of outdoor air for ventilation. Symptoms and symptom patterns may be assessed by personal interviews or by self-administered questionnaires (see Appendix A).

NIOSH staff conduct what they call Phase I environmental monitoring during the initial on-site survey. Typical measurements include carbon dioxide (CO_2), temperature, and relative humidity. Formaldehyde samples may be collected if the background assessment suggests that it may be a problem. In addition to these measurements, the operation of the HVAC system is evaluated to determine the adequacy of ventilation in complaint areas.

In a closing conference, NIOSH field investigators apprise building managers of the activities conducted, results obtained, and recommendations for corrective actions on problems identified. If no specific causal factor was determined or if the problem needs additional definition, a second more intensive site investigation may be conducted. This second investigation typically focuses on environmental monitoring of chemical and microbiological contaminants of interest.

Evaluation criteria used by NIOSH staff to interpret the results of environmental measurements include ASHRAE guidelines for acceptable IAQ[5] and the ASHRAE comfort guidelines for assessing thermal performance.[6]

For many (but not all) of the problem buildings investigated, NIOSH staff prepare a detailed report of their investigation including results of air testing, building inspection, and assessment of HVAC system operation. These reports are provided to building management requesting the investigation and are available to other interested parties on request.

USEPA/NIOSH Protocol for In-House Personnel

The U.S. Environmental Protection Agency in cooperation with NIOSH has developed a model protocol for investigating health and comfort complaints by building managers.[7] Though designed for in-house personnel, it contains many elements which make it appropriate for professional use as well.

The IAQ investigation is described as a cycle of information gathering, hypothesis formation, and hypothesis testing (Figure 7.1). An initial walk-through of problem areas is recommended to gather information on building occupants, HVAC systems, pollutant pathways, and contaminant sources. In the initial walk-through, the investigator collects easily available information about the history of the building and occupant complaints. Major components of the walk-through investigation include (1) notification of building occupants of the impending investigation, (2) identification of key individuals for access and information, (3) identification of known HVAC zones and complaint areas, (4) inspection of obvious pollutant sources, (5) evaluation of HVAC system deficiencies which serve the problem area or areas, and (6) evaluation of pathways and pressure differences which may cause cross-contamination.[7]

The acquisition of additional information is recommended if the walk-through does not identify the cause or causes of the problem. This would include information on occupant complaints, the HVAC system, pollutant sources, and pathways. Such information gathering is intended to provide a basis for developing a hypothesis for the identification of the problem. The use of forms and checklists developed by USEPA/NIOSH is recommended. Investigators are advised to review existing information about complaints including any records, complaint forms, and incident logs when they are available. They are also to collect additional information by means of in-person interviews and keeping of occupant diaries. Such data may be used to define complaint areas and evaluate symptom and timing patterns.

Since many IAQ complaints occur in poorly ventilated buildings and spaces, investigators are advised to inspect and evaluate the components of the HVAC system that serves complaint areas and other parts of the building. This inspection/evaluation is designed to determine whether (1) the ventilation system is functioning properly and is adequate for the current use of the building, (2) there are ventilation or thermal comfort deficiencies, and (3) the definition of the complaint area based on the HVAC system layout and operating characteristics is expanded. In collecting HVAC system information, investigators should review existing documents on HVAC design, installation, and operation; talk to facilities staff; and inspect system layout, condition, and operation. In the last case, measurements of temperature and relative humidity, pressure differentials, airflow patterns, air flow from diffusers, and CO_2 levels are recommended.

Investigators are also advised to collect information on potential pollutant pathways. These are determined by the use of chemical smoke tubes, release of odorants, such as peppermint oil, or tracer gases, such as sulfur hexafluoride

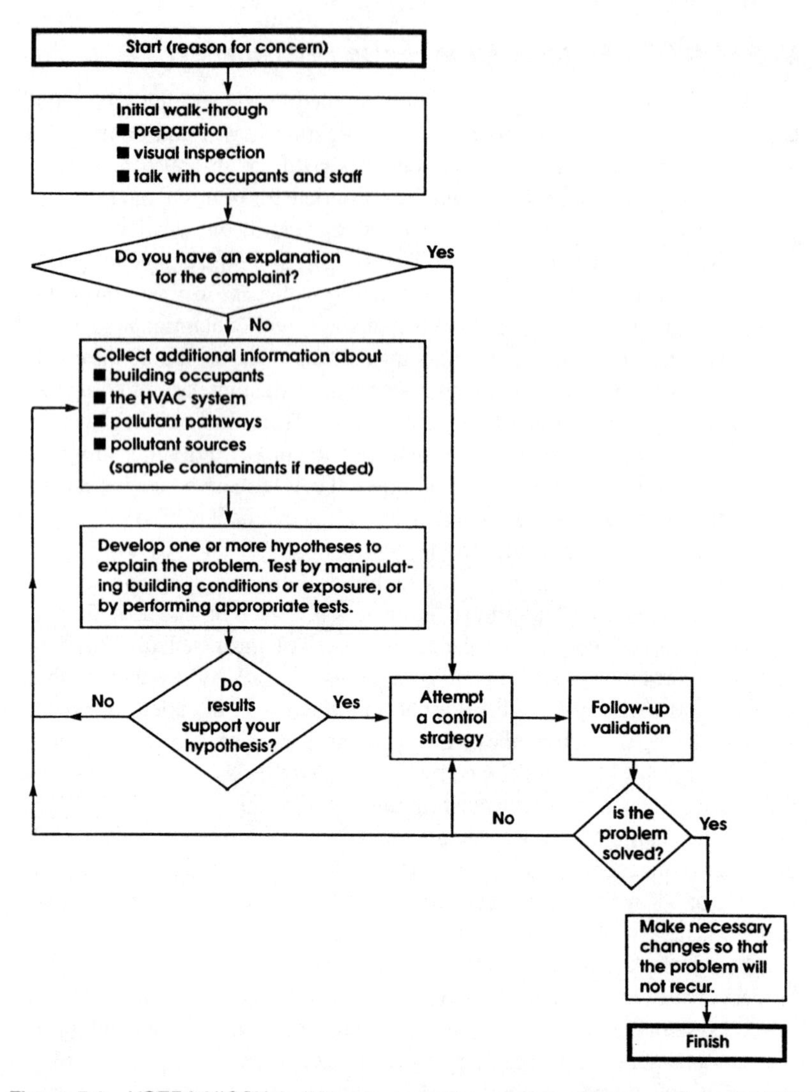

Figure 7.1 USEPA/NIOSH building investigation model sequence. (From USEPA/NIOSH, 1991. EPA/400/1–91/003. DDHS Publication no. 91–114.)

(SF_6). Pollutant pathway data are used to understand airflow patterns in and around complaint areas. They may indicate a need to enlarge the focus of the investigation or may direct attention to potential contaminant sources.

Lastly, investigators are advised to collect information on pollutant sources that may be responsible for reported complaints. During the building inspection, an inventory should be taken of potential pollution problems associated with outdoor sources (e.g., idling motor vehicles near fresh air intakes, soil-gas entry points, etc.), office/building equipment such as wet-process copiers, laser printers, building components and furnishings, and miscellaneous sources such as smoking lounges, print shops, and laboratories. An inventory of activities is also recommended. This includes a review of smoking policy, identification of areas of overcrowding, a review of products used in housekeeping and their application schedule, identification of supply storage areas, etc. Pollutant source data may be used to identify patterns linking emissions to complaints and evaluating potentially unrelated sources.

The USEPA/NIOSH investigative protocol gives limited attention to air sampling, suggesting that air sampling may not be required to solve the problem and may even be misleading. Air sampling is only recommended after all other investigative activities have been employed. Exceptions are tests which are indicative of very common IAQ concerns such as ventilation adequacy (CO_2 measurements) and thermal comfort (temperature, relative humidity, and air movement).

Investigators are to take into account concerns that IAQ complaints may be caused or complicated by other factors including a variety of environmental (lighting, noise, and vibration), ergonomic (fatigue, circulatory problems, and musculoskeletal aches caused by a mismatch between furniture/equipment and work task), and job-related psychosocial stressors (excessive workload and work pressure, lack of clarity in what is expected, poor interpersonal relations, little participation in decision-making, etc.).

As the investigation progresses, sufficient information may be available to develop one or more hypotheses. The accumulated information is to be reviewed to determine whether it is consistent with the proposed hypothesis. If it is not, then another hypothesis should be considered. The hypothesis may then be tested. Testing may include changing ventilation rates or pressure relationships between spaces, removing suspected sources, etc.

California Protocol

Quinlan et al.,[8] in association with the Occupational Health Clinic of San Francisco General Hospital and the California Indoor Air Quality Program, have developed a very comprehensive investigative protocol for use by public agencies and private consultants. It includes elements of the NIOSH protocol[3] and the protocol developed by Ontario Interministerial Committee on Indoor Air Quality[9] combined with the experience of the California Indoor Air Quality Program and the Occupational Health Clinic at San Francisco General Hospital. It is designed to be multidisciplinary in approach. The California protocol describes methods to obtain background information and to conduct site visits. The initial site visit usually leads to a more extensive medical investigation and environmental evaluation. The protocol describes how the results of both medical and environmental evaluations can be best analyzed to implement remedial measures.

A significant amount of background information is collected by telephone. This includes information about the occupants (Table 7.1), the general indoor environment of the building (Table 7.2), and the ventilation system.

Table 7.1 Background information related to affected individuals in a problem building.

- Population of the building
- Differences in building occupancy
- Relative percentage of building occupants with complaints
- Relationship of complaints to job activities, building location, etc.
- Nature of complaint initiation
- Consistency of symptoms/complaints
- Nature of symptom resolution away from the work site
- Patterns of symptom onset
- Onset of symptoms/complaints in time
- Medical treatment of affected individuals
- Availability of absence/sick-leave records
- Temporal history of specific illness among building population
- History of building, its newness, acceptance of relocation by occupants, etc.
- Recent employee/management relations
- Availability of a control group nearby

From Quinlan, P. et al. 1989. 771–796. In: Cone, J.E. and M.J. Hodgson (Eds.) *Problem Buildings: Building-Associated Illness and the Sick Building Syndrome. Occupational Medicine: State of the Art Reviews.* Hanley and Belfus, Inc. Philadelphia.

Table 7.2 Background information on the building environment.

- Type of building and use
- Building size with emphasis on number of stories, rooms per story, floor area, ceiling height, etc.
- Building age
- History of recent construction or renovation
- Building activities/operations
- Indoor sources of contamination
- Nature of office equipment operation
- Location of complaint areas
- Occupant satisfaction with general cleaning and maintenance
- Building smoking policy

From Quinlan, P. et al. 1989. 771–796. In: Cone, J.E. and M.J. Hodgson (Eds.) *Problem Buildings: Building-Associated Ilness and the Sick Building Syndrome. Occupational Medicine: State of the Art Reviews.* Hanley and Belfus, Inc. Philadelphia.

After background information is collected, a site visit is scheduled. The site visit is designed to be an initial walk-through inspection by an investigative team. Table 7.3 provides a suggested list of people to contact during the conduct of the investigation, documents to request prior to the site visit, and suggested equipment needs. The walk-through inspection includes (1) obtaining information about the building including anything unusual such as odors, water damage to ceilings, walls, floors, etc.; (2) interviewing occupants about their symptoms/complaints and any associated patterns; and (3) reviewing the operation and maintenance of the HVAC system and inspecting HVAC system components. The investigative team may include an industrial hygienist, epidemiologist, and a ventilation system engineer. The initial site visit is designed to provide valuable first impressions and clues to the nature of the problem.

Table 7.3 Contact people, requested documents, and suggested equipment for a site visit.

Individuals to contact
- Building manager
- HVAC system operator
- Office manager
- Employee representative

Documents to be reviewed
- Employee list by job category and location
- Absence/sick-leave records
- Blueprints or building floor plan
- HVAC system blueprints
- Maintenance records

Equipment for site inspection
- CO_2 monitor
- Air flow meter
- Psychrometer
- Thermometer
- Smoke tubes
- Tool kit
- Camera
- Measuring tape
- Flashlight
- Ladder
- Calculator
- Coveralls

From Quinlan, P. et al. 1989. 771–796. In: Cone, J.E. and M.J. Hodgson (Eds.) *Problem Buildings: Building-Associated Illness and the Sick Building Syndrome. Occupational Medicine: State of the Art Reviews.* Hanley and Belfus, Inc. Philadelphia.

If the site visit fails to identify and remediate the cause of the problem, a more comprehensive investigation involving medical, environmental, epidemiological, and engineering approaches is conducted. The goal of this comprehensive investigation is to develop and test an increasingly precise set of hypotheses about the causation of symptoms/health problems. These hypotheses are usually suggested from preliminary medical and environmental surveys.

One of the major steps in conducting a comprehensive building evaluation is designing an appropriate clinical epidemiological study. The purpose of this epidemiological investigation is to (1) determine the likelihood of specific types of building- associated illnesses or simply to confirm the existence of a problem, (2) to identify predominant patterns of symptoms and diagnosed illness, (3) to develop testable hypotheses concerning the nature of the problem based on symptom patterns, and (4) to identify individuals most affected and those in need for immediate removal or referral for further diagnosis and treatment.

Both subjective and objective data are collected in the epidemiological investigation. These include anecdotal case reports, semi-structured interviews conducted during the initial walk-through, structured questionnaires, and medical interviews with standard medical histories. By obtaining such data, investigators may be able to group symptoms around logical organ systems or potential etiological syndromes and begin the process of structuring an appropriate questionnaire. Objective data may be obtained by case verification involving medical records and discussions with treating clinicians, physical exams of affected individuals, routine clinical and specialized or research laboratory tests. Specialized laboratory testing may be valuable in confirming cases of hypersensitivity pneumonitis and identifying other potential etiological agents. The health evaluation attempts to identify high-risk groups and compare them to low-risk nonexposed populations.

The California protocol recommends that environmental measurements be conducted when potential exposures warrant verification, or documentation, or when potential problems can be best understood from test results. Measurements are considered appropriate where specific compounds or environmental parameters can be readily identified or measured. If outdoor air flow rates are suspected to be inadequate, actual flow rates are measured and compared to design or ASHRAE recommended outdoor air supply rates.

Collected health and environmental data are then analyzed in a variety of ways. These include an initial assembly of information from the site visit, questionnaire, health/medical investigation, and environmental monitoring, followed by a comparison of subjective health information obtained from interviews, case reports, health interviews, and structured questionnaires with any objective findings. Analyses of symptom prevalence by areas may provide clues in identifying the nature of the problem present. A variety of risk factors (smoking, odors, drafts, wet areas) and environmental contaminants present are evaluated by mapping out workers' locations in the building.

Cross-sectional analyses are applied to the data to define the frequency of outcomes and risk factors, including exposure variables, to determine whether a problem indeed exists, and if so, to define the nature of a case (a person with building-related symptoms). A comparison population is used to identify the symptoms and objective findings that distinguish the exposed population. Cross-sectional analyses may elucidate potential dose-response relationships and identify areas or jobs with the greatest risk. In case-control analyses, data are cross-tabulated by job title, location in building, and ventilation unit and correlated with environmental data to provide an estimate of risk and evidence for and against specific hypotheses about causes. Statistical analyses may also be applied to on-going surveillance programs to measure the effectiveness of mitigation measures which have been implemented.

When the initial and comprehensive investigations have been completed, the California protocol concludes with the team preparation of a final report. This report includes investigator observations, a summarization of questionnaire results, and an interpretation of the significance of environmental measurements. It also includes specific recommendations to remediate the problem.

AIHA Protocol

A detailed investigative protocol designed to meet the needs of industrial hygienists who conduct problem building investigations has been recently developed by the American Industrial Hygiene Association.[4] Though including elements of other protocols, it for the most part reflects the experiences of industrial hygienists who have conducted problem-building investigations.

The initial focus of the AIHA protocol is to conduct an intensive problem review. The goal of this review is to develop information to support specific hypotheses which provide a focus for the investigation. Problem review is, in theory, intended to develop the following information: (1) specific health and comfort complaints reported and proportion of occupants reporting each complaint; (2) time patterns of the onset of health and comfort complaints; (3) location patterns of affected occupants; (4) relationship between predominant health and comfort complaints and potential source materials; (5) time and location patterns of potential contaminant source introductions; (6) potential deficiencies in providing adequate ventilation air; (7) understanding the design, installation, operation, and manufacture of HVAC systems with concern relative to potential contaminant pathways; and (8) available medical opinions about causes of health complaints attributed to IAQ.

The problem-review stage includes (1) obtaining background information, (2) conducting an initial walk-through survey, and (3) gaining an overview of HVAC system design, operation, and maintenance. Information gathering includes discussing the problem by phone with building management and reviewing engineering reports, environmental audits, previous IAQ studies, building plans and specifications. The investigator is advised to identify and interview individuals familiar with the history of the building's construction, renovation, operation, and maintenance; patterns of past and current occupancy; and specific health and comfort problems experienced by occupants.

A checklist is used to obtain background information. It is designed to obtain and record information systematically relative to the identity of the facility, contact persons, facility description, facility maintenance, and the surrounding neighborhood. Factors which would describe the facility are summarized in Table 7.4. Typical or atypical activities or products related to the maintenance of the facility are summarized in Table 7.5. In addition, a review is conducted of chemical use including cleaners, disinfectants, boiler scale inhibitors, copier and printer materials, and specialty chemicals used in darkrooms, graphic arts, etc. The potential relationship between complaints and time and location of chemical use is to be evaluated and all sources of combustion identified and characterized. In the last case, this would include smoking policies and practices.

An initial walk-through survey is conducted for purposes of inspecting the problem area, non-problem control areas, major components of HVAC systems, and the building's immediate outdoor environment. This survey is de-

Table 7.4 Facility description.

- Nature of business/activities conducted by tenants/occupants
- Number of occupants in each space
- Activities conducted in or near the building (e.g., laboratory, manufacturing, cafeteria, parking garage, etc.)
- Time of day/days of week employees are on duty
- Construction — Includes number of floors, area per floor, age of building, intended use, and date of major renovations
- Furnishings — Includes floor and wall coverings, partitions, and ceilings

Rafferty, P.J. (Ed.) 1993. *The Industrial Hygienist's Guide to Indoor Air Quality Investigations.* American Industrial Hygiene Association. Fairfax, VA. With permission.

Table 7.5 Facility maintenance activities and products.

- Indoor and outdoor pesticide use
- Removal of asbestos and other insulation
- Leaking water pipes
- Repair or renovation of electrical, plumbing, or HVAC systems
- Installation of new partitions, wall covering, flooring
- Wax stripping
- Flooding
- Lead-base paint removal
- Cleaning upholstery, wall and floor coverings
- Insulation

Rafferty, P.J. (Ed.) 1993. *The Industrial Hygienist's Guide to Indoor Air Quality Investigations.* American Industrial Hygiene Association. Fairfax, VA. With permission.

signed to identify potential problems using the investigator's observational skills and, where possible, sensory perceptions. It focuses on areas and sources where contamination problems may be anticipated such as print shops, photo laboratories, kitchens, copy machines, sewage/drain odor problems, water-damaged materials, etc. Preliminary environmental and semiquantitative IAQ measurements are to be made in a second, more in-depth, walk-through survey. These may include temperature, relative humidity, carbon monoxide (CO), CO_2, airflow, and in some cases, total or specific volatile organic compounds (VOCs).

The AIHA protocol focuses on the need to evaluate various aspects of HVAC system design, operation, and maintenance relative to its potential for contributing to air quality problems. HVACs receive special attention in the background review and in more intensive site investigations.

All the activities described above are considered to comprise Phase I and are suggested to be sufficient to diagnose many if not most IAQ problems. If such investigations are not successful, more intensive Phase II studies are conducted based on working hypotheses developed in Phase I. Phase II studies may include sampling for VOCs, particles, combustion products, and biological aerosols in air and, in some cases, on materials. They may also include more intensive evaluations of the HVAC system and epidemiological studies including questionnaires and medical evaluations.

A questionnaire study is only conducted when it is deemed to be needed. A model health and comfort questionnaire (see Appendix B) is distributed to building occupants in complaint and control areas. Completed questionnaires are reviewed to identify patterns of health and comfort complaints for purposes of further hypothesis development.

The AIHA protocol makes claim that a thorough medical evaluation of individuals reporting health complaints is a vital element of problem-building investigation. It, nevertheless, acknowledges that such evaluations are rarely practical and actually conducted.

Investigative Protocols Used by Private Consultants/ Consulting Firms

The investigative protocols previously described have been, for the most part, developed to provide for the needs of public agency personnel in conducting problem-building investigations or recommendations for the conduct of such investigations by others. In the United States, problem-building investigations are very commonly conducted by private consultants associated with firms that provide specialized IAQ or IH services. In the latter case, IAQ investigations are, in many instances, conducted using traditional IH approaches, or they may be based on the recent AIHA protocol,[4] elements of the NIOSH protocol,[3] or the USEPA/NIOSH protocol developed for in-house personnel.[7] Most firms which provide IAQ services exclusively have developed investigative protocols which best serve their needs, philosophy, and professional strengths. These firms provide services nationally, and in some cases internationally as well. They typically utilize a team of investigators. Because services are often provided by multiple staff members, there is an obvious need to use a standardized investigative protocol to provide a consistent level of performance in identifying and solving IAQ problems.

The Building Diagnostics Protocol

Woods et al.[10] and Lane et al.[11] have described an investigative approach which can be used to assess IAQ in problem buildings and to assure acceptable air quality in buildings in which no dissatisfaction has been expressed. The approach is described as "Building Diagnostics". The Building Diagnostics (BD) protocol was developed for use by Honeywell Indoor Air Quality Diagnostics, a consulting firm in the United States specializing in the provision of IAQ services on a national basis. The BD approach is also, for the most, part used by Clayton Environmental Consultants[12] and Healthy Buildings International,[13] two major consulting firms which provide similar services.

The BD approach differs significantly from those discussed previously in that emphasis is placed on evaluating building performance rather than attempting to identify specific causal factors. It assumes that if the building and its systems are performing as designed or meet generally accepted performance criteria for comfort such as ASHRAE standards,[5,6] a substantial majority (≥80%) of building occupants will not express annoyance or dissatisfaction with IAQ and climate. The BD approach places a heavy emphasis on the evaluation of the performance of the building and its systems and the quantitative assessment of airborne contaminants and a variety of environmental parameters. It is described as using engineering principles for system analysis and scientific principles for measurements of contaminant concentrations and human response. It is based on an understanding of what is required to evaluate compliance with prescriptive and performance criteria for acceptable IAQ. As a consequence, considerable knowledge is required of the relationships which exist among the building, its systems, and its occupants.

The BD approach seeks to identify those factors which deviate from acceptable criteria so that recommendations can be made to improve system performance. As a consequence, it is important to know what to measure, to select appropriate instrumentation, interpret results, and predict building performance. The actual investigation itself is based on a three-stage approach including (1) consultation, (2) qualitative diagnostics, and (3) quantitative diagnostics.

The overall objective of the consultation phase is to determine the general scope of IAQ problems at the client's premises. A flow chart for the consultation phase can be seen in Figure 7.2. Preliminary information on problems/complaints is obtained from confidential meetings between members of the investigative team with building management and/or health and safety officers. Additional information on system characteristics is obtained in consultation with facilities management and staff. This is followed by a tour of facilities which includes an inspection of typical occupied spaces and the mechanical equipment that serves them. This tour or walk-through is designed to provide the investigative team with a first-hand exposure to the conditions in the space, communication with occupants, and observation of the performance of HVAC systems. After the desired information is obtained, the BD team develops what is described as a cost-effective diagnostic procedure by focusing on one or more hypotheses which are formulated in terms of suspected causes, manifestations, and system characteristics. These hypotheses are developed in preliminary form with the intent that they focus on possible recommendations or on a need for additional diagnosis.

The on-site consultation phase is concluded by a closing meeting with building management/health and safety officers where preliminary recommendations for mitigation are presented and, if necessary, proposals for additional diagnostic activities.

In the qualitative diagnostics phase engineering analysis techniques are employed to test the preliminary hypothesis or to validate initial recommendations. Performance criteria for interior environments of the building are established. These include chemical, physical, and microbial agents. This is fol-

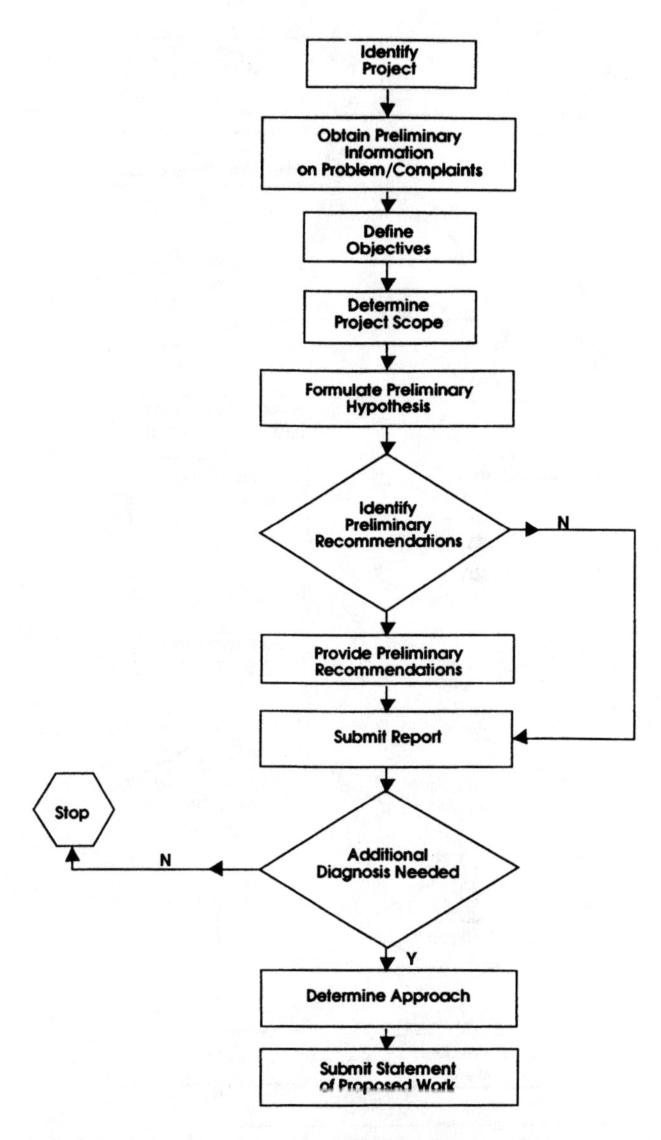

Figure 7.2 Building diagnostics consultation phase flowchart. (From Woods, J.E. et al. 1989. 80–98. In: Nagda, N.L. and J.P. Harper (Eds.) *Design and Protocol for Monitoring Indoor Air Quality.* ASTM STP 1002. American Society for Testing and Materials. Philadelphia. With permission.)

lowed by the characterization of occupant problems/complaints by interviewing individuals who have health and safety responsibility (Figure 7.3). If the problem appears to be one of annoyance/discomfort, an engineering systems analysis is conducted. If a specific type of building-related illness is apparent such as hypersensitivity pneumonitis, fiber glass dermatitis, or CO poisoning, immediate medical attention for the occupants is recommended, and the initiation

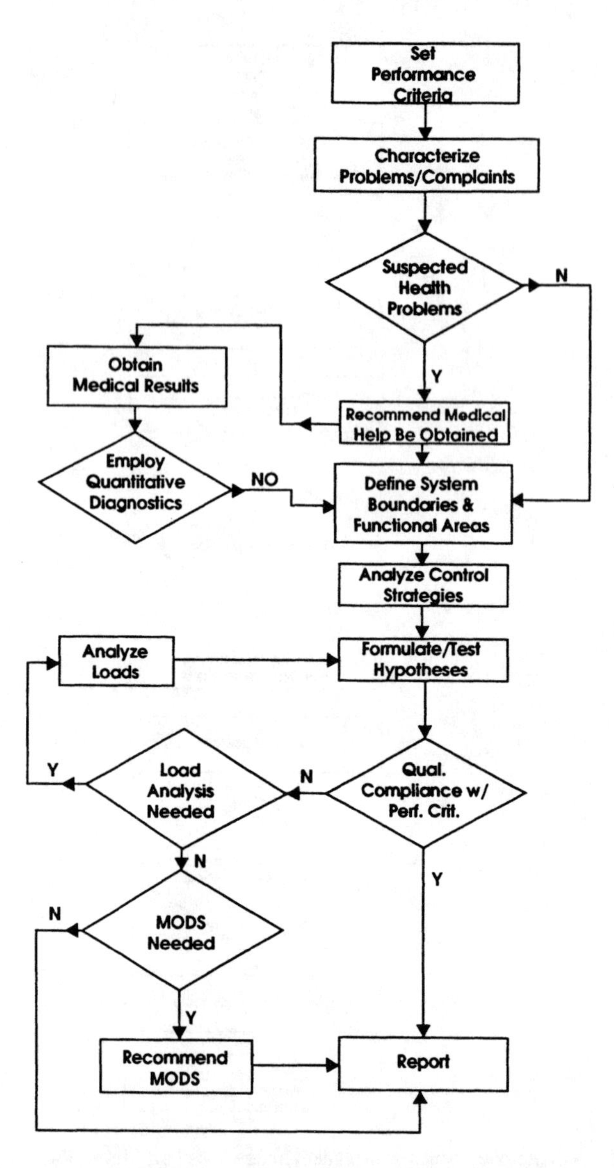

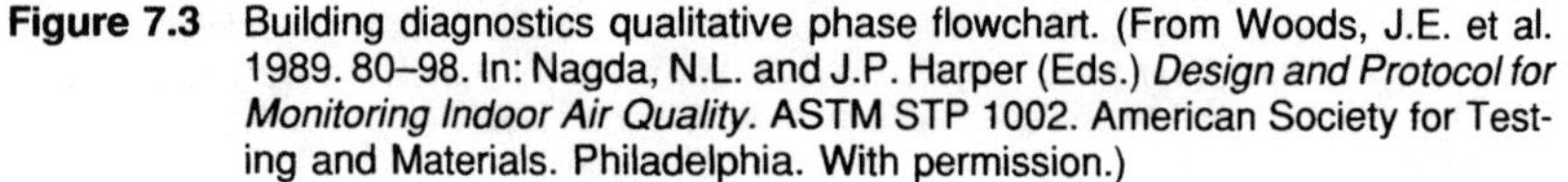

Figure 7.3 Building diagnostics qualitative phase flowchart. (From Woods, J.E. et al. 1989. 80–98. In: Nagda, N.L. and J.P. Harper (Eds.) *Design and Protocol for Monitoring Indoor Air Quality*. ASTM STP 1002. American Society for Testing and Materials. Philadelphia. With permission.)

of appropriate environmental measurements (quantitative diagnostics) is initiated.

Problem characterization is followed by a definition of system boundaries, that is, identifying the HVAC system or systems that are serving the complaint area. Control strategies are then reviewed and analyzed. These include those originally designed for the facility and actually installed, and those which currently exist. System loads are analyzed by comparing HVAC system capacities to the environmental requirements of the space. This analysis focuses on thermal loads and a consideration of outdoor air ventilation requirements.

After a hypothesis has been formulated and tested (e.g., by the application of nondynamic, simplified models), the results are evaluated for qualitative compliance with performance criteria. A written report is then provided to the client. This report describes the objectives, hypothesis tested, description of procedures used, results, conclusions, and recommendations.

The quantitative diagnostics phase is implemented if the qualitative diagnostics phase indicates a need for measurements of environmental parameters. In this phase, objective measurements are made of chemical, physical, and microbiological parameters as well as subjective responses of occupants to their environment. Objective measurements of thermal, lighting, acoustic, and air quality conditions in different functional areas of the building are conducted. A questionnaire is administered which attempts to assess subjective responses of occupants to their indoor environment. This questionnaire is putatively designed to elicit any real-time relationships with environmental factors which may affect occupant comfort. It includes an environmental section which asks the occupant to rate the acceptability of a number of environmental parameters in their space, a limited amount of demographic data, and a short symptoms checklist (Appendix C). If 80% or more of the occupants rate the overall environment as acceptable or better, the indoor environment is considered to be acceptable, and, in theory, no further investigative or remediation measures are necessary.

After the quantitative measurements have been made, the results are analyzed and interpreted. This is followed by preparation of a report which attempts to identify the problem and make recommendations for its mitigation.

Environmental Health and Engineering Protocol

The investigative protocol described here has been developed and used for private consulting purposes by staff of the consulting firm Environmental Health and Engineering (EH & E), many of whom have been or are associated with the IAQ program at Harvard University. Like investigative protocols previously described, the EH & E protocol includes three stages of investigation: problem description, survey monitoring, and health and comfort assessment.[14]

In the first stage, the scope and nature of the problem is determined by means of a detailed walk-through survey of the building, the administration of a screening questionnaire to the occupants of the space and interviews with management, employees, etc. Initial attention is given to the health and comfort concerns of building occupants by characterizing populations at risk by location, employment status, age, gender, or other potentially important factors. A guide used for the characterization of complaints can be found in Table 7.6.

Table 7.6 Guide for the initial characterization of complaints.

- Nature of complaints/symptoms
  - Site/organ affected (e.g., respiratory)
  - Severity
  - Duration
  - Associations
  - Treatment/confirmations
- Timing of complaints
  - Long term (continuing, periodic, seasonal, weekly, daily)
  - Short term (isolated events)
- Location of affected and nonaffected groups
- Numbers affected
- Demographics of occupants with and without complaints
  - Age
  - Gender
  - Employment status

From McCarthy, J.F. et al. 1991. 82–108. In: Samet, J.M. and J.D. Spengler (Eds.). *Indoor Air Pollution: A Health Perspective.* The Johns Hopkins University Press. Baltimore. With permission.

Such information may show evidence of linkages between changes in HVAC system operation, renovations, relocations, equipment use, and a variety of other factors.

In addition to the assessment of complaints, potential sources of contaminants are surveyed and evaluated. This source assessment takes into consideration whether contaminants are continuously or intermittently present, whether contaminants are produced from point or area sources, and whether contaminants originate from indoors or outdoors. Potential sources are identified in Table 7.7.

Finally, in determining the scope and nature of the problem, a building history is developed. This includes a review of the history of the building HVAC system and its operation, followed by an initial evaluation of the system's performance relative to its design goals. Principal factors/systems of building facilities inspected/surveyed are identified in Table 7.8.

After the conduct of the initial stage an experienced investigator may, in many cases, easily determine the cause and source of the problem. If the causal factor or factors can not be identified from the initial assessment, more detailed investigations are conducted. Information gathered in the initial assessment is used to develop a hypothesis and for the specification of protocols for further investigations.

The second stage includes an evaluation of systems, a definition of microenvironments, and environmental screening. The evaluation of systems initially involves a review of specifications, interviews with relevant personnel, and an inspection of the facility. It also includes an inspection and review of comfort controls in terms of sensor location, set points and access, determination of personnel responsibilities for comfort conditions, and responses to occupant requests. Systems evaluation also includes a thorough analysis of air

Table 7.7 Potential sources of significant indoor air contamination.

- Exterior sources affecting air intake
  - Loading docks
  - Roadways
  - Cooling towers
  - Other external sources
  - Locations of intake and exhaust vents
- Interior sources
  - Insulation (man-made mineral fibers, asbestos)
  - Equipment
  - Cleaning compounds
  - Office furnishings
  - Areas damaged by floods, high humidity, or other agents

From McCarthy, J.F. et al. 1991. 82–108. In: Samet, J.M. and J.D. Spengler (Eds.). *Indoor Air Pollution: A Health Perspective.* The Johns Hopkins University Press. Baltimore. With permission.

Table 7.8 Factors evaluated/inspected in surveys of building facilities.

- Physical layout of building
- Workstations
- Location of air supplies and returns
- HVAC zones
- HVAC inspection
  - Coils
  - Filters
  - Fans
  - Drip pans
  - Condensers
  - Air intakes
  - Distribution ductwork
- HVAC operation
  - Load analysis
  - Control analysis
  - Assessment of modifications

From McCarthy, J.F. et al. 1991. 82–108. In: Samet, J.M. and J.D. Spengler (Eds.). *Indoor Air Pollution: A Health Perspective.* The Johns Hopkins University Press. Baltimore. With permission.

flows through diffusers and other mechanical devices; an inspection of heating and cooling coils, drip pans, and filters; a review of maintenance schedules; and an assessment of thermal loads as originally designed and subsequent changes that may have affected them.

Microenvironments are identified in the second step of the survey monitoring stage. These may be defined as spaces sharing a common HVAC system, a ducted air supply, a zone, or a floor of a building. They may also be defined as locations where employees experience similar complaints. The identification of microenvironments is deemed essential for determining sampling locations. Microenvironments (for purposes of environmental monitoring) are defined as specific situations of exposure; that is, they are locations in space and time over which pollutant concentrations are assumed to be uniform and constant.[15]

The last step of survey monitoring is the conduct of environmental screening which attempts to determine potential contaminant levels or physical conditions of the space. It includes the collection of data by means of real-time survey monitors for measurement of CO, CO_2, temperature, relative humidity, light levels, or integrated measurements for contaminants such as RSP and VOCs.

A third-stage investigation is conducted when environmental monitoring fails to provide sufficient evidence to implicate a causal agent or condition related to the symptoms or complaints. This stage involves a more thorough inquiry on symptoms and complaints. A detailed occupant evaluation is conducted in order to correlate the subjective responses of building occupants with the objective measurements made in the investigation. This evaluation may include (1) collection of health and symptom information related to the workplace and to all other environments, (2) use of appropriate clinical tests such as serum precipitins against antigens associated with hypersensitivity pneumonitis, and (3) documentation of symptom rates. In the last case, self-administered questionnaires may be used for identifying symptom clustering by location or in groups of employees categorized by time of occurrence and other factors.

Canadian Investigative Protocols

At the Canadian federal level, investigative services for problem buildings are provided by Health and Welfare and Public Works Canada. Indoor air quality investigations are conducted by federal environmental health officers and their staff in the various provincial regions. Health and Welfare Canada investigations have, for the most part, been conducted on an *ad hoc* basis with little attempt at standardization.[16] Public Works Canada (PWC), on the other hand, has developed and uses standardized building investigative procedures. Public Works Canada is responsible for the construction and management of all federal government buildings. Its mandate is to provide and maintain working conditions and environments conducive to safety, good health, and comfort for all governmental employees, commercial tenants, and the general public who either occupy or visit federal buildings.[17] At the provincial level, Ontario has developed an investigative protocol for problem buildings under its jurisdiction.

Public Works Canada Protocol

The PWC IAQ investigative protocol was developed in response to investigation requests involving complaints in federal buildings. It consists of an IAQ assessment strategy and an IAQ test kit and manual which are used by PWC investigators.[18]

The PWC assessment strategy employs a two-part approach.[17,18] In the preliminary assessment stage, investigators collect information for the purposes of locating and identifying contaminant sources and defining the nature and severity of the problem. Such information is derived from a review of building plans; complaint records; interviews with occupants, managers, and operating personnel; and a walk-through inspection of the building. The walk-through is designed to identify potential contaminant sources such as wet-process photocopiers, loading docks, print shops, kitchens, laboratories, furnishings, paints, and carpeting. The building walk-through is thought to be sufficient in many cases to identify the problem, with relatively simple measurements employed to confirm the hypothesis.

In the second stage of the assessment, a variety of IAQ parameters are measured. Typical measurements include levels of CO_2, CO, respirable suspended particles (RSP), and VOCs. Formaldehyde, mold, temperature, and relative humidity may also be measured.

Contaminant measurement is a standard practice in the PWC protocol. Test results are used to confirm or exclude what are considered to be commonly experienced problems. Carbon dioxide is used as a indicator of ventilation adequacy; CO, the presence of vehicle exhaust migrating from parking garages and loading docks; RSP, the effectiveness of the air filtration system; temperature and relative humidity for thermal comfort; and mold for microbial contamination.

The second assessment stage also includes an inspection and evaluation of HVAC systems. This involves the determination of system type, control logic, operating schedule, general condition of filters, humidifiers, mixing plenums, operation of dampers, and location of exhaust and outdoor air intakes.

Public Works Canada investigators utilize a standard test kit[19] to facilitate the use of the assessment strategy in responding to complaints in federal buildings. It includes equipment to measure CO_2, CO, RSP, VOCs, temperature, relative humidity, and mold. A users manual which comes with the kit provides a checklist for conducting the building inspection as well as detailed instructions on the use and maintenance of instruments. These kits are provided to six regional maintenance sections along with a training program. Public Works Canada headquarters provides an IAQ advisory service which (1) offers guidance and assistance in the interpretation of data, (2) analyzes VOC samples by GC/mass spectrometry, and (3) lends and calibrates equipment.

The PWC IAQ assessment strategy does not place any significant emphasis on official or semi-official IAQ guidelines. It does seriously consider office worker exposures to a broad spectrum of contaminants which are present in very low levels relative to occupational exposures and the many contaminants in indoor air which have no occupational exposure limit and whose relative hazard or safety is unknown. The PWC assessment also takes note of the special case of the hypersensitive individual and the difference in expectations between industrial and office workers.

Ontario Interministerial Committee Protocol

The Ontario Interministerial Committee's (OIC) protocol[9] for conducting IAQ investigations is based on the premise that the problem building should be investigated first. Then, if necessary, symptoms and exposures of occupants are evaluated and measurements of contaminants and physical factors conducted. It consists of four stages: Stage 1. Preliminary assessment; Stage 2. Questionnaire; Stage 3. Simple Measurements; Stage 4. Complex Measurements.

In the preliminary assessment, a building inspection is conducted and information on the building is collected. This includes observations made in the building, details of HVAC system operation, and the nature of occupant complaints. A building checklist (developed by OIC) is often used by investigators. This checklist focuses on CO, other pollutant sources, HVAC system operation, building maintenance and design, and complaint area observations. Once this information is collected, the investigator attempts to interpret the results from an assessment summary page which is specifically designed to assist investigators in interpreting data collected.

If the problem is not identified, the investigation proceeds to a second stage which focuses on the use of a "comprehensive" self-administered questionnaire (see Appendix D) developed to gather information from the occupants on problems being experienced. In theory, information obtained from the questionnaire survey should allow the investigator to establish whether and to what extent IAQ problems exist and possible sources of contamination. The questionnaire focuses on symptoms which are typically associated with IAQ complaints as well as general complaints associated with the physical environment of the building. It is structured so that both symptom and exposure scores can be determined. These scores are then compared to determine whether high exposure and symptom scores can be linked to a particular contaminant or physical agent. A pattern of high symptom scores associated with high exposure is interpreted to mean that there may be a widespread problem with a particular contaminant or physical factor. Symptom-exposure evaluations are conducted for VOCs, formaldehyde, ozone (O_3), CO, CO_2, relative humidity, temperature, air movement, noise/light, biological contaminants, particulate matter, and second-hand smoke.

A series of worksheets are provided to the investigator for the purpose of assisting in the analysis of the questionnaire as well as a "Lotus" software package and summary tables for individual questionnaires.

In the third stage, measurements are conducted of building air contaminants or physical factors which require relatively simple instruments. These include temperature, relative humidity, air movement, CO_2, CO, and formaldehyde. Guidance is given for suitable locations and time for making measurements.

Stage four measurements focus on contaminants which require relatively complex instrumentation and trained personnel. These include microorganisms, RSP, VOCs, O_3, nitrogen oxides, and asbestos. Both stage three and four measurement results are compared with guideline values provided by OIC. These include acceptable/unacceptable values for CO_2, CO, formaldehyde, nitrogen dioxide (NO_2), O_3, microorganisms, and physical factors.

European Investigative Protocols

In Europe, problem-building complaints are typically conducted by public agencies. A model investigative protocol[20] has been developed for such use by the Danish Building Research Institute (DBRI) and the Nordic Ventilation Group.[21] A protocol based on the DBRI model has been developed as a guide for investigating problem buildings in countries of the European Economic Community.[22]

Danish Building Research Institute Protocol

The Danish Building Research Institute has developed a stepwise investigative protocol[20] based on the principle that obvious IAQ problems (identified from a questionnaire investigation) are solved first. If the problem's cause is uncertain, more detailed investigations are then conducted.

The DBRI protocol includes five stages. In the first stage described as operational control, building operation and maintenance personnel are contacted to determine if the ventilation system is operating properly and whether individual thermal comfort needs are being addressed. If operational conditions are considered to be satisfactory, a technical and employee hygiene questionnaire investigation (second stage) is initiated to determine the extent and nature of IAQ problems. The questionnaire (see Appendix E) is submitted to all employees or a representative group. It is designed to elicit responses about occupant symptoms and problems associated with the indoor environment.

A technical investigation of the building and its use is then conducted and documented on a standard form or checklist. The results of the employee and technical surveys are compared and evaluated to identify probable causes which can be either dealt with immediately or further analyzed in the next stage of the investigation. In evaluating symptoms, only those prevalence rates which are statistically above 20–25% are considered to be unusual and suggestive of a need for further investigation of the building or parts of it. Indoor air quality and comfort are documented and mapped in a report before proceeding to the third stage. This report includes statistics of symptoms and complaints and a description of technical conditions in the building. It may include other observations, including when complaints arose, if certain rooms caused specific complaints, seasonal patterns, etc. Based on this report, corrective actions are recommended and expected to be carried out.

In the third stage, a building inspection is conducted and simple environmental measurements are made. A number of standard criteria are evaluated (Table 7.9) in this inspection for purposes of identifying sources/causes of

Table 7.9 Danish building research institute building inspection criteria.

- Placing of outdoor air intakes
- Potential for re-entry/entrainment/cross-contamination
- Presence and sources of odors
- Potential for indoor quality problems from tobacco smoke
- Presence, location, and ventilation exhaust conditions for copying machines and laser printers
- Extent of paper handling and supplies of newly printed paper
- Presence and maintenance conditions of large indoor plants
- Presence and maintenance conditions of local humidifiers
- Presence of water damage or mold
- Adequacy of building maintenance
- Visual conditions associated with video-display units
- Noise levels and tonal quality
- Effect of furnishings/building materials on difficulty of cleaning
- Presence of dust on horizontal surfaces
- Cleanliness of ventilation inlets/exhaust openings
- Use of sunshielding
- Employee population relative to design capacity

From SBI Report 212 Indoor Climate Problems: Investigation and Remedy. 1990. Denver Buiding Research Institute. Denver.

employee complaints. Random measurements of CO_2, temperature, and air flows are conducted. Odors are evaluated, as are other factors such as noise and lighting. Measurements and evaluations are conducted in areas with and without problems.

If actions taken in preceding steps have not reduced occupant complaints, a thorough analysis of the ventilation system and IAQ and environmental factors is conducted (stage four). Ventilation systems are inspected for mechanical defects, damper system adjustment, and humidifier condition. Supply and exhaust air flows and their temperatures are measured to determine the need for commissioning and balancing. Air exchange rates are determined if volumetric air flows indicate that they may not be adequate, and ventilation efficiency is measured in "typical rooms" if outdoor air supply is close to the lowest acceptable limit. Smoke tubes are used to evaluate air flow in and between rooms. The percentage of recirculated air is determined. Automatic control systems are evaluated to determine whether they are in service and working properly. Maintenance procedures are reviewed. Air quality at the outdoor air intake is evaluated. Carbon dioxide and CO measurements are made in the work area, at air inlets, at exhaust openings and the outdoor air intake. Variations in CO_2 and CO concentrations are determined during the course of the day in "typical rooms".

Based on observations made in these investigations, a variety of potential contaminants are measured. They may include total and individual VOCs, if significant odors of an irritative nature are present in new or refurnished buildings; formaldehyde, if building materials and furniture are possible sources; mineral fibers, if sound absorption lining of HVAC components is found to be unprotected or damaged; dust levels in air including microbial and organic components of dust where poor cleaning and excessive paper handling occurs;

lighting measurement and evaluation, particularly for video display users; sound measurement; and supplementary measurements of thermal radiation asymmetry, air velocity, etc.

Additional evaluations are conducted if problems persist after remedial measures (based on stage four evaluations) have been implemented. These include a clinical assessment of affected individuals and supplemental investigations of building air quality and environmental factors. Such investigations would be carried out on the basis of a hypothesis of associated causes.

Under the DBRI protocol a number of environmental factors and contaminants have been identified as causing SBS symptoms directly or contributing to health and comfort complaints. Many of these are described in the Nordtest protocol which follows.

Nordtest Protocol

A problem-building investigative protocol has been developed and published by the Nordic Ventilation Group.[21] It not surprisingly contains may elements of the DBRI protocol. It includes, however, eight investigative steps.

The Nordtest protocol document initially describes effects of the indoor climate on occupant comfort and health symptoms and provides a statistical approach (similar to the DBRI protocol) to determine whether the prevalence of symptoms in a problem building is higher than "normal" and thus in need of intensive investigation. A large number of risk factors are described. These include those related to personal characteristics (gender, illness, smoking), psychosocial factors (position in workplace, social relations) and working conditions (influence/pace of work, amount of work, and type of work.) Notably, working with VDTs, photocopying, and using carbonless copy paper are considered to be major risk factors for SBS symptoms.

Risk factors for health and comfort complaints associated with the building environment are summarized in Table 7.10; contaminants as risk factors, in Table 7.11. In both cases, risk factors are characterized by their relative level of risk based on measured values or observations. These risk factor tables provide major guideposts for both the conduct of problem-building investigations and interpretation of results. Many of these risk factors are also described in the DBRI Protocol.

The Nordtest protocol is a step-by-step investigative procedure. In order, these include (1) operational checks, (2) questionnaire survey or interviews, (3) technical description, (4) inspection, (5) measurement of indicator values, (6) measurement of ventilation rate and performance of ventilation, (7) measurement or assessment of indoor climate effects and sources, and (8) specific examination of persons and effects.

In the first step, individuals responsible for the operation of the building and technical services are asked whether operational conditions are normal, for example, whether the ventilation system works as designed. Set values are

Table 7.10 Environmental risk factors for health and comfort complaints in buildings.

	Level of risk			
Factor	**Low**	**Medium**	**High**	**Effects**
Air temperature, °C	21–23	21–22 23–24	<20 >24 >26 (summer)	Draft, cold, hot dryness, SBS
Daily temperature rise, °C	<2	2–3	>3	Dryness, SBS
Air velocity, m/s	<0.15	0.15–0.20	>0.20	Drafts
Noise, dBA				
Average	<60	60–65	>65	Noise complaints
Background	<35	35–40	>40	General symptoms
Low frequency noise, dB	<20	20–25	>25	General symptoms
Lighting				
General	Suitable	Suitable	Poor	General symptoms,
Individual	Yes	No	No	eye complaints
Glare	No	Control	Glare	General symptoms, eye complaints
Contrast	Good	Control	Too much/ little	General symptoms, eye complaints
Static electricity, kV	<1	1–2	>2	Complaints—shocks
Ventilation/person, L/s	>14	14–8	<8	Bad air, SBS
Ventilation/area, L/s/m^2	>2	2–1	<1	Bad air, SBS
Ventilation system	Natural exhaust, well-working supply system	Supply with heating or cooling	Humidification, badly monitored	SBS, allergy
Size of room; number of workplaces/room	<3	3–7	>7	General symptoms
Office machinery and processes	In separate room		Several in same room	SBS
Cleaning program	Thorough (daily)	Suitable 3–4/week	Unsuitable superficial (<2/week)	SBS, allergy
Cleaning, tidiness, accessibility	Good	Average	Poor	SBS, allergy
Floor coverings	Medium hard	Carpet	Carpet >10 years service	SBS, allergy
Fleece factor, m^2/m^3	<0.35	0.35–0.70	>0.70	SBS, allergy
Shelf factor, m/m^2	<0.2	0.2–0.5	>0.5	SBS, allergy
Odor assessment, decipols	<0.8	0.8–1.5	>1.5	Bad air
Moisture damage	None	Minor, short duration	Major, long duration	SBS

From Kukkonen, E. et al. 1993. Nordtest Report NT Tech. Rep. 204.

checked. The distribution of air flow may also be checked. If problems are found which may have contributed to complaints, they are to be resolved before further investigations are made. Complainants are informed of these initial actions.

Table 7.11 Contaminant risk factors for health and comfort complaints in buildings.

Contaminant	Level of risk			Effects
	Low	Medium	High	
Formaldehyde				
mg/m³	<0.05	0.05–0.10	>0.10	Mucous membrane
ppm	0.04	0.04–0.08	>0.08	irritation, SBS
VOCs, mg/m³				SBS
Ozone				
mg/m³	<0.05	0.05–0.10	>0.10	Mucous membrane
ppm	<0.03	0.03–0.05	>0.05	irritation
Hydrogen chloride				
mg/m³	<1.4	1.4–4	>4	Mucous membrane
ppm	<1	1–3	>3	irritation
Nitrogen dioxide				
mg/m³	<0.2	0.2–0.5	>0.5	Mucous membrane
ppm	<0.1	0.1–0.3	>0.3	irritation, asthma
Carbon monoxide				
mg/m³			>10	General symptoms
ppm			>9	
Carbon dioxide, ppm	<700	700–1000	>1000	Stale air
Mineral fibers				
Air, f/m³	<200	200–1000	>1000	Mucous membrane
Surfaces, f/m²	<10	10–30	>30	and skin irritation
Bacteria in air CFU/m³				Allergy, SBS, respiratory complaints
Fungi in air CFU/m³				Allergy, respiratory complaints
Tobacco smoke	None	Sometimes	Constant	Eye irritation, SBS
Dust (air)				
mg/m³	<0.1	0.1–0.3	>0.3	SBS, mucous membrane irritation
Floor dust, g/m²	<0.2	0.2–0.5	>0.5	SBS
Bacteria in floor dust, CFU/g	<6 × 10³	6–10 × 10³	>10 × 10³	SBS ?
Fungal spores in floor dust, CFU/g	<1000	1000–3000	>3000	SBS ?, allergy
Macromolecular organic dust, mg MOD/g dust	<1	1–3	>3	SBS
Dust mites				
Allergen/g dust	<100 ng	100–2000 ng	>2000 ng	Allergy, asthma
Mites/g dust	<5	5–100	>100	

From Kukkonen, E. et al. 1993. Nordtest Report NT Tech. Rep. 204.

A questionnaire survey is carried out if operational controls are thought to be normal and complaints are not related to any observed abnormal condition. In some cases, systematic interviews are conducted of small groups of building occupants. The purpose of the survey is to determine the extent and nature of problems. It also serves as a basis for assessing whether problems are purely of a technical nature or whether there are hygienic or psychological considerations involved. The questionnaire is administered to all employees or to a representative sample. It includes questions relative to symptoms and specific complaints which can be attributed to the effects of factors in the indoor environment. The results of the survey questionnaire are evaluated in a systematic way to separate those problems which are of general occurrence, and thus considered "normal," from those which are more prevalent and in need of further investigation.

In the third step, a technical description of the building is made with the assistance of operational personnel and design architects and engineers. The technical description includes a variety of building-associated details such as materials and their condition, mechanical services and their operational condition and strategy, the rate of ventilation and its distribution per unit area and person, cleaning conditions, various service loads, etc. It is designed to provide a basis for an assessment of risk factors inherent in the building, its use, and operation. It also provides a basis for an inspection and any later measurements which may be made.

Steps four and five include an on-site building inspection and measurement of indicator values. They typically are carried out at the same time. Major elements of the building inspection include (1) building use, fittings, and furnishings; (2) cleaning, (3) moisture damage, (4) external design of the building, and (5) odors. In the first case, major attention is given to tobacco smoking, noise from ventilation systems, quantity and relative age of paper handled, presence of green plants and chemicals used in their care, and vision or noise problems associated with VDTs. Cleaning considerations include the standard of cleaning, factors which make cleaning difficult, and presence of dirt in ventilation openings. Evaluative factors associated with the building exterior include siting of air intakes; presence of laboratories, restaurants, etc. which pose a risk of entrainment or cross-contamination; noise from other activities; and state of building maintenance.

Indicator measurements are collected on a random basis in problem and nonproblem areas. Recommended measurements include air temperature, air movement patterns using smoke tubes, room CO_2, and CO levels either in the supply or exhaust air or at the outside air intake. In step six, a careful analysis is made of ventilation system performance.

Additional measurements are to be made in step seven. These include (1) the concentration of individual organic gases/vapors, and possibly formaldehyde, when the presence of irritant substances is suspected, but the sources cannot be identified; (2) concentrations of airborne mineral fibers when the ceiling surface is lined with, for example, mineral wool which is unprotected and damaged; (3) the concentration of airborne dust or dust on floor surfaces when cleaning is considered to be unsatisfactory, and large quantities of paper are being handled; (4) additional measurements of radiation asymmetry, head radiation from the ceiling, air velocity, etc. when thermal load is suspected; and (5) checking sound and light conditions in typical rooms if there are specific complaints or where these are suspected to be problems. Assessments may also be made to determine whether (1) sufficient air is supplied to occupied spaces, (2) air flow between rooms is such that air passes from cleaner to less clean rooms, (3) ventilation efficiency is satisfactory, and (4) outdoor air quality is satisfactory.

In step eight, a medical examination is conducted. This examination may involve all individuals or a selected sample. Additional investigations of a hygienic nature may be carried out based on the results of the medical examination.

The Nordtest protocol is intended to be used in a step-wise fashion until the causal nature of problems have been identified and remedial measures recommended and implemented. The entire eight-step process would only be conducted in those cases where the identification and resolution of the building-related health and comfort problem would prove to be relatively difficult.

FEATURES OF INVESTIGATIVE PROTOCOLS

The investigative protocols described above represent efforts by a variety of public agencies, investigators, and private consultants to standardize procedures for conducting problem-building investigations. The goal of any investigative protocol is to achieve a relatively high rate of problem identification and resolution. Differences evident in the investigative protocols described suggest that success rates in solving IAQ problems will depend in good measure on the protocol employed. There is, unfortunately, no evidence to indicate the relative success rates of individual protocols since systematic follow-up studies of the efficacy of recommended mitigation measures have not been conducted and reported.

Investigative protocols described above are characterized by common elements, the emphasis they place on certain aspects of the investigation, and elements which are unique to each.

Multiple Stages of Investigation

Most protocols include two or more stages or phases of investigation. These multi-phasal investigative approaches reflect the fact that many problems are easily identified and resolved from initial information gathering and/or a walk-through investigation. They also reflect the fact that problems in some buildings are not amenable to easy identification and require more intensive assessment which focuses on environmental monitoring and building systems evaluation.

Investigative protocols tend to differ as to factors which receive emphasis at each stage. In some, the first or early stages are limited to information gathering and an initial walk-through (PWC, OIC, DBRI, Nordtest, California, EH & E). NIOSH also includes limited environmental monitoring. In some protocols, the assessment of occupant symptoms/complaints is conducted during the initial stage (NIOSH, EH & E, Nordtest, OIC, DBRI, California), while in others symptom/complaint assessments are made in the advanced stages of the survey (AIHA, BD), or more intensive investigations of symptoms/complaints are made in the advanced stages (EH & E, DBRI, California). Assessments of HVAC system operation and maintenance vary as to when they are conducted. In general, some preliminary HVAC system assessments are made in initial investigative activities with intensive evaluations at advanced stages.

Personnel Conducting Investigations

Many protocols recognize that a variety of skills are required to conduct a problem-building investigation, and single investigators do not generally have all the necessary skills. A multidisciplinary team approach appears to be a feature of the NIOSH, California, DBRI, Nordtest, EH & E, and BD investigative protocols.[3,8,12,14,20,21] The PWC, OIC, and AIHA approaches appear to depend on a single investigator with training in IH. The multidisciplinary approach requires at least two and in most cases three specially trained individuals to conduct an investigation. As such, it is both labor and resource intensive. At the state level in the United States, where many problem-building investigations are initially conducted, resources for such multidisciplinary teams are not generally available. In the private sector, investigative teams may be comprised of highly specialized professionals. As a consequence, fees for services rendered may be relatively high, and many potential clients may find the service unaffordable.

Site Visits

Site visits are an integral part of all investigative protocols. These typically include discussions with building management and occupants, a walk-through inspection, a limited evaluation of building HVAC systems and their performance, and additional site visits if initial investigations fail to identify the problem or when the hypothesis formulated on the basis of information gathered in the initial phase of the investigation is to be tested.

Assessment of Occupant Symptoms/Complaints

Most investigative protocols recognize to various degrees that health and comfort problems putatively being experienced by building occupants need to be assessed. The exceptions are those protocols which give little consideration to assessing complaints directly (AIHA and BD) or assume that building management prefers employees be kept uninformed of the investigation and its nature.[13] Those protocols that include the assessment of occupant symptoms/complaints as a major focus of the investigation recognize that such an assessment provides important clues to the nature of the problem and its resolution. These assessments represent the epidemiological component of problem-building investigations.

Significant attention is given to epidemiological assessments in the California, EH & E, DBRI, and Nordtest protocols.[8,14,20,21] Though an epidemiologist is typically a part of the NIOSH investigative team, the role of epidemiology appears to be less clearly defined in the NIOSH approach than it is in the California, DBRI, and EH & E protocols. The EH & E protocol emphasizes epidemiological evaluations in both stage one and stage three investigations. The former is designed to characterize the nature of symptoms/complaints, their relationship in time, and the location and numbers of affected individuals. This stage one epidemiological assessment appears to be particularly appropriate in identifying causal factors because it focuses on the specific nature of the health and/or comfort problems. It seeks to identify the problem based on the specific health effects and the epidemiological patterns which are associated with them.

The questionnaire survey in the Nordtest protocol[21] also serves as a rudimentary epidemiological assessment. The California, DBRI, and Nordtest protocols[8,20,21] involve significant epidemiological assessments at advanced stages of the investigation as does the EH & E protocol. The epidemiological assessments included in California and DBRI protocols are intensive and require considerable statistical analyses. Such assessments require specialists trained in epidemiological surveys and statistical analyses with considerable staff time and resources allocated to them. Such resource demands would not be practical in most problem-building investigations.

Assessment of building occupant symptoms/complaints are conducted in the second stage of the OIC protocol. It employs a comprehensive questionnaire and detailed instructions to the investigator on how to summarize collected data and interpret results. It also provides detailed summary sheets, a software package, and instructions so that investigators not trained in epidemiology can interpret results and identify the cause of the problem.

In the BD and AIHA protocols,[4,10,11] an assessment of occupant symptoms is only conducted in the last stage of the investigation. In the former, a human resources questionnaire (see Appendix B) is distributed to building occupants to obtain a limited amount of demographic data, occupant symptoms, and a rating of the acceptability of environmental parameters in the building. Implicit in the BD approach is that one can solve IAQ problems, including health-related complaints, without any special knowledge of epidemiology or toxicology, that these are comfort complaints which are amenable to resolution if performance factors for building systems are being met. It assumes, for the most part, that health-related complaints are due to the failure of building systems to perform their jobs as designed or should have been designed. It is an investigative protocol that depends on engineering assessments of building systems and IH measurement of contaminant concentrations. Because the BD approach[10,11] relies almost exclusively on system performance as the key to addressing most IAQ problems, it will fail to resolve those problems where there is no apparent association between system performance and building-related health and comfort complaints. Such cases are unfortunately quite common.

The AIHA protocol describes goals which require significant epidemiological assessment. These goals are set forth in the section, AIHA Protocol, this chapter. Unfortunately, the protocol itself contains few elements of investigative conduct by which those goals can be achieved. The protocol appears to eschew contact with building occupants which unfortunately is not consistent with its stated goals of epidemiological assessment.

Use of Questionnaires/Checklists

Questionnaires are used in most IAQ investigative protocols. They may be designed to elicit demographic information about the occupants, their perception about environmental and work conditions in the building, the nature of their symptoms/complaints, etc. The value of questionnaires lies in the fact that they standardize data collection relative to factors which may explain the nature and causes of complaints. Questionnaires may be self-administered or be administered by investigators. Self-administered questionnaires are more commonly used because they reduce staff time required. The nature and use of questionnaires is quite varied. In the NIOSH protocol,[3] a questionnaire (see Appendix A) is used to conduct personal interviews during the initial site visit. It may be either self-administered or serve as a guide for occupant interviews. A comprehensive self-administered questionnaire is used in the OIC protocol (see Appendix D). It is notable for the detailed instructions provided to the investigator for summarizing and interpreting data results and attempting to correlate symptoms with exposures. In the DBRI and Nordtest protocols,[20,21] a questionnaire (see Appendix E) designed to elicit information from employees, their demographic characteristics, environmental and work conditions, and symptoms, is administered in the early stages of the investigation. In the BD and AIHA protocol,[4,10,11] questionnaires are only administered in the last stage when all other actions have failed to identify and resolve the problem. In the California protocol,[8] a relatively simple, one-page questionnaire is used in the early stages of the investigation, and a more extensive questionnaire is used in the later comprehensive stage.

In addition to questionnaires, a variety of checklists are used. Checklists are designed to assure standardized/consistent collection of information about the building environment, HVAC system operation and maintenance, and potential contaminant sources. Typical types of information gathered and recorded in problem-building investigations can be found in Tables 7.1–7.9. Checklists used to characterize/assess the problem-building environment, the HVAC system, and potential sources are available in published form in the protocols of the Danish Building Research Institute,[20] the California program,[8] and USEPA/NIOSH.[7]

Assessment of HVAC System Operation and Maintenance

Evaluations of the operation and maintenance of HVAC systems in buildings under investigation are common to all investigative protocols. The reason for this consistency is a shared belief that a large percentage of IAQ problems are due to inadequate ventilation or ventilation system deficiencies, or that if the causal factor or factors cannot be identified, the problem can nevertheless be mitigated by increasing ventilation rates. This is evident in NIOSH[1] and Health and Welfare Canada[16] reports of the results of their IAQ investigations.

In the BD protocol,[10,11,12,13] which is widely used by private consulting firms in the United States, investigations are conducted on the premise that most IAQ problems are due to inadequacies of building system design, operation, and maintenance. As a consequence, much investigative energy is directed to the evaluation of building HVAC systems. With few exceptions (e.g., hypersensitivity pneumonitis, fiberglass dermatitis, and CO exposures), the BD protocol appears to treat IAQ problems in a generic way; that is, they are comfort problems addressable within the context of building systems being designed and operated to meet accepted performance criteria.

The strength of the BD approach[10,11] and other protocols that emphasize assessment of HVAC system performance is that on a theoretical basis ventilation can be used to reduce contaminant exposures even when the causal agent cannot be identified. If increased ventilation does significantly reduce the concentration of the unknown contaminant or contaminants, then a reduction in symptom prevalence may be anticipated. Building investigators who recommend the use of increased outdoor ventilation rates for complaint mitigation report anecdotal testimony from building managers and occupants that symptoms/complaints have decreased as a consequence of increasing ventilation rates. There have, however, been few systematic studies to scientifically verify the effectiveness of increasing outdoor ventilation rates for the resolution of problem-building complaints. Discussions of this problem can be found in Chapters 3 and 10.

Environmental Measurements

The use of environmental measurements varies widely across investigative protocols described. Some protocols (NIOSH, USEPA/NIOSH) deemphasize environmental measurements, particularly contaminant measurements, reasoning that in most cases no single contaminant will have a sufficiently high concentration to explain observed symptoms.[3,7] The exception is CO_2. In all investigative protocols, CO_2 measurements are recommended for the purpose of determining ventilation adequacy. Only limited environmental measurements are made in the early stages of most problem-building investigations. They typically include CO_2 and environmental parameters such as temperature, relative humidity, and airflow. Significant quantitative environmental measurements are usually reserved for advanced investigative stages as initial efforts fail to identify the problem or when environmental measurements are necessary to test a hypothesis relative to a potential causal factor or agent.

In theory, it would be desirable to conduct measurements of only those contaminants which have a reasonable chance of being a potential cause of complaints. However, since symptoms are usually subjective and nonspecific, it is often difficult (with few exceptions) to identify a causal factor by means of symptom patterns. Consequently, many investigators conduct screening measurements of contaminants (1) which have been previously associated with health complaints at relatively low levels of exposure, (2) for which there is an IAQ guideline, and (3) for which sampling/analytical procedures are available and relatively easy to use. In the second stage of the PWC protocol,[18,19] routine measurements are made of CO_2, CO, RSP, and TVOCs. Screening measurements of a variety of indoor air contaminants are also made in the advanced stages of the NIOSH, AIHA, California, OIC, BD, and EH & E protocols.[3,4,8,9,10,11,14]

Source Assessments/Contaminant Considerations

Inspections or walk-through investigations are conducted in all protocols. These serve a number of purposes. Among the more important of these is the identification of potential contaminant sources which may be responsible for the problem or problems. These source assessments are of a qualitative nature. They are based on the assumption that identifiable contaminant sources are responsible for building air contamination and occupant health complaints. These include, for example, contaminant problems associated with entrainment, cross-contamination, and re-entry, contaminants generated from office equipment and supplies, microbial infestation of building materials and systems, pressed wood furnishings which may emit significant quantities of formaldehyde, etc. Such contaminant/source problems may have been identified in previous investigations, known from experience, drawn from the reports of others, or have been the subject of intensive research investigations. As a consequence, many investigators have a sense (or should have a sense) of the kind of contamination problems which have the potential for occurring in a building and how such problems can be evaluated.

It is apparent from a review of the various investigative protocols that some consider only a relatively narrow range of potential problems, whereas others are more inclusive. TVOC levels are given significant consideration as a causal factor in both the PWC and DBRI protocols and less so in the Nordtest protocol.[18,19,21] Other protocols do not appear to recognize TVOCs as a potential cause of building-related health complaints. The different treatment of TVOCs as a potential SBS causal factor can be seen in the comparative results of NIOSH[1] and PWC[17] investigations. In approximately 50% of 30 PWC investigations, occupant complaints were determined to be due to excessive exposures to TVOCs and/or formaldehyde. In many cases, TVOCs were in excess of 5 mg/m^3 (1 ppm), a guideline value based on the studies of Molhave and his co-workers.[23,24] This is in contrast to NIOSH investigations[1] where excessive TVOC levels appear not to have been seriously considered as a potential cause of sick building complaints. Not surprisingly, inadequate ventilation is less frequently identified as the primary cause of building-related health complaints in PWC investigations compared to those of NIOSH (43 to 53%).

As seen from the above discussion, the outcome of an investigation is dependent on what investigators and the agencies which support them consider to be the cause of building-related health complaints. Nowhere is this more evident than in DBRI and Nordtest protocols.[20,21] Potential causal factors evaluated by DBRI and Nordtest investigators include contaminants/sources (see Table 7.10 and 7.11). which are not commonly considered to be problems by investigators in the United States.

In North America, many investigators are generally unaware that many of these factors (Table 7.10 and 7.11) have been reported to be associated with problem-building complaints. If they are, they are not certain of their potential significance because of the limited research and/or investigations on which they are based. North American investigators appear to be relatively conservative in accepting the various concepts and findings of European researchers and investigators in conducting problem-building investigations. What one considers to be a potential problem will, of course, significantly affect the outcome of an investigation. It is evident from these differences that the probability of identifying the wrong casual factor or no causal factor at all is quite high.

Use of IAQ/Comfort Guidelines

As previously described, many of the early problem-building investigations were conducted by industrial hygienists who employed traditional IH concepts and methodologies. Measurements of various IAQ parameters were often made and compared to occupational PELs or TLVs. In almost all instances, contaminant levels were several orders of magnitude lower than occupational standards and guidelines. Based on these comparisons, many investigators concluded that there was either no problem or that it could not be explained from chemical or physical measurements. Though the days of using industrial PELs and TLVs for purposes of interpreting the results of IAQ investigations are over for the most part, there is nevertheless considerable interest in using guidelines for the purpose of interpreting test results and determining whether building problems are due to one or more indoor contaminants or other environmental parameters. This is true for most investigative protocols. The use of IAQ guidelines reflects an inherent desire on the part of investigators for simplicity and security as well. By using guidelines, an investigator can safely (from a professional standpoint) make conclusions relative to causal factors based on some established set of criteria which have official or quasi-official acceptance or recognition. The ASHRAE IAQ and thermal comfort guidelines[5,6] and the WHO IAQ guidelines[25] are examples of this.

Because of the historical reliance on occupational standards and guidelines, the availability of IAQ guidelines is of considerable importance to industrial hygienists who conduct IAQ investigations. It is important to note in this regard that the AIHA protocol specifically avoids the use of guidelines and advises extreme caution in interpreting air test results. Guidelines are important in the NIOSH, OIC, BD, DBRI, and Nordtest protocols.[3,9,10,11,20,21] NIOSH investigators typically use ASHRAE guidelines[5,6] for both air contaminants and thermal comfort. The BD approach[10,11] places heavy emphasis on air quality and thermal comfort guidelines using these as performance criteria. Thermal and air quality guidelines used by BD investigators are summarized in Tables 7.12 and 7.13. Guidelines (Tables 7.10 and 7.11) for a number of indoor air contaminants are used to interpret results of technical investigations conducted under the Nordtest protocol[21] and in the DBRI protocol.[20] Guidelines used for interpreting results of investigations conducted by the OIC protocol are summarized in Table 7.14.

Guidelines are developed by panels employing a consensus process. In the United States, the American Society of Heating, Refrigerating, and Air-Conditioning Engineers (ASHRAE) has been a leader in formulating IAQ and thermal comfort guidelines. In Europe, the World Health Organization (WHO) has exercised a leadership role in developing IAQ guidelines. Guidelines for individual contaminants such as formaldehyde have been developed by a variety of European countries and governmental entities in the United States and Canada (Table 7.15).

The use of guidelines in IAQ investigations has both advantages and disadvantages. On the positive side, they assist investigators in interpreting the results of environmental measurements. If measured results are clearly above guideline values, an investigator can feel reasonably confident that the contaminant or contaminants were in whole or in part responsible for causing the problem. If measured results are clearly below guideline values (particularly considerably lower), an investigator can be reasonably confident that the contaminant or contaminants were not responsible.

On the negative side, many investigators treat guidelines as if they are magic numbers, that is, dividing lines between illness and safety. No guidelines developed by a consensus process, or any process, can provide such a clear-cut delineation. Guidelines are developed by panels who represent a variety of fields of investigation and interests. They are often subject to considerable pressure and lobbying efforts from interests who may be adversely affected by the outcome of their deliberations. Particularly notable has been the case of the ASHRAE guideline for formaldehyde[26] which was deleted from the 1989 ASHRAE IAQ guidelines allegedly because of threatened litigation.

Table 7.12 Thermal environment performance criteria.

Parameter	Guideline
Operative temperature (winter)	68.5–76°F (20.3–24.4°C) (at 30% RH)
Operative temperature (summer)	73–79°F (22.8–26.1°C) (at 50%)
Dew point	>35 F (1.7°C) (winter); 62°F 16.7°C) (summer)
Air movement	≤30 FPM (9.1 MPM) (winter); ≤50 FPM (15.2 MPM) (summer)
Vertical temperature gradient	Shall not exceed 5°F (2.8°C) at 4 in. (10.2 cm) and 67 in. (170.1 cm) levels
Plane radiant symmetry	<18°F (10°C) in horizontal direction; <9°F (5°C) in vertical direction

From Woods, J.E. et al. 1989. 80–98. In: Nagda, N.L. and J.P. Harper (Eds.). *Design and Protocol for Monitoring Indoor Air Quality.* ASTM STP 1002. American Society for Testing and Materials. Philadelphia. With permission.

Table 7.13 Air quality guidelines used in Building Diagnostic Protocol.

Substance threshold	Odor recognition	Occupational TLV	Protection factor over TVL by use of odor annoyance threshold	Recommended AQ standard criteria
Butyric acid	0.001 ppm	N.A.	N.A.	0.002 ppm c
Carbon dioxide	N.A.	5000 ppm	N.A.	1000 ppm c
Suspended particles	N.A.	10 mg/m^3	N.A.	0.05 mg/m^3 y (≤10 μm)
Pyridine	0.02 ppm	5 ppm	100	0.05 ppm c
Furfural	0.002 ppm	2 ppm	500	0.004 ppm c
Toluene	2 ppm	100 ppm	33	3 ppm c

Note: N.A. = not applicable; c = ceiling concentration; y = annual average.

From Woods, J.E. et al. 1989. 80–98. In: Nagda, N.L. and J.P. Harper (Eds.). *Design and Protocol for Monitoring Indoor Air Quality.* ASTM STP 1002. American Society for Testing and Materials. Philadelphia. With permission.

Table 7.14 Guidelines for the interpretation of measured contaminant levels in indoor spaces.

	Measured concentrations (ppm)			
Contaminant	Normal outdoor	Normal indoor	Possible problem	Do not exceed
Carbon dioxide	330–400	330–800	800–1000	1000
Carbon monoxide	0–4	0–4	4–11	11
Formaldehyde	0–0.02	0–0.04	0.03–0.05	0.05
Ozone	0–0.03	0–0.01	0.05–0.10	0.10

From Occupational Health and Safety Division, Ontario Ministry of Labor, Health and Safety Branch. 1988. Ontario.

In many cases, the scientific evidence to support a guideline is meager. As a consequence, guideline development is often associated with considerable uncertainty. Consensus panels convened to develop guidelines must often decide whether to err on the side of safety or take a more conservative approach

Table 7.15 Guidelines for acceptable formaldehyde levels in indoor air.

Agency/government	Permissible level (ppm)	Status
HUD, USA	0.40 target	Recommended
ASHRAE, USA	0.10	Recommended (withdrawn)
California Department of Health, USA	0.10 (action level) 0.05 (target level)	Recommended
Canada	0.10 (action level) 0.05 (target level)	Recommended
Sweden	0.20	Promulgated
Denmark	0.12	Promulgated
Finland	0.12	Promulgated
Netherlands	0.10	Promulgated
Germany	0.10	Promulgated
Italy	0.10	Promulgated
WHO	0.08	Recommended

(less stringent). The latter course appears often to be the case. This is evident in the relatively high initial ASHRAE guideline[26] and WHO guidelines[25] for formaldehyde (120 μg/m^3, 0.10 ppm; and 96 μg/m^3, 0.08 ppm; respectively) given the fact that there is considerable evidence that formaldehyde may cause symptoms below these levels.[27-29]

The potential for formaldehyde to cause symptoms at levels below 0.08 ppm is addressed in OIC[9] and Nordtest risk guidelines.[21] A similar concern may be expressed for TVOCs in the PWC and DBRI guidelines.[18,20] In each case, the guideline is 5 mg/m^3 (1 ppm), initially the lowest level at which Molhave and his co-workers[23,24] observed symptoms in human subjects exposed to a mixture of 22 VOCs. Because exposures to 5 mg/m^3 TVOCs cause symptoms, the strict application of this guideline makes no provision for symptom occurrence at lower TVOC exposure levels. The 5 mg/m^3 guideline does not take into account the recent dose-response relationships for TVOCs developed by Molhave[30] in which concentrations in excess of 3 mg/m^3 are described as causing human discomfort, and concentrations in the range of 0.2–3.0 mg/m^3, which may cause irritation and discomfort when other exposures interact. It also does not take into account the studies of Norback et al.[31,32] who observed dose-response relationships between SBS symptoms in 11 buildings at TVOC concentrations in the range of 0.05–1.38 mg/m^3 and in problem school buildings in the range of 0.07–0.18 mg/m^3.

Used properly, guidelines can serve as useful tools for investigators to evaluate the probability that building complaints are related to the factor under measurement. As such, they are not to be used as discrete dividing lines between safe and unsafe levels of exposure. They should be used as a reference value from which judgments can be made relative to the role of contaminants in the observed problem and what mitigation measures should be implemented. These judgments should include both a consideration of the guideline and the literature base associated with the contaminant. One should also use good sense as well, particularly when the facts of the exposure strongly suggest that the contaminant under question is the cause of complaints. The Nordtest and OIC guidelines found in Tables 7.10, 7.11, and 7.14 are notable since they provide values which indicate an elevated risk of symptoms and a range of values which indicate that possible problems may occur.

The use of guidelines poses other difficulties. These include the unavailability of guidelines for some exposures, the limitations of air testing involved in one-time sampling, and the selection of contaminants for screening measurements because guidelines are available. In the absence of guidelines, investigators must attempt to relate the symptoms/complaints to an exposure which is consistent with toxicological and epidemiological evidence. For example, when symptoms of a neurotoxic nature coincide with solvent exposures, it would make sense to test a hypothesis that would evaluate the solvent as the potential causal factor. This could include removing the source or controlling contaminant levels in some other way. This approach, of course, makes sense even if a guideline were available and if contaminant levels were observed to be below guideline values based on one-time sampling.

Since exposures to a particular gas or particulate-phase contaminant vary considerably over time, there is a risk that the results of one-time sampling (which is common in IAQ investigations) may be misinterpreted when compared to guideline values without an understanding of the range of contaminant variation which has been or is occurring. Such misinterpretation is likely to occur when one-time test results are in the low end of the range of exposure variation.

It is unfortunate that the availability of guidelines often directs the process of selecting contaminants for screening measurements. In theory, measurements should only be made of those contaminants which can reasonably be expected to be potential causal factors in a particular problem building. Unfortunately, contaminants are often measured simply because there is a guideline to which results can be compared.

As can be seen by the above discussion, IAQ guidelines pose a significant dilemma. On both theoretical and practical grounds, they are important tools in conducting problem-building investigations. On the other hand, they are often misused and thus may be an impediment to the successful conduct of problem-building investigations.

PROCEDURES FOR EVALUATING PROBLEMS WITH HVAC SYSTEMS

All investigative protocols place some to considerable emphasis on evaluating building HVAC systems relative to contributing to or mitigating IAQ health and comfort problems. The evaluation of HVAC systems and their potential role in occupant health and comfort complaints requires a technical understanding of their design, operation, maintenance, and the problems which may occur as a result of system deficiencies. Systematic protocols are important in evaluating HVAC systems just as they are in the larger investigative protocols of which they are components.

Before conducting HVAC system evaluations, building investigators need to have a working knowledge of the basic types of HVAC systems used in buildings, their design and operating characteristics, and the kinds of problems that can be expected to be associated with their operation.

A variety of HVAC system-associated problems have been linked to poor IAQ health and comfort complaints. These include (1) insufficient outdoor air for the effective control of bioeffluent and other contaminants generated in buildings; (2) migration of contaminants from one building zone or space to another (cross-contamination); (3) re-entry of building exhausts; (4) entrainment (in intake air) of contaminants generated outdoors; (5) generation of man-made mineral fibers (MMMF) from the disintegration of sound liners in air-handling units (AHUs); (6) microorganisms and organic dust contamination in condensate drip pans, humidifiers, filters, and porous thermal/acoustical insulation; (7) inadequate dust control; (8) inadequate control of temperature, relative humidity, and air velocity; and (9) inadequate air flows into building spaces due to system imbalances.

Inadequate outdoor air supply has been a commonly reported problem in building investigations. This problem has been associated with such factors as (1) the provision of outdoor air for ventilation purposes was not designed into the system, (2) design outdoor air capacity is inadequate for building needs, (3) increase in building occupancy beyond initial design, (4) malfunction of system inlet dampers, (5) system operating practices which minimize the provision of outdoor air for purposes of energy conservation and minimizing operating costs, and (6) reduced air flows due to poor maintenance of filters, fans, and other HVAC system components.

The HVAC system may contribute to IAQ complaints because it can serve as a pathway for contaminants in moving from a source of generation to other building spaces and for outdoor contaminants to enter the building. Cross-contamination often results as a consequence of pressure imbalances between building zones served by different AHUs. Pressure imbalances can also occur in a single air-handling unit if room supply and exhausts are not properly balanced.[33] Outdoor contaminants may enter the building air supply as a result of the improper location of air intakes near loading docks, or at street level near high traffic areas, or in close proximity to cooling towers. Parking garages located on lower levels of high-rise buildings may also serve as a source of building air contamination. Contaminants may re-enter the building from exhaust vents or boiler chimneys because of the poor design and location of both exhausts and outdoor air intakes and by infiltration due to pressure imbalances in the building.[34]

The HVAC system may also be a source of indoor air contaminants. The most notable are microorganisms which may proliferate in poorly maintained condensate drip pans,[35] contaminated air filters,[36] and porous acoustical duct and air-handler insulation.[37] Such contamination has been associated with outbreaks of hypersensitivity pneumonitis[35] and linked with allergic symptoms.[36] The generation of airborne MMMF from sound/thermal liners has also been reported.[38]

The HVAC system may contribute to occupant complaints because of inadequacies associated with its operation and maintenance. In many cases, dust control is limited to low efficiency dust stop filters which are designed to protect mechanical equipment only. In such cases, system filters may not be adequate.

The management of thermal comfort is always a major problem for building managers. Thermal comfort problems associated with temperature, relative humidity, and air movement are commonly associated with HVAC system design, operation, and/or maintenance deficiencies.

Types of HVAC Systems

A variety of HVAC systems are employed to climate control buildings. The three basic types include the all-air system, the air-water system, and the all-water system.[33] One or more of these basic systems may be employed in an individual building.

All-Air Systems

All-air systems include large central AHUs which serve an entire building or AHUs which serve small areas or zones. In such systems, air is conditioned as it passes over heating and cooling coils. Conditioned air may be delivered to a space through a single duct or through hot and cold ducts and mixed in a box prior to entering building spaces. These systems may provide a constant (CAV) or variable air volume (VAV) to conditioned spaces (Figures 7.4 and 7.5). The former is operated at a fixed flow rate, and the temperature is varied to meet heating/cooling needs. The VAV system has the capability of varying the air volume as well as temperature in response to the heating/cooling demands of spaces. Both systems utilize outdoor air for ventilation and can control the relative percentages of outdoor and recirculated air which may vary from 10–100%.[33] Variable air volume systems are both more complicated and difficult to operate. Because of greater technical demands for proper operation, VAV systems have acquired a reputation for being problem plagued. A major problem occurs when VAV valves are designed to close when thermal requirements are met.[39] As a consequence, little or no ventilation air may be provided to the occupied space.

Air-Water Systems

Air and water systems differ from all-air systems by the fact that before supply air is discharged to a space, it passes by a fan coil unit supplied with heated or cooled water. These are described as terminal reheat systems. Other than this difference, they operate similarly to single-duct all-air systems.[33]

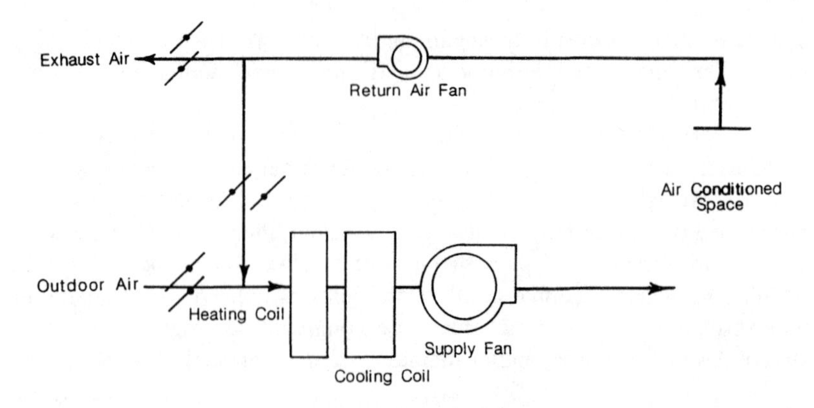

Figure 7.4 Constant-volume HVAC system design. (From McNall, P.E. and A.K. Persily. 1984. *Ann. ACGIH: Evaluating Office Environmental Problems.* 10:49–58. With permission.)

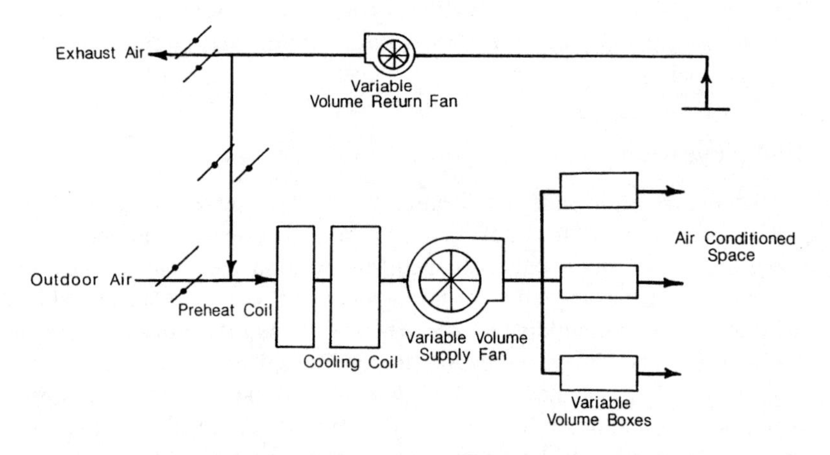

Figure 7.5 Variable air volume HVAC system design. (From McNall, P.E. and A.K. Persily. 1984. *Ann. ACGIH: Evaluating Office Environmental Problems.* 10:49–58. With permission.)

All-Water Systems

In all-water systems, heating and cooling occurs through terminal reheat units which are located in each occupied space. Unlike all-air or air-water systems, the terminal reheat unit only provides air circulation in a space. A separate duct system must be utilized to provide outdoor air for ventilation.

HVAC System Evaluations

Procedures used to evaluate HVAC systems in NIOSH investigations have been described by Salisbury.[40] These initially include discussions with building personnel familiar with HVAC system operation and maintenance, followed by a walk-through inspection. During the walk-through, investigators identify the HVAC system type, locate its major components, and determine air flow paths. Major components inspected include AHUs, fan coil units, condenser units, heating systems, cooling systems, air supply inlets, and outdoor air intakes. The basic operation of temperature control systems and air control dampers is reviewed, as well as HVAC system start-up and shutdown times. Outside air dampers are checked to determine whether they are functional; fans are checked for proper operation.

The potential for entrainment of outside air contaminants is considered and contaminant migration pathways evaluated. Interiors of AHUs where standing water or moisture is likely to be found (cooling coils, condensation pans, and drains) are inspected. The potential for humidification systems to cause air contamination problems is also evaluated.

If available, HVAC system specifications, blueprints, or HVAC system test and balance data are reviewed to determine (1) return, outdoor, and supply air volumes; (2) number and location of HVAC zones; (3) air-duct configuration; (4) location of air supply diffusers and return air grills; and (5) the theory of operation of the HVAC system and controls. If possible, the outdoor air ventilation rate is measured directly using a pitot tube, air flow hood, or some similar air flow measuring device. If available, direct air hoods are used to measure supply and return air flow rates. Carbon dioxide levels are measured as an indirect indicator of building ventilation when direct measurement of outdoor air flows is not possible or practical.

In addition to these assessments, the AIHA protocol[4] recommends the evaluation of occupancy and space-use patterns, inspecting and evaluating chemicals present in mechanical rooms, the type and condition of HVAC system filters, differences in air flow patterns from selected supply air diffusers and return air grilles in problem and nonproblem areas, and the potential for air stratification and poor air distribution.

In the Nordtest protocol,[21] the assessment of ventilation system performance includes an inspection of AHUs for mechanical problems, blockage of filters, visual examination of heating surfaces, heat exchangers, humidifiers, and damper settings. It also includes measurements of supply and exhaust volume flow rates; air change rates, if the measurement of volume flow rates does not provide satisfactory answers; the efficiency of ventilation in "typical" rooms, if the supply of outdoor air is near the lower limit; and the proportion of recirculated air. In addition to these measurements, random sampling of air flows between rooms is recommended.

Assessing Building Ventilation

Outdoor air ventilation rates can, in theory, be determined directly by the use of air flow instruments (assuming that infiltration air inflows are low). In many cases, it may neither be feasible nor desirable to conduct such measurements. Instead, outdoor air flows may be determined indirectly using a variety of techniques. These include measurements of CO_2, conducting enthalpy balances, and using tracer gases. These techniques can be used to determine the effectiveness of the ventilation system in removing indoor air contaminants.

CO_2 Techniques

Outdoor air flow rates or effective ventilation rates (EVRs)[41] can be determined from measurements of CO_2 in return, supply, and outdoor air.

The percentage of outdoor air can be determined from Equation 7.1.

$$\text{Outdoor air (\%)} = \frac{C_S - C_R}{C_O - C_R} \times 100 \tag{7.1}$$

where C_S = ppm CO_2 in supply or mixed air in the AHU
C_R = ppm CO_2 in return air
C_O = ppm CO_2 in outdoor air.

The percentage of outdoor air can be converted to an outdoor air flow rate by means of Equation 7.2.

$$\text{Outdoor air flow rate (L/s)} = \frac{\text{\% Outdoor air} \times \text{Total air flow (L/s)}}{100} \tag{7.2}$$

The value used for the total air flow may be that provided to a room or zone, the capacity of the AHU, or the total air flow of the HVAC system as determined from actual measurements.[7]

Outdoor air flow rates can also be determined by graphic procedures. Assuming peak CO_2 at the time of measurement, ventilation rates in L/s/person can be determined from Figure 7.6. A peak CO_2 level of 800 ppm would be equal to a ventilation rate of 10 L/s/person; a peak CO_2 level of 1000 ppm would result in a ventilation rate of 8 L/s/person.[40]

Using Figure 7.6 for determining ventilation rates assumes that peak CO_2 levels have been measured. Obtaining peak levels requires knowledge of the appropriate time to measure CO_2 in a building.

CO_2 levels can also be used to determine outdoor air exchange rates, air changes/hour (ACH). Such determinations can be made from graphs derived from the following equation:

$$C_t - C_i = \frac{qa}{V} \cdot \frac{1}{n}(1 - e^{-nt}) \tag{7.3}$$

where C_t = CO_2 at time t, m^3/m^3
C_i = CO_2 ventilation air concentration, m^3/m^3
q = emission of CO_2 in m^3/hour/person

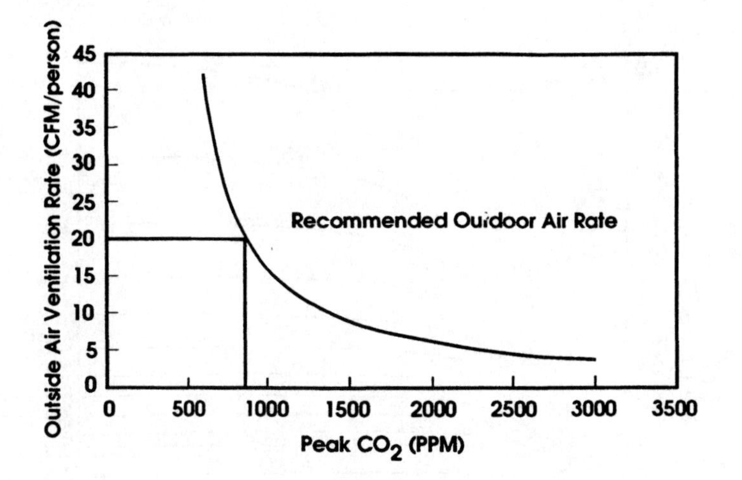

Figure 7.6 Outdoor ventilation requirements as a function of peak indoor CO_2 concentrations. (From Salisbury, S.A., 1990. 87–98. In: Weeks, D.M. and R.B. Gammage (Eds.). *Proceedings Indoor Air Quality International Symposium: The Practitioner's Approach to Indoor Air Quality Investigations.* American Hygiene Association. Akron, OH. With permission.)

a = number of occupants
V = volume of room air, m^3
n = air exchange per hour.

For purposes of these calculations, the per person CO_2 emission rate is set to $18 \times 10^{-3} m^3/hour$ and the outdoor CO_2 level to $330 \times 10^{-6} m^3/m^3$.[20]

By setting y equal to $1/n\ (1\text{-}e^{-nt})$, Equation 7.3 can be reduced to:

$$y = \frac{(C_t - C_i)V}{qa} \tag{7.4}$$

By solving for *y*, the air change rates for various time periods after building occupancy can be determined from Figures 7.7 and 7.8.[20]

This method assumes that the same number of individuals are in the occupied space at all times, the CO_2 is equal to the outdoor value C_i at time 0, and the results express an air change rate based on CO_2 concentration. It also assumes a ventilation efficiency consistent with complete mixing.

The curves in Figures 7.7 and 7.8 can be used to calculate air exchange rates at almost any time after building occupancy begins and do not require the measurement of peak levels of CO_2 as indicated in the technique previously described. This method does require a knowledge of the numbers of individuals present and the volume of the ventilated space.

The curves in Figures 7.7 and 7.8 illustrate how rapidly CO_2 levels reach equilibrium under different air exchange conditions. At high outdoor air exchange rates (2–5 ACH), CO_2 levels reach equilibrium in 1–2 hours, whereas

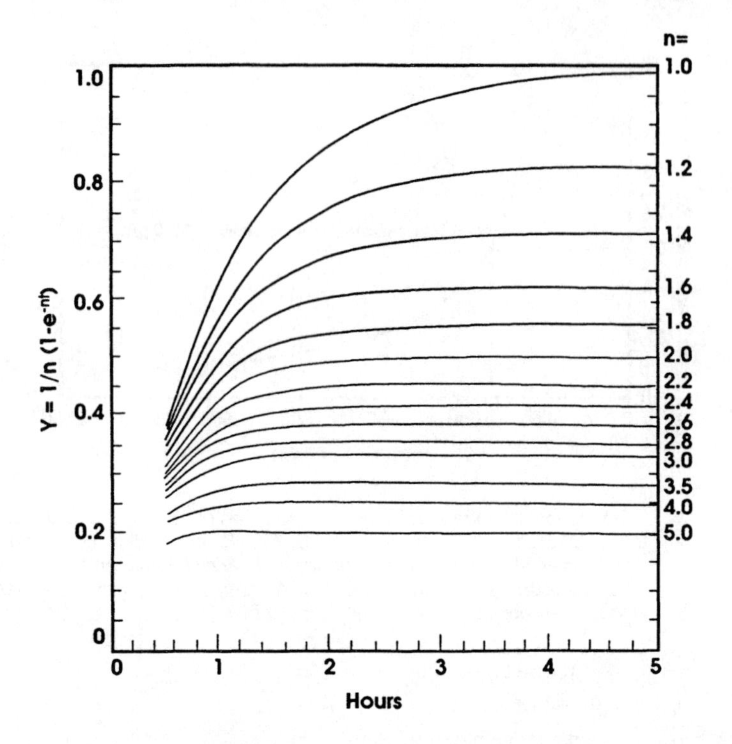

Figure 7.7 Calculation of building air change rates from changes in CO_2 concentrations, Y = <1. (From SBI Report 212 Indoor Climate and Air Quality Problems. 1990. Danish Building Research Institute.)

equilibrium or peak CO_2 levels do not occur until about the fifth hour at one ACH, with even longer periods for air exchange rates less than one.

The ventilation rate per person can be determined from the following equation:

$$VR = \frac{nV(1000)}{a(3600)} \tag{7.5}$$

where VR = ventilation rate/person (L/s/person).

Thermal Balance

The percentage of outdoor air can also be determined from temperature measurements of airstreams representing outdoor, return, and mixed air (supply air before it has been heated or cooled). Access to the return/outdoor air mixing box (for purposes of temperature measurements) is, of course, essential. This may be difficult or impossible in some systems. A large number of temperature measurements of the mixed airstream is recommended.[7]

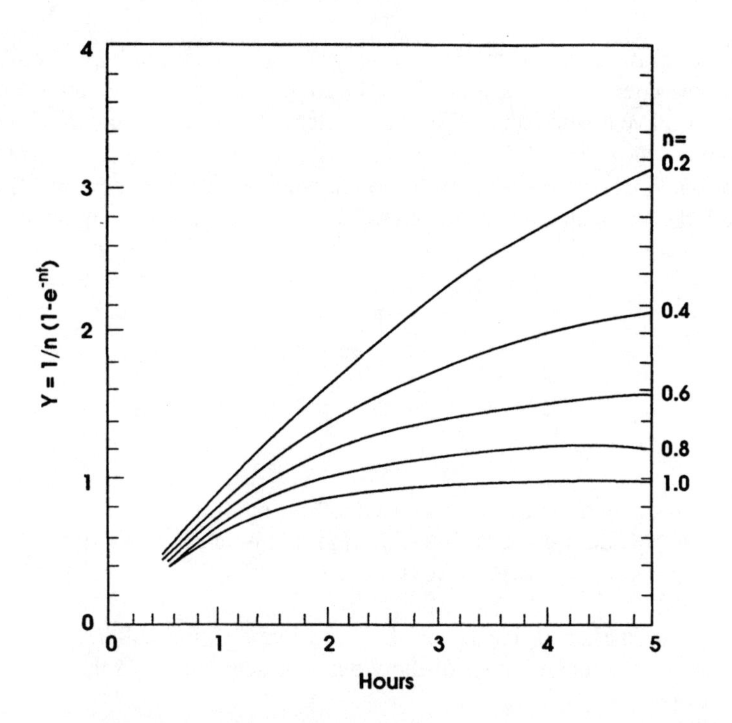

Figure 7.8 Calculation of building air change rates from changes in CO_2 concentration, Y = >1. (From SBI Report 212 Indoor Climate and Air Quality Problems. 1990. Danish Building Research Institute.)

The percentage of outside air can be calculated from Equation 7.6:

$$\text{Outdoor air (\%)} = \frac{T_R - T_M}{T_R - T_O} \times 100 \tag{7.6}$$

where T_R = temperature of return air
T_M = temperature of mixed air
T_O = temperature of outdoor air

The percentage of outdoor air can be converted to a volumetric outdoor air flow rate by means of Equation 7.2.

The major disadvantage of the thermal balance method is that it only assesses the amount of outdoor air introduced into the HVAC system and ignores infiltration. This is not true for CO_2 methods.

Using Tracer Gases

The outdoor air ventilation rate can be determined by the use of tracer gases such as sulfur hexafluoride (SF_6).[42] Tracer gas use is based on the mass balance assumption that the change in tracer gas concentrations in building spaces is equal to the amount introduced minus the amount removed by mechanical ventilation and exfiltration. After its introduction and mixing with indoor air, the decay or decrease of the tracer gas with time is directly related to the outdoor air ventilation rate. From measurements of tracer gas levels over time,[43] the actual air exchange rate (ACH) can be calculated using the following equation:

$$n = \left[\frac{\ln \frac{C_O}{C_t}}{t} \right] \tag{7.7}$$

where C_O = initial tracer case concentration (ppm)
C_t = tracer gas concentration (ppm) at time t (hours), ppm
n = air changes/hr (ACH)

Multiplying the air exchange rate by the volume of the building space and dividing by the total number of individuals present will yield the ventilation rate per person.

The use of tracer gases[43] such as SF_6 to determine building ventilation rates is technically more demanding than the measurements of CO_2 and thermal balance techniques previously described. It requires systems for tracer gas injection, sampling, and analysis. Sulfur hexafluoride analysis is conducted by gas chromatography using an electron capture detector.

An alternative to SF_6 is to monitor the decay of CO_2 levels after hours, when the building is no longer occupied. The limitations of this method are obvious; it conflicts with the normal work hours of building investigators. Nevertheless, its value lies in the fact that it is not as technically demanding as other tracer gas procedures.

RECAPITULATION

Diagnosing problem buildings is in many cases a challenging and complex task. Early efforts based on *ad hoc* responses to complaint requests and the use of traditional industrial hygiene practices were not adequate in diagnosing and mitigating building-related health complaints. As a result of increasing needs, a number of public and private groups have developed standardized protocols for conducting problem-building investigations. These include those developed by NIOSH, an in-house protocol developed jointly by USEPA and NIOSH, a comprehensive protocol developed by state health programs and groups in California in the United States, the protocol developed by AIHA to be used by industrial hygienists, protocols developed by PWC for Canadian federal buildings and the Ontario Interministerial Committee for provincial use, protocols developed by the DBRI and the Nordic Ventilation Group for use by European investigators, and protocols developed by private consulting companies in the United States. Though most investigative protocols contain common elements (multiple stages of investigation, on-site investigations, HVAC system evaluations), they differ widely in terms of investigative philosophies, emphasis on interviewing building occupants and use of questionnaires, approaches to conducting IAQ and environmental measurements, factors which are considered to be causal, and the use of guideline values for interpreting measured contaminant levels. Because of the great diversity in the various investigative protocols which are commonly used in North America and Europe, it would not be surprising for different groups of investigators to come to different conclusions relative to causal factors in a particular problem building. Such potential disparities represent, in part, our relatively limited understanding of the true causes of problem-building complaints which we often ascribe to sick building syndrome and differences in training and understanding of the problem among building investigators.

HVAC systems and their operation are often implicated in problem-building complaints, and increased outdoor ventilation rates are recommended to reduce building-related health and comfort complaints in the absence of any identified causal factors. The assessment of HVAC system performance is therefore an inherent part of most problem-building investigations. Assessment techniques of HVAC system operation and performance are discussed. This includes measurement techniques used to determine ventilation rates from measurements of CO_2 levels, enthalpy balance, and use of tracer gases.

APPENDIX A.
NIOSH IAQ SURVEY QUESTIONNAIRE

1. Complaints Yes ____ No ____
   (If yes, please check)

   ____ Temperature too cold
   ____ Temperature too hot
   ____ Lack of air circulation (stuffy feeling)
   ____ Noticeable odors
   ____ Dust in air
   ____ Disturbing noises
   ____ Others (specify)

2. When do these problems occur?
   ____ Morning ____ Daily
   ____ Afternoon ____ Specific day(s) of the week
   ____ All day Specify which day(s):____________
   ____ No noticeable trend ______________________

3. Health Problems or Symptoms
   Describe in three words or less each symptom or adverse health effect you experience more than two times per week.

   Example: runny nose
   Symptom 1 ______________________________
   Symptom 2 ______________________________
   Symptom 3 ______________________________
   Symptom 4 ______________________________
   Symptom 5 ______________________________
   Symptom 6 ______________________________

   Do **all** of the above symptoms clear up within 1 hour after leaving work?
   Yes ____ No ____

   If no, which symptom or symptoms persist (noted at home or at work) throughout the week? Please indicate by drawing a circle around the symptom number below.
   Symptom: **1** **2** **3** **4** **5** **6**

   Do you have any health problems or allergies which might account for any of the above symptoms?
   Yes ____ No ____

   If yes, please describe.______________________________
   ____
   ____

4. Do any of the following apply to you?
   ___ Wear contact lenses.
   ___ Operate video display terminals at least 10% of the workday.
   ___ Operate photocopier machines at least 10% of the workday.
   ___ Use or operate other special office machines or equipment.
   (Specify) ______________________________

5. Do you smoke? Yes ____ No ____

6. Do others in your immediate work area smoke? Yes ___ No ___

7. Your office or suite number is ______________________

8. What is your job title or position? ______________________

9. Briefly describe your primary job tasks. ______________________

____

____

10. Can you offer any other comments or observations concerning your office environment? (Optional)

____

____

____

____

____

11. Your name? (Optional)______________________________

12. Your office phone number? (Optional) ______________________

From Gorman, R.W. and K.M. Wallingford. 1989. 63–72. In: Nagda, N.L. and J.P. Harper (Eds.). *Design and Protocol for Monitoring Indoor Air Quality.* ASTM STP 1002. American Society for Testing and Materials. Philadelphia.

APPENDIX B. AIHA OCCUPANT HEALTH AND COMFORT QUESTIONNAIRE

1. (Optional) Name: ______________________________
   Job Title: ______________________________
   Department: ______________________________
   Phone: ______________________________

2. Area or room where you spend the most time in the building:

3. Do any of your work activities produce dust or odor?
   ____ YES ____ NO
   Describe: ______________________________
   ____

4. Gender ____ Male ____ Female
   Age ____ Under 25
   ____ 25–34
   ____ 35–44
   ____ 45–54
   ____ 55 and over

5. Do you smoke? ____ YES ____ NO
   Have hay fever/pollen allergies? ____ YES ____ NO
   Have skin allergies/dermatitis? ____ YES ____ NO
   Have a cold/flu? ____ YES ____ NO
   Have sinus problems? ____ YES ____ NO
   Have other allergies? ____ YES ____ NO
   Wear contact lenses? ____ YES ____ NO
   Operate video display terminals? ____ YES ____ NO
   Operate photocopiers 10% of the time? ____ YES ____ NO
   Use other special office machines? ____ YES ____ NO
   Specify: ______________________________
   Take medication currently? ____ YES ____ NO
   Reason: ______________________________

6. Office characteristics: ______________________________
   Number of persons sharing same room/work area: ____________
   Number of windows in room/work area: ____________
   Do windows open? ____ YES ____ NO

   Please rate adequacy of work space per person:

Poor		Average		Excellent
1	2	3	4	5

   Please rate room temperature:

Poor		Average		Excellent
1	2	3	4	5

Do others smoke in your work area? ____ YES ____ NO

7. How long have you worked?
   In this room/area? ____________________
   In this building? ____________________

8. Symptoms: Select symptoms you have experienced in this building. This is a random list — not all symptoms listed have been noted in this building.

Symptom	Occasionally	Frequently	Not related to building	Appeared after arrival	Increased after arrival
Difficulty in concentrating					
Aching joints					
Muscle twitching					
Back pain					
Hearing problems					
Dizziness					
Dry, flaking skin					
Discolored skin					
Skin irritation					
Itching					
Heartburn					
Nausea					
Noticeable odors					
Sinus congestion					
Sneezing					
High stress levels					
Chest tightness					
Eye irritation					
Fainting					
Hyperventilation					
Problems with contacts					
Headache					
Fatigue/drowsiness					
Temperature too hot					
Temperature too cold					
Other (specify)					

Have you seen a doctor for any or all of these symptoms?
____ YES ____ NO
When do you experience relief from these symptoms? ____________
____

9. When do these problems usually occur?
   Time of day: Morning Afternoon Evening

Day of week: S M T W TH F S

Month: J F M A May June J Aug. S O N D

Season: Spring Summer Fall Winter

10. Do symptoms disappear? ____ YES ____ NO
    When? ________________________________

11. In your opinion, what is the cause of perceived indoor air quality problems? ________________________________

____

12. Comments: Please take this opportunity to comment on any factors you consider to be important concerning the quality of your work environment.

____

____

____

____

____

From Raferty, P.J. (Ed.). 1993. *The Industrial Hygienist's Guide to Indoor Air Quality Investigations*. American Industrial Hygiene Association. Fairfax, VA.

APPENDIX C. BUILDING DIAGNOSTICS HUMAN RESOURCE QUESTIONNAIRE

1. Environmental Factors

   Listed below are 12 items related to the environment of the area in which you work. In front of each item, enter the number from the following scale that best describes the acceptability of your work area at this time.

6 = Very acceptable	3 = Somewhat unacceptable
5 = Acceptable	2 = Unacceptable
4 = Somewhat acceptable	1 = Very unacceptable

____ Temperature	____ Humidity
____ Air movement	____ Odor (smell)
____ Amount of dust	____ Loudness of sounds
____ Number of noisy distractions	____ Pitch/frequency of sounds
____ Brightness of lighting	____ Glare
____ Shadows	____ Overall quality

2. Demographic Data & Symptoms

   Please answer the following questions by filling in the blank or checking the correct response:

What is your age? _____		
What is your gender?	____ Male	____ Female
Are you currently a tobacco smoker?	____ Yes	____ No
Have you smoked at your workstation during the past hour?	____ Yes	____No

   Please circle any of the following symptoms that you may have at this moment:

Headache	Unexplained memory loss
Backache	Stiff arm
Drowsiness	Nose irritation
Hand cramps	Itchy skin or rash
Eye irritation	Neck pain
Dry mucous membranes	Mental fatigue
Sore throat	Leg cramps
Itchy foot	Dry skin

   Other current symptoms that may be related to your work environment:______________________________

From Woods, J.E. et al. 1990. 80–98. In: Nagda and J.P. Harper (Eds.) *Design and Protocol for Monitoring Indoor Air Quality.* ASTM STP 1002. American Society for Testing and Materials. Philadelphia.

APPENDIX D. ONTARIO INTERMINISTERIAL COMMITTEE IAQ SURVEY QUESTIONNAIRE

HEALTH SURVEY

During the Previous Week While Working in Your Area

Check "YES" if symptoms interfered with work.	Check symptoms which have given you trouble.
(1) Nasal symptoms	
YES ____	Nosebleeds ____
NO ____	Congestion ____
	Sinus problems ____
	Sneezing ____
	Runny nose ____
	Dry nose ____
	Other (specify)________________
(2) Throat symptoms	Sore throat ____
YES ____	Dry cough ____
NO ____	Other (specify________________
(3) Eye symptoms	Redness ____
YES ____	Watering ____
NO ____	Burning ____
	Puffiness ____
	Dryness ____
	Irritation ____
	Blurred vision ____
	Other (specify)________________
(4) Contact lens wearer	
YES ____ NO ____	
Problems related to wearing contact lenses	Problems with
	Cleaning ____
YES ____ NO ____	Deposits ____
	Discomfort ____
	Pain ____
	Other (specify)________________

(5) Skin problems	Dryness ____
YES ____	Flaking ____
NO ____	Rash ____
	Irritation ____
	Other (specify)________________

---

(6) Aches and pains	Headache ____
YES ____	Backache ____
NO ____	Muscle/Joint pain ____
	Other (specify)________________

---

(7) General complaints	Drowsiness ____
YES ____	Dizziness ____
NO ____	Faintness ____
	Difficulty in concentration ____
	Other (specify)________________

---

(8) Other symptoms	Breathing ____
YES ____	Digestive ____
NO ____	Menstrual ____
	Other (specify)________________

---

(9) Were you ever absent because of any health problem(s) that you feel may have been caused or aggravated by working at your present location?
YES ____ NO ____
If YES, state the health problem: ______________________________
____
____
____

(10) Did you seek medical treatment because of any health problem(s) caused by working at your present location?
YES ____ NO ____

(11) Do you have allergies?
YES ____ NO ____
If YES, allergic to: ______________________________
____
____

(12) Are you taking any prescribed medication for any symptoms you mentioned?
YES ____ NO ____

General Information
(21) Age (in years) _______
(22) Sex: Male ____
Female ____

Are there any further comments which you would like to make?
____
____
____
____

Description of Job and Working Environment

(1) My employment status:
Full time ____
Part time ____
Temporary ____
Other (specify) ______________________________

(2) I have worked in this area since (year)_______, (month)_____, (day)____.

(3) I work at the present location _____ (hours) a day.

(4) I am in:
A closed office ____
My own cubicle ____
An open area shared with others ____
Other (specify) ______________________________

(5) What is the geographic location of your desk within the building?
East side ____ Northeast corner ____
West side ____ Northwest corner ____
South side ____ Southeast corner ____
North side ____ Southwest corner ____

(6) Are you sitting within 3.7 meters (12 feet) of a window?
YES ____ NO ____

(7) Can the window be opened?
YES ____ NO ____

(8) Within 10 meters (approximately 33 feet) of your work location, is there:

A typewriter?	YES ____	NO ____
A photocopying machine?	YES ____	NO ____
A keyboard with a video display screen (e.g., VDT, CRT, data or word processor)?	YES ____	NO ____
A printer?	YES ____	NO ____
A teletype or fax machine?	YES ____	NO ____
A posting machine?	YES ____	NO ____

Other (specify)________________________________

Working Environment

Please record your general assessment of the working environment at your present location.

Check "YES" if disturbing	Check aspects which are disturbing
(1) Noise YES ____ NO ____	Nearby conversation ____ Lighting ____ Ventilation system ____ Office equipment ____ Other (specify)________________
(2) Ventilation YES ____ NO ____	Temperature ____ Humidity ____ Air movement ____ Other (specify)________________
(3) Lighting YES ____ NO ____	Too bright ____ Not bright enough ____ Glare, flicker ____ Other (specify)________________
(4) Others YES ____ NO ____	Specify: ________________________ ________________________ ________________________ ________________________

(5) If there is a smell in your area, how would you describe the smell:

(a) The smell resembles:
- Glue ____
- Vinegar ____
- Alcohol ____
- Ammonia ____
- Propane ____
- Gasoline ____
- Perfume ____
- Other (specify)____________________

(b) It smells:
- Smoky ____
- Dusty ____
- Musty ____
- Stale ____
- Other (specify)____________________

(c) In your opinion, where is the smell coming from?______________

____

____

____

---

(6) Do you use any of the following in your work location?

(a) A desk lamp ____
(b) A fan ____
(c) A heater ____
(d) A humidifier ____
(e) An ion generator ____
(f) An air cleaner ____
(g) Personal care products (e.g., hand cream, hairspray) (Please specify what products) ____
(h) No items ____

---

(7) Have you any control over your work location?

(a) Ventilation ____
(b) Temperature ____
(c) Humidity ____
(d) Lighting ____

---

(8) Is there a smoking room near your working location?

YES ____ NO ____

(9) If YES, how far are you away from smoking room? _____ (feet)

GENERAL INFORMATION

(18) Age (in years) _____

(19) Sex: Male _____

Female _____

Are there any further comments which you would like to make?

____

____

____

____

____

From Occupational Health and Safety Division, Ontario Ministry of Labor, Health and Safety Branch. 1988. Ontario.

APPENDIX E. DANISH BUILDING RESEARCH INSTITUTE INDOOR CLIMATE SURVEY QUESTIONNAIRE

Background Factors

Year of birth 19____

Sex Male ____ Female ____

Do you smoke? Yes ____ No ____

Occupation________________

How long have you been at your present place of work?

________ years

WORK ENVIRONMENT

Have you been bothered during the last 3 months by any of the following factors at your work place?

	Yes, often (every week)	Yes sometimes	No, never
Draft	____	____	____
Room temperature too high	____	____	____
Varying room temperature	____	____	____
Room temperature too low	____	____	____
Stuffy "bad" air	____	____	____
Dry air	____	____	____
Unpleasant odor	____	____	____
Static electricity, often causing shocks	____	____	____
Passive smoking	____	____	____
Noise	____	____	____
Light that is dim or causes glare and/or reflections	____	____	____
Dust and dirt	____	____	____

Work Conditions

	Yes, often	Yes, sometimes	No, seldom	No, never
Do you regard your work as interesting and stimulating?	____	____	____	____
Do you have too much work to do?	____	____	____	____
Do you have any opportunity to influence your working conditions?	____	____	____	____

Do your fellow workers help you with problems you may have in your work?	____	____	____	____

Past/Present Diseases/Symptoms

	YES	NO
Have you ever had asthmatic problems?	____	____
Have you ever suffered from hayfever?	____	____
Have you ever suffered from eczema?	____	____
Does anybody else in your family suffer from allergies (e.g., asthma, hayfever, eczema?)	____	____

Present Symptoms

During the last 3 months, have you had any of the following symptoms?

	Yes, often (every week)	Yes, sometimes	No, never	If YES: Do you believe that it is due to your work environment? Yes	No
Fatigue	____	____	____	____	____
Feeling heavy-headed	____	____	____	____	____
Headache	____	____	____	____	____
Nausea/dizziness	____	____	____	____	____
Difficulty concentrating	____	____	____	____	____
Itching, burning, or irritation of the eyes	____	____	____	____	____
Irritated, stuffy, or runny nose	____	____	____	____	____
Hoarse, dry throat	____	____	____	____	____
Cough	____	____	____	____	____
Dry or flushed facial skin	____	____	____	____	____
Scaling/itching scalp or ears	____	____	____	____	____

Hands dry, itching red skin	___	___	___	___	___
Other	___	___	___	___	___

Further Comments

From SBI Report 212 Indoor Climate and Air Quality Problems. 1990. Danish Building Research Institute.

REFERENCES

1. Seitz, T.A. 1990. "NIOSH Indoor Air Quality Investigations: 1971 through 1988." 163–171. In: Weekes, D.M. and R.B. Gammage (Eds.). *Proceedings of Indoor Air Quality International Symposium: The Practitioner's Approach to Indoor Air Quality Investigations.* American Industrial Hygiene Association. Akron, OH.
2. Woods, J.E. et al. "Indoor Air Quality and the Sick Building Syndrome: A Diagnostic Approach." Unpublished manuscript.
3. Gorman, R.W. and K.M. Wallingford. 1989. "The NIOSH Approach to Conducting Air Quality Investigations in Office Buildings." 63–72. In: Nagda, N.L. and J.P. Harper (Eds.). *Design and Protocol for Monitoring Indoor Air Quality.* ASTM STP 1002. American Society for Testing and Materials. Philadelphia.
4. Rafferty, P.J. (Ed.). 1993. *The Industrial Hygienist's Guide to Indoor Air Quality Investigations.* American Industrial Hygiene Association Technical Committee on Indoor Environmental Quality. American Industrial Hygiene Association. Fairfax, VA.
5. ASHRAE Standard 62–1981R. 1989. "Ventilation for Acceptable Indoor Air Quality." American Society for Heating, Refrigerating and Air-Conditioning Engineers. Atlanta.
6. ANSI/ASHRAE Standard 55–1981. 1981. "Thermal Environmental Conditions for Human Occupancy." American Society for Heating, Refrigerating and Air-Conditioning Engineers. Atlanta.
7. USEPA/NIOSH. 1991. "Building Air Quality: A Guide for Building Owners and Facility Managers." EPA/400/1–91/003. DDHS (NIOSH) Publication Number 91–114.
8. Quinlan, P. et al. 1989. "Protocol for the Comprehensive Evaluation of Building-Associated Illness." 771–797. In: Cone, J.E. and M.J. Hodgson (Eds.). *Problem Buildings: Building-Associated Illness and the Sick Building Syndrome. Occupational Medicine: State of the Art Reviews.* Hanley and Belfus, Inc. Philadelphia.

9. Occupational Health and Safety Division, Ontario Ministry of Labor, Health and Safety Support Branch. 1988. "Report of the Inter-Ministerial Committee on Indoor Air Quality." Ontario.
10. Woods, J.E. et al. 1989. "Indoor Air Quality Diagnostics: Qualitative and Quantitative Procedures to Improve Environmental Conditions." 80–98. In: Nagda, N.L. and J.P. Harper (Eds.). *Design and Protocol for Monitoring Indoor Air Quality*. ASTM STP 1002. American Society for Testing and Materials. Philadelphia.
11. Lane, C.A. et al. 1989. "Indoor Air Quality Diagnostic Procedures for Sick and Healthy Buildings." 237–240. In: *Proceedings of IAQ '89: The Human Equation: Health and Comfort*. American Society of Heating, Refrigerating and Air-Conditioning Engineers. Atlanta.
12. MacPhaul, D.E. et al. 1990. Database on Indoor Air Quality Evaluations. (unpublished manuscript). Clayton Environmental Consultants, Inc., Wayne, PA.
13. Turner, S. and P.W.H. Binnie. 1990. "An Indoor Air Quality Survey of Twenty-Six Office Buildings." 27–32. In: *Proceedings of the Fifth International Conference on Indoor Air Quality and Climate*. Vol. 4. Toronto.
14. McCarthy, J.F. et al. 1991. "Assessment of Indoor Air Quality." 82–108. In: Samet, J.M. and J.D. Spengler (Eds.). *Indoor Air Pollution: A Health Perspective*. Johns Hopkins University Press. Baltimore.
15. Ryan, P.B. and W.E. Lambert. 1991. "Personal Exposure to Indoor Air Pollution." 109–127. In: Samet, J.M. and J.D. Spengler (Eds.). *Indoor Air Pollution: A Health Perspective*. Johns Hopkins University Press. Baltimore.
16. Kirkbride, J. et al. 1990. "Health and Welfare Canada's Experience in Indoor Air Quality Investigations." 99–106. In: *Proceedings of the Fifth International Conference on Indoor Air Quality and Climate*. Vol. 5. Toronto.
17. Nathanson, T. 1990. "Building Investigations: An Assessment Strategy." 107–110. In: *Proceedings of the Fifth International Conference on Indoor Air Quality and Climate*. Vol. 5. Toronto.
18. Davidge, R. et al. 1989. *Indoor Air Quality Assessment Strategy*. Building Performance Division, Public Works Canada. Ottawa.
19. Public Works Canada. 1984. *Indoor Air Quality Test Kit: User Manual*. Building Performance Division, Public Works Canada, Ottawa.
20. Valbjorn, O. et al. 1990. "Indoor Climate and Air Quality Problems: Investigation and Remedy." SBI Report 212. Danish Building Research Institute.
21. Kukkonen, E. et al. 1993. "Indoor Climate Problems—Investigation and Remedial Measures." Nordtest Report NT Tech. Rep. 204.
22. Molina, C. et al. 1989. "Sick Building Syndrome: A Practical Guide." Rep. No. 4. Commission of the European Communities. Brussels-Luxembourg.
23. Molhave, L. et al. 1986. "Human Reactions to Low Concentrations of Volatile Organic Compounds." *Environ. Int.* 12:167–175.
24. Kjaergaard, S.K. et al. 1990. "Human Reactions to a Mixture of Indoor Air Volatile Organic Compounds." *Atmos. Environ.* 25A:1417–1426.
25. World Health Organization. 1987. "Air Quality Guidelines for Europe." World Health Organization Regional Publications. European Series No. 23. Copenhagen.

26. Janssen, J. 1983. "Ventilation for Acceptable Indoor Air Quality: ASHRAE Standard 62–1981." In: *Ann. ACGIH: Evaluating Office Environmental Problems*. 10:77–92.
27. Broder, I. et al. 1988. "Comparison of Health and Home Characteristics of Houses Among Control Homes and Homes Insulated with Urea-formaldehyde Foam, II. Initial Health and House Variables and Exposure Response Relationships." *Environ. Res.* 45:156–178.
28. Krzyzanowski, M. et al. 1990. "Chronic Respiratory Effects of Indoor Formaldehyde Exposure." *Environ. Res.* 52:117–125.
29. Liu, K.S. et al. 1987. "Irritant Effects of Formaldehyde in Mobile Homes." 610–614. In: *Proceedings of the Fourth International Conference on Indoor Air Quality and Climate*. Vol. 2. West Berlin.
30. Molhave, L. 1990. "Volatile Organic Compounds, Indoor Air Quality and Health." 15–33. In: *Proceedings of the Fifth International Conference on Indoor Air Quality and Climate*. Vol. 5. Toronto.
31. Norback, D. et al. 1990. "Volatile Organic Compounds, Respirable Dust, and Personal Factors Related to Prevalence and Incidence of Sick Building Syndrome in Primary Schools." *Br. J. Ind. Med.* 47:733–741.
32. Norback, D. et al. 1990. "Indoor Air Quality and Personal Factors Related to the Sick Building Syndrome." *Scand. J. Work Environ. Health*. 16:121–128.
33. Hughes, R.T. and D.M. O'Brien. 1986. "Evaluation of Building Ventilation Systems." *Am. Ind. Hyg. Assoc. J.* 47:207–213.
34. Gorman, R.W. 1984. "Cross Contamination and Entrainment." In: *Ann. ACGIH: Evaluating Office Environmental Problems*. 10:115–120.
35. Morey, P.R. et al. 1984. "Environmental Studies in Moldy Office Buildings: Biological Agents, Sources and Preventative Measures." In: *Ann. ACGIH: Evaluating Office Environmental Problems*. 10:21–35.
36. Elixmann, J.H. et al. 1990. "Fungi in Filters of Air-Conditioning Systems Cause the Building-Related Illness." 193–196. In: *Proceedings of the Fifth Internatinal Conference on Indoor Air Quality and Climate*. Vol. 1. Toronto.
37. Morey, P.R. and C. Williams. 1990. "Porous Insulation in Buildings: A Potential Source of Microorganisms." 529–534. In: *Proceedings of the Fifth International Conference on Indoor Air Quality and Climate*. Vol. 4. Toronto.
38. Samimi, B.S. 1990. "Contaminated Air in a Multi-Story Research Building Equipped with 100% Fresh Air Supply Ventilation Systems." 571–576. In: *Proceedings of the Fifth International Conference on Indoor Air Quality and Climate*. Vol. 4. Toronto.
39. McNall, P.E. and A.K. Persily. 1984. "Ventilation Concepts for Office Buildings." *Ann. ACGIH: Evaluating Office Environmental Problems*. 10:49–58.
40. Salisbury, S.A. 1990. "Evaluating Building Ventilation for Indoor Air Quality Investigations." 87–98. In: Weekes, D.M. and R.B. Gammage (Eds.). *Proceedings of the Indoor Air Quality International Symposium: The Practitioner's Approach to Indoor Air Quality Investigations*. American Industrial Hygiene Association. Akron, OH.
41. Turner, W.A. and D.W. Bearg. 1986. "Evaluation Procedures for Measuring Indoor Air Quality." *Building Operating Management*. November, 38–46.

42. Persily, A.K. et al. 1989. "Investigation of a Washington, D.C. Office Building." 35–50. In: Nagda, N.L. and J.P. Harper (Eds.). *Design and Protocol for Monitoring Indoor Air Quality*. ASTM STP 1002. American Society for Testing and Materials. Philadelphia.
43. Grimsrud, D.T. 1984. "Tracer Gas Measurements of Ventilation in Occupied Spaces." *Ann. ACGIH: Evaluating Office Environmental Problems*. 10:69–76.

8 MEASUREMENT OF INDOOR AIR CONTAMINANTS

Though significant differences exist in investigative philosophies relative to the appropriateness of testing for air contaminants in problem-building investigations, most investigators will at some time conduct one or more air quality measurements. These may range from relatively limited sampling of CO_2 levels for the purpose of determining the adequacy of ventilation, to screening for a range of common contaminants, to suspected causal contaminants.

As with other aspects of indoor air quality (IAQ) investigations, measurements of contaminant levels need to be conducted in a thoughtful and systematic way. The conduct of air testing without specific objectives and use of good sampling practice contributes little to the identification and resolution of building contamination problems under investigation. It is imperative that investigators conduct measurements using appropriate sampling and analytical procedures.

CONTAMINANT MEASUREMENT CONSIDERATIONS

Thorsen and Molhave[1] have described elements of a standardized measurement protocol. They propose that the first step should be to define sampling objectives or hypothesis to be tested. Sampling objectives depend on a number of variables including the perceived nature of the problem and investigator philosophies relative to the appropriateness of air testing. Once sampling objectives have been defined, investigators are advised to establish a list of relevant sampling variables and ranges of variation.[1] These include the many types of biological, chemical, and physical exposures and emission controlling factors which affect qualitative and quantitative variation of concentrations in time and space. Variables which may affect dose-response relationships in a direct or indirect way are listed in Table 8.1. For each selected variable or covariable, a sampling/measuring specification is to be established. These include (1) the selection of appropriate sampling methods or equipment, (2) calibration of sampling instruments, and (3) determination of sampling location,

Table 8.1 Major classes of variables which may affect indoor air quality, climate, and human responses to them.

Variables
The indoor atmospheric environment
Biological exposure
Microorganisms
Allergens
Chemical exposure
Gases, vapors, particulate-phase materials
Physical exposure
Temperature, humidity, air movement, light, sound, dust/fibers, ions
Emission controlling covariables
Qualitative variation in time and space
Building site and type
Materials
Processes
Type of ventilation system
Biological contaminant sources
Outdoor contaminants
Quantitative variation in time and space
Emission rates
Elimination rates
Covariables for human reactions
Symptoms involving
Mucous membranes
Pulmonary system
Cardiovascular system
Digestive system
Central nervous system
Skin reactions
Heat balance
Hyper-reactivity
Psychological effects
Changes in human activity patterns
Other human reactions
Nonhuman reactions
Effects on animals or plants
Effects on building and other properties

From Thorsen, M. A., and L. Molhave, 1987. *Atmos. Environ.,* 21:1411–1416. With permission from Elsevier Science Ltd.

time, duration, and the number of samples or measurements to be made and how the sampling and analysis are to be administered.

Sampling Methods

It is essential that sampling and instrumental measuring techniques appropriate to the task are selected. Sampling procedures should be sufficiently sensitive to measure the contaminant or contaminants in the range expected in the sampled environment. The use of sampling techniques/instruments which have a lower limit of detection (LOD) above the range which can be expected for indoor environments is wasteful of both resources and time. Unfortunately, the use of measurement techniques with LODs in excess of normal ranges of contaminant variation in indoor environments has been a much too common practice among investigators in the recent past, particularly in measuring formaldehyde levels. As a general rule, McCarthy et al.[2] recommend that the LOD of a sampling/measuring procedure be 1% of the industrial threshold limit values (TLV) or permissible exposure limits (PEL).

In addition to the LOD, sampling methods should be selected on the basis of their specificity and accuracy. Procedures which have gained acceptability as a consequence of systematic evaluation by testing/evaluating organizations such as the National Institute of Occupational Safety and Health (NIOSH), the U.S. Environmental Protection Agency (USEPA), and the American Society for Testing and Materials (ASTM), or similar agencies/bodies in other countries are in most cases more appropriate than those for which there is little knowledge of performance and limitations of use. The use of reference methods or approved methods may give investigators a relative degree of confidence about the validity of a chosen method and its application.

In addition to performance characteristics, the selection of a contaminant measuring method will depend on available resources. Despite significant limitations, less expensive methodologies are, in many cases, preferred to more costly ones. This is particularly true for gas sampling tubes (GSTs) which have been widely used for carbon monoxide (CO), carbon dioxide (CO_2), and formaldehyde. In addition to cost, other factors may affect the selection of sampling procedures or instruments. These include equipment availability, portability, and degree of obtrusiveness.[2]

A number of sampling/measuring procedures are available for use in IAQ investigations. These include one-time direct-read devices such as GSTs, direct-read continuous sampling instruments, integrating-active sampling devices, and integrating-passive sampling monitors.

Gas Sampling Tubes

Gas sampling tubes or gas detector tubes are widely used for contaminant measurement in industrial workspaces and less commonly in IAQ investigations. They are designed to provide quick and simple real-time measurement of contaminant concentrations. They typically sample a small quantity of air (minimum of 0.1 L) over a short period of time (seconds to minutes) by means of a specially designed gas syringe or bellows (Figure 8.1). Gas sampling tubes have also been designed for longer sampling times (hours) based on contaminant collection by diffusion.

Gas sampling tubes are comprised of hermetically sealed glass that contains a granular sorbent such as silica gel, alumina, or pumice impregnated with substances which react on contact with a specific contaminant or group of contaminants with similar chemical properties. Reactions with a contaminant result in a color change in the sorbent. The length of the stain or color change on the sorbent is directly proportional to the concentration. The concentration can be read directly on a calibrated scale on the side of the tube.

Gas sampling tubes are commercially available for the measurement of a large number of gases and vapors (100+) and concentration ranges. They are

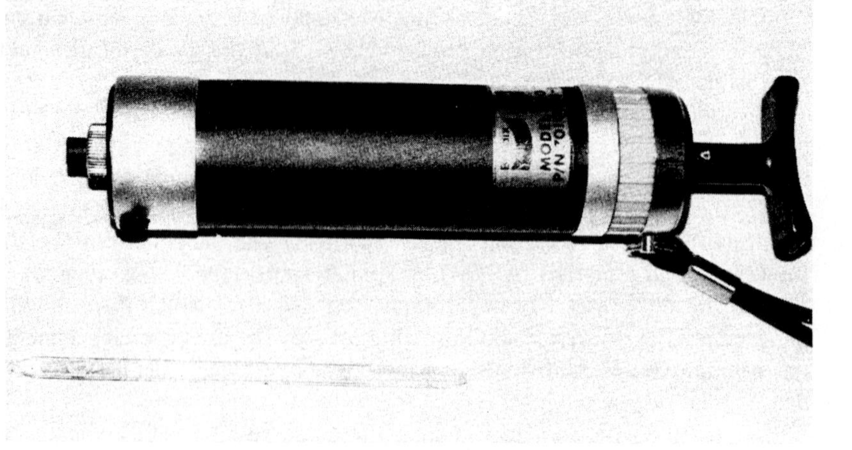

Figure 8.1 Gas syringe with gas sampling tube.

typically designed to provide an accuracy of ±25%. They are attractive as sampling devices because of their relatively low cost, simplicity of use, and near real-time test results. They unfortunately have significant limitations. These include, in many cases, a lack of specificity, high LODs, and limited shelf life (1 year or so). Because the measurement technique is based on a color reaction, sampled gases and vapors cannot be differentiated in a sample environment where several or more chemically similar contaminants are present. (This problem, of course, is not just limited to GSTs.) This lack of specificity can seriously compromise the validity of test results. Many GSTs are designed for use in industrial environments or for contaminant measurement associated with chemical spills. As a consequence, many have high LODs and are not suitable for IAQ measurements.

Not withstanding these limitations, GSTs have been and continue to be widely used in IAQ investigations, most notably for measurements of CO and CO_2. Poor results, however, have been associated with the use of GSTs in measuring indoor air levels of formaldehyde.[3]

Electronic Direct-Read Instruments

A variety of electronic direct-read instruments are available for measuring the concentrations of indoor air contaminants. Such instruments allow the operator to obtain real-time or quasi-real-time measurements of gas, vapor, or particulate matter concentrations. They are usually pump-driven devices which draw air into a chamber where contaminant concentrations are determined by chemical or physical sensors, employing such principles as electrochemistry, photometry, infrared analysis, and chemiluminescence. Specific contaminant concentrations can be read from the deflection of a meter needle on a concentration scale or a digital output. Portable electronic direct-read instruments are available and widely used for the measurement of CO_2, CO, and respirable particles (RSP) in indoor environments. Portable direct-read instruments are also available for ozone (O_3), a contaminant associated with office copiers.

Active Integrated Sampling

Though direct-read real-time sampling devices are especially attractive for use in IAQ investigations, they are not always available or best suited for the collection of a variety of contaminants found in indoor air. This is especially true for formaldehyde and VOCs. In such cases, samples are usually collected by dynamic sampling procedures wherein air from the test environment is drawn through a liquid or solid sorbent medium over a sampling duration ranging from minutes to hours depending on the LOD for the contaminants under investigation. As a consequence, measured concentrations are average or integrated values for the sample collection period. In addition to formaldehyde and VOCs, integrated sampling procedures are widely used for sampling asbestos, RSP, and biological aerosols. Principles of contaminant collection, include absorption, adsorption, impaction, and filtration.

Passive Integrated Sampling

A number of contaminants can be measured by passive sampling devices which collect contaminants by diffusion onto solid sorbents (e.g., charcoal) or in liquid media. Since passive devices have very low sampling rates, the concentration is an average value for an exposure period of several or more hours to seven days. Passive devices have the advantage of low cost and simplicity of use and the disadvantage of relatively long averaging times. Their low cost and reliability make passive sampling devices attractive for IAQ measurements, particularly in research studies where contaminant concentrations are being determined in a large population of buildings. They are, however, less useful in problem building investigations. Passive sampling techniques are commonly used to measure formaldehyde, nitrogen dioxide (NO_2), and radon in indoor environments.

Quality Assurance/Calibration

Contaminant measurements, whether conducted in systematic research studies or as one-time tests in problem-building investigations, need to be quality assured to provide confidence in measured values. The calibration of instruments used in collecting air samples or making contaminant measurements is the primary focus of quality assurance programs. Calibration is a process whereby measured values of air flow or contaminant levels are compared to a standard. The standards used may be primary or secondary with the latter traceable to the former. Sampling equipment should be calibrated in the anticipated range of field measurements.[2] It is customary to calibrate field instruments both before and after sampling.

In addition to calibration, good quality assurance programs will also include the use of field blanks, media blanks, replicate samples, and split or spiked samples as appropriate. Replicate samples may be used to assess the precision of sampling and analytical methods. An appropriate number of field blanks should be employed so that extraction efficiency or potential media contamination can be assessed by laboratory personnel conducting sample analyses. Two field blanks for each of ten active samples is recommended as a general guide.[2]

Sampling Decisions

Once sampling objectives have been defined and measurement methods selected, the investigator must determine where samples are to be taken, when the sampling is to be done, the duration of sampling, the number of samples to be collected, and how the sampling and analysis is to be administered.

Location

Selection of sampling locations should reflect the objectives of the contaminant measurement effort. McCarthy et al.[2] suggest that five types of areas should be considered in a sampling strategy. These include areas with (1) the most susceptible occupants, (2) potential sources, and (3) the least effective ventilation. They should also include outdoor air and control areas having the least problems. In theory, the primary objective of contaminant measurement is to identify the contaminant or contaminants responsible for the problem. As a consequence, priority should be placed on those areas where complaints are the most prevalent and where the highest concentrations of contaminants can be expected based on observations made relative to sources in the building walk-through and the perceived or measured adequacy of ventilation. In many investigations, samples may also be collected from so-called control or noncomplaint areas for purposes of determining differences in both the types of contaminants and their levels and to determine whether complaint rates are consistent with exposures. Carbon dioxide sampling should be conducted in return air, supply air, and outdoors in determining the adequacy and effectiveness of ventilation.

Time

In general, building investigators place little importance on the time period when measurements are conducted. They are typically made after arrival at the site and instruments have been set up. Because concentrations of contaminants vary considerably with time, it is important that time-dependent contaminant variations are considered before sampling is commenced. This is particularly true for such contaminants as formaldehyde, CO_2, and CO. Over the course of the seasons, formaldehyde levels in buildings may vary by a factor of two to three. In residences, formaldehyde levels are at their lowest values during cold winter months and highest values during warm humid weather.[4] Carbon dioxide values in poorly ventilated buildings increase in concentration with time, reaching peak values four or more hours after building occupancy.[5] Carbon monoxide levels are highly episodic with peak values occurring in time with such transient phenomenon as flue gas spillage, entrainment in outside air supplies, etc. The identification of CO as the cause of occupant complaints from contaminant sampling efforts may be particularly difficult due to the episodic nature of CO contamination of building spaces. It is almost axiomatic that the time period selected for CO measurements (when CO is actually the problem) is the wrong time.

The selection of a sampling time for contaminant measurements should be made with some knowledge of the range of variation which may occur with time. Thorsen and Molhave[1] suggest that a sampling time be chosen to minimize the influence of such variations, by sampling when potential cofactors are expected to be at a constant and average level. However, since such a choice of sampling time is not often practical, the sampling program should allow estimates of the range of variation of relevant covariables. If resources are limited, the sampling strategy should focus on the highest concentrations or exposures. This is critical when IAQ guidelines are being used to determine whether exposures are within acceptable limits or not.

Duration

In addition to the selection of a time when contaminant measurements are to be made, investigators must determine the time period (sampling duration) over which samples are to be collected. The sampling duration should reflect the toxic nature of the contaminant and actual exposures. McCarthy et al.[2] suggest that if there are short-term exposure limits for specific pollutants, they may have a significant toxic effect from short-term exposures and therefore should be evaluated with this concern in mind. Long-term integrated sampling is recommended for the characterization of exposures to pollutants that cause toxic effects on chronic exposure. In most problem-building complaints, however, symptoms appear to be of an acute nature. As a consequence, short-term sampling appears to be appropriate and is typically the norm.

What is meant by short term is highly relative. In practice, this may mean the several seconds-minutes or so necessary to obtain a measured value from direct-read sampling devices to the several hours required commonly in using active sampling procedures. In the latter case, the sampling duration is determined by time required to collect sufficient sample to reflect the LOD needs of the analytical technique. The sampling duration will also be determined by the sampling rate necessary to achieve the desired collection efficiency.

Number

A sufficient number of samples should be collected to provide reasonable confidence that measured values are reflective of exposure conditions. Though there is no prescribed rule for the number of samples to be collected, it is advisable that multiple samples be collected to avoid problems associated with bad samples or sample loss and to adequately characterize the exposure problem. Thorsen and Molhave[1] suggest that the number of samples be determined statistically from estimates of the distribution of environmental concentrations or from pilot studies. With the possible exception of studies of a large population of buildings, most investigators would probably find this to be impractical. Investigators will, in most cases, determine the number of samples to be collected based on their judgment of the minimum number of samples required to obtain meaningful results within the resource constraints involved. Though increasing the number of samples increases confidence in interpreting sampling results, it also significantly increases time spent in sample collection and analysis and costs of the investigation.

Administrative Practices

Administrative practices which assure the accuracy and validity of test results are vital to all contaminant efforts. These include the use of an unambiguous numbering system for each sample collected as well as blanks used, sample data sheets which document the history of each sample, the application of strict sample transport and storage requirements, and the maintenance of adequate laboratory analysis records.

PROCEDURES FOR COMMONLY MEASURED CHEMICAL/PHYSICAL CONTAMINANTS

The concentrations of gas, vapor, or particulate-phase contaminants are often measured during problem-building investigations based on a hypothesis that a particular contaminant or contaminants may be a problem or for screening purposes. These often include CO_2, CO, formaldehyde, total volatile organic compounds (TVOCs), and specific VOCs. Less commonly, particulate matter (TSP or RSP) levels are measured.

Carbon Dioxide

Because it is used as an indicator of ventilation adequacy, CO_2 is the most commonly measured contaminant in IAQ investigations. Such measurements are exclusively made by means of direct-read sampling devices. These include grab samples collected by GSTs and real-time measurements with continuous sampling instruments.

Because of low cost, GSTs are widely used for CO_2 measurements in problem building investigations ($2 per GST with a kit cost of approximately $300 U.S.). The CO_2 concentration is read directly from the GST. Sampling accuracy is reported to be in the range of ± 25% minimum.

Despite their low cost and ease of use for CO_2 measurements, GSTs, not surprisingly, have significant limitations. These include a need to collect multiple samples in a variety of spaces over time. Multiple sampling not only reduces the cost advantage over real-time direct-read electronic instruments but also may be relatively labor intensive. The accuracy of GSTs used for CO_2 measurements may also be a problem. Ancker et al.,[6] for example, evaluated the use of three major commercially available CO_2 GSTs for determining air recirculation rates in buildings. They observed significant variation and relatively poor precision for GSTs as compared to measurements of inlet, recirculated, and mixed air by an infrared spectrophotometer. In some cases, air recirculation rates calculated from means or medians of GST values were not believable with values below 0% and above 100%. They concluded that CO_2 GSTs currently commercially available are not suitable for making CO_2 measurements for purposes of determining building air recirculation rates. In addition, field observations made by the author indicate that at least some CO_2 GSTs produce unusually high readings when supply air temperatures are considerably above room temperatures.

The initial studies reported by Ancker et al.[6] were followed up by more extensive field evaluations of CO_2 detector tubes.[7] Two types (Draeger #3080 and Kitigawa 126B) were shown to be very strongly correlated ($r = 0.98$) with a reference method over the concentration range of 400–2200 ppm; a third (Gastec 2LL) was less strongly correlated ($r = 0.91$). Like Ancker et al.,[6] large errors in the range of 400–600 ppm were observed. However, because of the overall good accuracy and precision observed, Norback et al.[7] concluded that CO_2 detector tubes could be used both to estimate outdoor air exchange rates and to determine whether guideline values have been exceeded.

Portable battery-operated, real-time electronic direct-read instruments are widely used for CO_2 measurements in building investigations (Figure 8.2). The operation of these instruments is based on the principle of the differential absorption of infrared energy by different levels of CO_2. An internal pump draws air through the instrument allowing it to respond quickly to changes in CO_2 levels. These instruments may have one or two operating ranges: 0–2000 ppm and 0–5000 ppm. They are relatively easily calibrated from zero gases and CO_2 air standards.

Carbon dioxide direct-read monitors have the advantage of high accuracy (±10% or less) and the ability to make numerous measurements in time and space with considerable ease. These devices can easily be connected to recorders so that CO_2 level changes with time can be easily determined.

Direct-read CO_2 monitors may be more expensive than GSTs (in the range of $500–4000 U.S.). Despite the cost differential, such instruments are superior

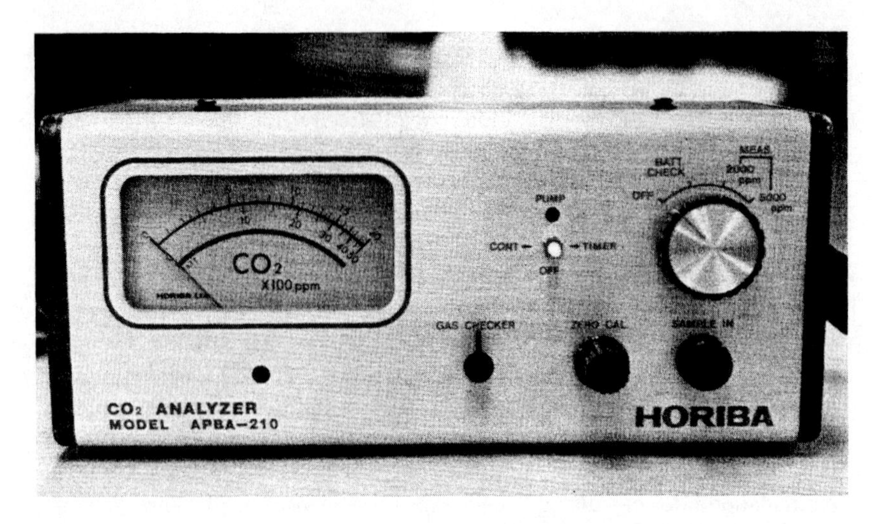

Figure 8.2 Real-time CO_2 monitor.

to GSTs and are the instruments of choice for such measurements. In this context, GSTs should be viewed as an interim measurement method, used only until a real-time electronic instrument can be acquired.

Carbon Monoxide

Carbon monoxide levels are often routinely measured in problem-building investigations or when there is evidence from occupant symptoms or the building inspection to suspect that CO may be a problem. Carbon monoxide levels, like CO_2, are commonly measured by means of GSTs or by direct-read real-time electronic instruments.

Carbon monoxide measurements are often made with GSTs because of their low cost, ease of use, and acceptable accuracy. Gas sampling tubes for CO are available in several concentration ranges with a range of 1–50 ppm used for IAQ investigations.

A variety of direct-read real-time CO measuring devices are available for CO monitoring. These instruments have concentration ranges up to 2000 ppm. Accuracy at the low end of the range may be on the order of several ppm. Carbon monoxide monitoring devices determine concentrations by electrochemically oxidizing CO to CO_2. As CO is oxidized, it produces an electrical signal proportional to the CO concentration in the airstream.[2] These instruments may be small hand-held devices in which CO enters the electrochemical cell by diffusion or larger pump-driven devices. The latter produces nearly instant real-time values, whereas the former is somewhat slow, taking a minute or so to respond.

Real-time or quasi-real-time CO monitors have advantages which are similar to those used for CO_2. They can, for example, be easily calibrated with standard gas mixtures providing relatively good accuracy, are portable, and, of course, provide instantaneous or nearly instantaneous readings. The electrochemical cells unfortunately have a limited life-time and must be periodically replaced at significant cost.

Formaldehyde

Though formaldehyde is rarely found by itself in sufficient concentrations in offices and other public access buildings to be causally connected to building-/work-related complaints, it is nevertheless commonly measured in such investigations. The most widely used formaldehyde sampling technique is the NIOSH bubbler chromotropic acid method.[8] In this procedure, samples are collected by a sampling pump which draws air through 15–20 ml of a 1% sodium bisulfite solution. The sampling rate is typically 1 L/min with a sampling duration of one to several hours. Collected samples are analyzed colorimetrically. The reported accuracy is approximately ±10%.[8] The NIOSH chromotropic acid method has a long history of use and is considered to be very reliable.

Formaldehyde sampling is typically conducted on a one-time basis. Because concentrations vary widely in buildings (due to the response of source materials to changes in temperature and humidity), it is important that environmental factors be taken into consideration when formaldehyde sampling is conducted. This is particularly important in the interpretation of sampling results.

Formaldehyde measurements are less frequently made in problem buildings by means of passive sampling devices. Such devices may include tubes into which formaldehyde diffuses through an open end to a solid sorbent on the other end or through a gas-permeable disc to an absorbing liquid or solid medium inside the tube. They may include badge-type monitors in which formaldehyde diffuses through a membrane to react with a solid or liquid sorbent inside the device. Historically, devices based on absorption on a 1% sodium bisulfite medium with subsequent analysis with the chromotropic acid method have dominated the market for such devices. Though widely used, they have been problem-plagued, most notably the loss of sampling efficiency if samples have been exposed to low humidity conditions.

Though passive samplers based on the chromotropic acid method[9] have dominated the commercial market for formaldehyde samplers during the 1980s, their limitations have encouraged the development of new devices based on different analytical principles. One of the most popular of these (at the time of this writing) is samplers which collect formaldehyde and other aldehydes on 2,4-dintrophenylhydrazine (DNPH)-coated glass fiber filters. Collected aldehydes are converted to stable hydrozones which are analyzed by high-performance liquid chromatography (HPLC).[9,10] The DNPH method has a number of advantages. These include specificity for different aldehydes including

Figure 8.3 DNPH passive formaldehyde sampler.

formaldehyde and high accuracy and sensitivity. It can detect a concentration of 0.009 ppm in a 20-liter sample volume[11] using a 37-mm sampling cassette. Dingle et al.[12] reported that such monitors had a LOD of 1 ppb in a 24-hour sample and 3 ppb in an 8-hour sample. Given these low LODs, DNPH passive monitors (Figure 8.3) can be exposed for as little as a few hours to measure formaldehyde concentrations which are typically reported in building environments. They are, as a consequence, suitable for use in problem-building investigations. DNPH analysis requires the use of expensive equipment. However, samplers and associated analytical services are commercially available at relatively low cost ($30–50 U.S./sample).

Other passive formaldehyde sampling devices have also been recently developed. One of these uses what is described as a biosensor in which enzymes react with formaldehyde to produce a colored molecule.[11] It is a semi-quantitative method with concentrations determined by the naked eye using a color code. The method has the advantage of being easy to use by the average layman and has high selectivity. The technique has a reported sensitivity of 0.03 ppm for a 2-hour exposure and would appear to be suitable for measurements in problem-building investigations. Its semi-quantitative nature, however, suggests concerns about its accuracy and precision. The device is commercially available.

Osada et al.[13] developed a passive sampler for formaldehyde using triethanolamine-impregnated silica gel packed in a porous tube made from tetrafluroethylene resins with analysis by the 4-amino-3-hydrazino-5-mercapto-1,2,4 triazol method. The reported measurable range is 0.008 to 6 ppm over an exposure period of 24 hours. This sensitivity would make it suitable for the short-term (2 to 3 hour) measurements which are common in IAQ investigations. Its commercial availability is not known.

The passive sampling techniques described above are both specific for formaldehyde and have very low LODs. A passive sampler developed by Miksch[14] measures total aldehydes by means of the MBTH method. Though its sensitivity is sufficient for short-term sampling (2 to 3 hours), its lack of specificity for formaldehyde has limited its usefulness for IAQ investigations.

Since a number of passive samplers are commercially available at relatively low cost ($30–60 U.S. per sampler including analysis), they represent an attractive approach to conducting formaldehyde measurements in buildings. Passive samplers which have previously dominated the market had significant limitations relative to their use in problem-building investigations. The most significant of these were low sensitivity with exposure durations of 1 to 7 days required to collect sufficient formaldehyde for analysis. The higher sensitivity of the methods described above makes them suitable for conducting measurements that reflect exposure conditions of building occupants and time constraints associated with conducting building investigations. Such devices, though providing integrated results, have the potential for determining peak levels of formaldehyde over a period of one or several hours which may be important in causing the typical acute symptoms associated with this substance.

Formaldehyde can also be determined by means of automated devices which provide quasi-real-time results. These devices are based on the absorption of formaldehyde in a reagent with subsequent analysis by means of the pararosaniline colorimetric method.[15] This method has the advantage of both specificity and sensitivity with an LOD significantly lower than the chromotropic acid method. The detection range is from 0.002 to 5 or 10 ppm, depending on the range setting selected. These devices have the disadvantage of relatively high cost ($6000+ U.S.), a readout which must be converted to formaldehyde concentration, and laborious calibration, set-up, and cleanup procedures.

Volatile Organic Compounds

Though technically demanding and expensive, volatile organic compound (VOC) measurements are nevertheless fairly commonly conducted in problem-building investigations. These include the measurement of TVOC concentrations as a screening tool as suggested by Gammage et al.[16] for the identification and quantification of target VOCs and the use of the spectrum of identified and unidentified VOC compounds for purposes of fingerprinting by which specific VOC sources may be identified. The desirability of measuring TVOCs is based on research studies and TVOC theories of Molhave[17] which implicate TVOC exposures as a causal factor in mucous membrane and general symptoms at relatively low TVOC concentrations (circa 2–5 mg/m^3). It is also based on the intuitive judgment of investigators that in the absence of any other known specific causal factor, it is likely to be one or more of the VOCs ubiquitously present in indoor spaces which are responsible for IAQ complaints.

Because of the diversity of their chemical and physical properties, VOC measurement poses unique challenges to IAQ investigators. The most significant of these is sample collection. In general practice, VOCs are collected on a solid sorbent medium employing dynamic or passive sampling procedures or by means of evacuated metallic canisters.

VOC sampling is commonly conducted by drawing contaminated air through a glass or metal sampling tube which contains one or more solid sorbents. For typical indoor air applications, sampling rates in the range of 50–250 ml/minute are employed[18-20] with durations of 20–60 minutes.[18,21] The major problem with such sampling is the choice of collection media or sorbents. Sorbents used to sample VOCs in IAQ investigations have included charcoal and a variety of synthetic sorbents such as Tenax, Poropak Q, XAD-2, and other porous polymers. These sorbent media vary considerably in their ability to capture and retain specific VOCs. All have significant limitations in VOC collection and retention.

Collection of VOCs on Tenax has been reported by Wallace[22] to be the most popular and preferred method of VOC measurement in indoor environments. Tenax is a polymer of 2,6-diphenyl phenylene oxide. Advantageous properties include its hydrophobicity (making sampling possible under high humidity conditions), stability under high desorption temperatures, and its reusability. The disadvantage of Tenax is its relative inability to capture very volatile substances, such as vinyl and methylene chloride; formation of reaction products, such as acetophone, benzaldehyde, and phenol;[23,24] and high background levels of benzene, styrene, and possibly toluene.

Multisorbent sampling techniques have been developed for VOCs sampling in indoor environments. Multisorbent samplers have the advantage of collecting VOCs over a broad range of volatilities with high accuracy and precision for detecting small spatial and temporal changes.[19] Additionally, the low sampling volumes required make them adaptable for a variety of applications. Typically, multisorbent samplers include in series glass beads, Tenax, ambersorb, and charcoal. Black and Bayer[25] have reported the use of multisorbent sampling devices in IAQ investigations which have used in series glass beads, Tenax, followed by an additional porous sampling medium.

Less commonly used in IAQ investigations are sampling devices which collect VOCs by passive diffusion to an adsorbent surface. Such samplers are typically badge-like devices using charcoal as the sampling medium. Passive samplers are commonly exposed for a period of one to two weeks before they are analyzed. This extended sampling period collects sufficient sample to overcome the problem of high background concentrations on the collection medium and loss of sensitivity from solvent desorption due to the redilution of the collected sample. Unfortunately, the sampling duration is too long to reflect the nature of exposures which occur in problem buildings on a day-to-day basis.

In addition to the use of adsorption media in both dynamic and passive sampling methods, VOCs can be collected for analysis using evacuated metallic containers. The collected sample is directly injected into a gas chromatograph for analysis. The evacuated canister has several advantages over sorption methods. These include the avoidance of chemical reactions on the sorbent and low recoveries due to breakthrough or incomplete desorption. As a consequence, a wider variety of VOCs can be detected and measured. Disadvantages include the relatively small portion (1 ml) of the 1–10 liter sample which is analyzed and the contamination potential of the sample by pumps, tubes, and fittings used with the technique.[22]

Because of the low concentrations (ppb) which are typical of most VOCs in indoor spaces, the preferred method of analysis is to use thermal desorption of samples collected on synthetic sorbents such as Tenax.[19-22,25,26] Thermal desorption has several advantages over the commonly used solvent (CS_2) desorption. These include fewer analytical operations, shorter operating time, recovery of the entire collected sample for analysis rather than only the rediluted portion, and, of course, increased sensitivity to low VOC levels.

In such analyses, samples are thermally desorbed into a capillary gas chromatograph or a gas chromatograph interfaced with a mass spectrometer, with the latter employed to aid in the identification of individual VOC components. TVOC concentrations are usually quantified by the use of a gas chromatograph with a flame ionization detector (GC-FID) with toluene or hexane as the standard for determining concentrations.

TVOC determination procedures based on GC-FID and a cyclohexane standard have been reported by Tsuchuya and Kanabes-Kaminska.[21] TVOC concentrations are calculated from values determined from a variety of measured parameters based on the following equations. The calibrant gas is a mixture of cyclohexane and helium.

$$f = c_s/V_s/C_s \tag{8.1}$$

where f = response factor in counts/μg
c_s = FID count for a known volume of the cyclohexane/helium standard
V_s = volume of cyclohexane/helium standard, m^3
C_s = concentration of the cyclohexane/helium standard, $\mu g/m^3$

The TVOC concentration is then calculated as

$$T = c/f/V \tag{8.2}$$

where c = FID count for air sample
V = volume of air sampled, m^3
T = TVOC concentration, $\mu g/m^3$

The determination of TVOC concentrations has the advantage of taking less effort with a shorter turn-around time for air sampling results. Such measurements have the additional advantage that they can be compared to guideline values which have been suggested for purposes of assessing whether air contamination by VOCs may be responsible for causing or contributing to health complaints in a building. It may be noted that the magnitude of TVOC values will depend on sampling methods used. Tsuchuya and Kanabes-

Kaminska[21] report that TVOC values are higher when sampling is conducted with multi-sorbent tubes as compared to Tenax. TVOCs can also be determined by the use of flame ionization detector instruments which report TVOC levels in units based on methane.

It has been suggested[21] that the determination of TVOCs without further qualitative and quantitative analysis of individual components is both less informative and a waste of resources. Such further analysis has been reported to be used successively by Tsuchuya and Kanabes-Kaminska[21] as a finger-printing tool for the identification of major sources of VOCs such as wet-process photocopiers in office buildings. Brooks and Davis[27] have also reported the successful use of VOC fingerprinting in identifying the source or sources of VOCs suspected to be responsible for illness or odor complaints associated with office printers or computer systems.

Airborne and Settled Dusts

As indicated in Chapter 5, there is considerable evidence to implicate particulate-phase materials or "dusts" with SBS symptoms. Though both airborne and settled dusts have been associated with SBS symptoms, settled dust appears to be the greater risk factor. Methods to assess settled dust concentrations, fractional composition, and deposition rates have been reported for several systematic studies (see Chapter 5). Unfortunately, methodologies used have not been sufficiently detailed to be of practical use to IAQ investigators.

A variety of sampling and instrumental methods are available for measuring particulate matter in the respirable size range. These include gravimetric methods in which particles are collected on filters with concentrations reported in micrograms or milligrams per cubic meter. Respirable particles can also be measured by real-time or near real-time instruments using optical techniques.[2] These include instruments that measure particle concentrations in the aerodynamic size range of 0.1–10 μm by forward light scattering. Respirable particle levels are often measured using piezoelectric resonance. In piezoelectric devices, nonrespirable particles are removed by an impactor or cyclone. The respirable particles are then deposited electrostatically on a quartz crystal sensor. The difference in oscillating frequency between sensing and reference crystals is monitored and displayed as concentration in milligrams per cubic meter. Particle measuring devices range in cost from hundreds to several thousand dollars (U.S.).

SAMPLING BIOLOGICAL AEROSOLS

Bioaerosol sampling in problem buildings is commonly conducted by IAQ investigators. Bioaerosols consist of a variety of both viable and nonviable microorganisms, reproductive propagules, and microbial fragments. Organisms of particular concern include mold, bacteria, and actinomycetes.

The conduct of bioaerosol sampling in a problem building depends on what the investigator thinks may be a problem, the availability of sampling instrumentation and associated services, the experience of investigators in this type of sampling, and the needs or perceived needs of building management. Like any other type of sampling, the likelihood of obtaining meaningful results depends on clearly defined objectives. Many of the same sampling considerations which are applicable to gas/vapor- and particulate-phase substances apply to bioaerosol sampling as well. There are, of course, sampling considerations which are given special attention in bioaerosol measurements.

Sampling Approaches

How one measures bioaerosols depends on the organisms of interest and the availability of methods, collection instruments, and analytical capabilities. Over the past several decades, a need for bioaerosol sampling has developed in such areas as infection control, aeroallergen quantification and identification, and, increasingly, measurements of human exposures in both occupational and residential environments. A number of methods and sampling devices have been developed for the purposes of meeting these needs.

In conducting bioaerosol sampling, one has to make a determination as to how concentrations are determined. Many widely used techniques only quantify those organisms which can grow on the collection and/or incubation medium. Such concentrations are described as reflecting the "viable" fraction of collected bioaerosol particles. Devices used to make such measurements are often referred to as "viable" samplers. This is not totally accurate since growth of specific microorganisms depends on the sample medium. Increasingly, the term "culturable" is being used to indicate that such measurements only reflect the concentration of those microbial structures which have grown on the medium used in sample collection and/or incubation.[28] The concentration of such structures is expressed as the number of colonies which have developed on the growth medium per unit volume of air sampled, colony-forming units per cubic meter (CFU/m^3). Typically, all colonies are counted and expressed as a total count irrespective of species types present. When the analyst has sufficient skill in identification, concentrations may be expressed per individual species, or dominant species, or simply the dominant genera or species types are reported along with the total count.

Viable or culturable bioaerosol particles represent only a fraction of the total bioaerosol concentrations in air samples. After having been subject to environmental stresses, many organisms, reproductive propagules, and microbial fragments are no longer viable or culturable. Though they are nonliving, they may nevertheless have considerable biological and medical significance. In the case of mold spores and hyphal fragments, nonviable particles are similarly antigenic to those which are viable and grow in culture. In such instances, total mold spore/particle concentrations may be more important in assessing human exposures and health effects than viable/culturable concentrations.

Sampling Devices

A variety of sampling devices are available for conducting bioaerosol measurements. These include devices which are designed to collect and quantify viable/culturable microbial particles and those that quantify all microbial particles collected. The latter are used primarily for mold sampling.

Viable/Culturable Samplers

Viable/culturable particle sampling devices available for determining microbial concentrations in air are listed and partially described in Table 8.2. These devices vary considerably in terms of the principle of collection, sampling rates, and collection times employed. They are similar in that, in all cases, a solid culture medium is used to "grow out" collected viable/culturable particles for enumeration and identification.[29]

Samplers differ in their "efficiencies". Efficiency has several dimensions in bioaerosol sampling. These include (1) the efficiency with which particles are collected, (2) the efficiency with which the viability of collected microorganisms is preserved, and (3) overall efficiency which includes both collection efficiency and viable recovery. The overall collection efficiency determines the suitability of a device for a given application. Differences in efficiency are evident when filters are compared to impactor or impinger samplers. The former is reported to be very efficient in collecting small particles with lower viable recoveries. Similar results are reported for AGI-4 impinger and AGI-30 impinger comparisons.[29]

Both the multi-stage-sieve impactor and impingers have been suggested to be used as reference samplers to which all other sampler results can be compared.[30] Because of its widespread use in both IAQ research studies and problem-building investigations, the single-stage (Andersen N-6) sampler (Figure 8.4) has become the *de facto* reference sampler by which the performance of other bioaerosol samplers is compared.

Comparative Studies

Because of differences in collection principles and efficiencies, available bioaerosol sampling devices are unlikely to provide similar test results when samples are taken from the same environment. This is particularly true of samplers which characterize different properties of particles such as impingers, filters, and impactors. Chatigny et al.[29] suggest that reasonable comparisons can be made between different types of impactors. Sieve and slit samplers are described as being efficient for bioaerosol particles in the respirable size range

Table 8.2 Devices/techniques used for viable/culturable bioaerosol sampling.

Sampler type	Operating principle	Sampling rate (L/min)	Recommended sampling time (min)
Slit or slit-to-agar impactor	Impaction onto solid culture medium on rotating surface	30–700	1–60 Depending on model and sampling circumstances
Sieve impactors			
Single-stage, portable impactor	Impaction onto solid culture medium in a "rodac plate"	90 or 180	0.5–5
Single-stage (N-6) impactor	Impaction onto solid solid culture medium in a 10-cm plate	28.3	1–30[a]
Two-stage impactor	As above	28.3	1–30[a]
Multiple-stage impactor	As above	28.3	1–30[a]
Centrifugal	Impaction onto solid culture medium in plastic strips	40±	0.5
Impingers			
All glass/AGI-30	Impingement into liquid; jet 30 mm above impaction surface	12.5	1–30
All glass/AGI-4	As above; jet 4 mm above impaction surface	12.5	1–30
Filters			
Cassette	Filtration	1-2	5–60

[a] Contemporary practice is 1–5 minutes

From Chatigny, M.A. et al., 1989. *Air Sampling Instruments for Evaluation of Atmospheric Contaminants*. ACGIH. With permission.

(<3 μm) with the portable single-stage impactor[31] and centrifugal samplers[32,33] having relatively poor collection efficiencies in the respirable range.

Several investigators have attempted to compare the performance of a variety of samplers to the Andersen N-6 single-stage sampler. The studies of Verhoeff et al.[34] showed that the N-6 sampler collected higher numbers of viable mold spores than the portable sieve-impactor surface air sampler (SAS). The N-6 and membrane-filter samplers gave similar results in sampling culturable mold particles in the studies of Samimi et al.[35] Results reported for a centrifugal sampler (Biotest RCS Plus) were, however, considerably lower.

The development and use of bioaerosol samplers has, unfortunately, received little systematic evaluation of the various physical and biological parameters which affect their performance. Such studies are currently being conducted by a multi-disciplinary team coordinated by Klaus Willeke at the University of Cincinnati. Physical principles involved in bioaerosol collection have been described in detail by Nevalainen et al.[36] and Willeke et al.[37] Preliminary results reported from their studies indicate that bioaerosol

Figure 8.4 Andersen N-6 single-stage sampler.

concentrations determined from one sampling device cannot be easily compared to those of other devices. Because Andersen sieve impactors are commonly used as reference samplers, it is notable that Willeke[38] observed that both the Andersen single- and two-stage samplers collected significantly lower concentrations of bacterial aerosols than specially designed samplers developed at the University of Cincinnati.

That the Andersen N-6 sampler may have performance limitations is not surprising. Potential problems associated with its use were raised in the studies of Stanevich and Petersen[39] who reported anomalous results associated with different sampling durations (Table 8.3). Sampling durations of three and five minutes resulted in significantly lower total culturable mold particle counts than a sampling duration of 1 minute.

Given the widespread use of the Andersen N-6 sampler for both research studies and problem-building investigations and the availability and use of other devices which appear to perform less well, the reported results of Willeke[38] are disquieting. This would particularly be the case if similar results were observed for mold sampling as well. Willeke's results portend a considerable level of uncertainty related to performance and reliability of bioaerosol sampling devices currently being used. This uncertainty will continue for some time since Willeke's studies are still in their early stages and at this time, the results have not been published and, as such, are not available for scientific scrutiny.

Table 8.3 Variation in mold levels (CFU/m³) associated with different sampling durations using Andersen single-stage (N-6) samplers.

Sampling duration (min)	N	Colonies/ plate	CFU/m³	Coefficient of variation
1	50	6	200	.36
3	49	10	117	.47
5	48	14	100	.28

From Stanevich, R., and M. Petersen, 1990. *Proceedings Fifth International Conference on Indoor Air Quality and Climate.* Vol. 2. Toronto.

This uncertainty poses a major problem for users of the Andersen N-6 and other bioaerosol samplers. It raises a number of presently unanswerable questions. If the N-6 sampler has significant performance problems, how reliable and useful are results obtained by it. Should we continue to use the N-6 sampler (and other samplers) in research studies and problem-building investigations?

It is imperative that users recognize that bioaerosol sampling devices currently available have significant limitations which affect the accuracy of measured values and that these problems will be eventually sorted out by systematic scientific inquiry. In the interim, there is little reason not to continue to use the methods currently available. It is good practice, in general, not to treat a measured value in absolute terms. Additionally, if performance problems are shown to be systematic, that is, all measurements are affected more or less similarly, then such measurements may be used to determine relative differences in exposures in various environments. In addition, the Andersen N-6 and other samplers can be used to distinguish between the different forms of bioaerosols present, their relative abundance, and potential sources.[40] Despite potential performance problems, viable/culturable samplers can nevertheless provide useful information to both building investigators and research scientists.

Total Spore/Particle Samplers

Bioaerosol sampling methods/devices designed for the collection and enumeration of total microbial particle concentrations (both viable and non-viable) are less widely used than the viable/culturable methods described above. This reflects, in part, a historical emphasis on viable/culturable bioaerosol sampling methods and their amenability to organism identification. In addition, the recent studies of Willeke indicate that because of high cutoff diameters (circa 2.5 μm), total particle samplers are only suitable for the collection of organisms that produce large airborne structures, such as mold spores and pollen.[38]

In commercially available sampling devices, air is drawn through a slit opening. Particles are inertially collected on greased glass slides which can be subsequently examined microscopically to enumerate collected mold spores

Figure 8.5 Mini-Burkhard total spore sampler.

and hyphal fragments. Though many mold types can be identified to genera, this method generally cannot be used to identify species or genera with similar spores such as *Penicillium* and *Aspergillus*. A commonly used portable total mold spore sampling device can be seen in Figure 8.5.

Total mold spore samplers have one very significant advantage vis-a-vis viable/culturable samplers; that is, they are more likely to better reflect the antigenic potential of bioaerosols since both viable and nonviable fractions are quantified. Differences on the order of 2–10× (total to viable counts) have been reported.[41,42] Differences as high as 2×10^5 have been observed by the author in a mold-infested building.[43] These results suggest that investigators who conduct only viable/culturable sampling may at times fail to identify potentially significant airborne mold contamination problems when they are, in fact, occurring. They also indicate the need for the simultaneous conduct of both viable/culturable and total mold spore/particle sampling to better assess the nature of potential bioaerosol contamination of building environments.

Recommended Bioaerosol Sampling Practices

As with other contaminants, bioaerosol sampling should be conducted in a manner that provides consistently good results. In addition to the good practice principles described for indoor air sampling earlier in this chapter, bioaerosol sampling has some unique requirments. Recognizing the need for standardization, the Bioaerosols Committee of the American Conference of Governmental Industrial Hygienists (ACGIH) has attempted to develop guidelines for sampling and assessment of bioaerosols in workplace environments. Relatively detailed guidelines were published in 1986[44] and 1987[45] with more general guidelines in 1989.[40]

Sampling Objectives

The primary objective of bioaerosol sampling for the average investigator is to determine the relative concentrations of biological contaminants such as mold, actinomycetes, or bacteria in problem indoor environments in the hope of determining whether levels are excessive or within acceptable limits. Unfortunately, there is no consensus as to what excessive or acceptable levels are. Recognizing this problem, and the fact that the orders of magnitude fluctuate in bioaerosol levels which occur in buildings over periods of days to months (and in the outdoor environment), the ACGIH Bioaerosols Committee has increasingly downplayed the use of bioaerosol sampling for quantitative determinations of the prevalence of airborne biological contaminants. Rather, they emphasize that the primary objective of bioaerosol sampling should be the identification of the source or sources of airborne microorganisms so that effective remediation may be undertaken.[40]

Given the limitations of the various bioaerosol sampling devices previously discussed, problems in interpreting sampling results due to temporal fluctuations in concentrations[46] and prevalence rates of different organisms and the ACGIH Bioaerosols Committee's deemphasis of quantitative assessments of air concentrations, the author nevertheless believes that conducting bioaerosol sampling in problem-building investigations can be of considerable value. Such sampling should not be done routinely. As with other contaminants, sampling should be conducted to evaluate a hypothesis based on other information available to the investigator. This may include observations of water-damaged interior materials, excessive accumulations of "slime" in condensate drip pans, pervasive mold/microbial odors, symptom/illness patterns suggestive of exposure to biological contaminants, etc.

As indicated in Chapter 6, mold levels in mechanically ventilated buildings are usually relatively low (circa 50–300 CFU/m^3). Orders of magnitude higher levels would be suggestive of a significant mold contamination problem and a need for more intensive investigation. Quantitative air sampling may be useful in confirming the presence of a biological contamination problem. It can also be used to determine the prevalence of specific organisms which may more likely be suspected to cause potential illness problems, such as species of *Aspergillus*. On the other hand, low measured levels should not be construed to mean that a biological contamination problem does not exist. Because of the snapshot nature of one-time sampling and because of temporal variations, one may fail to observe a contamination problem which may, in fact, be occurring or the transient high-level releases associated with disturbance of infested porous insulation associated with HVAC systems.[47]

In addition to the objectives described above, quantitative bioaerosol sampling can be useful in assessing the effectiveness of remedial measures and for recommendations for reoccupancy of buildings or spaces where remediation measures have been implemented.[48]

Sample Collection—Viable/Culturable Bioaerosol Sampling

A variety of factors have to be taken into consideration when conducting air sampling for viable/culturable contaminants. These include sampling approaches, media selection, sampling locations/sample numbers, sampling volume/duration, sampler disinfection, sampler calibration, etc.

Area vs. Aggressive Sampling

As suggested previously, one's approach to sampling will depend on sampling objectives. This may include the collection of area samples for the purpose of determining whether occupied spaces are excessively contaminated by biological contaminants and the type of biological contamination which is occurring such as fungi, actinomycetes, or bacteria and particular genera or species. Area sampling may also be conducted to determine the efficacy of remedial measures. As recommended by the ACGIH committee, sampling may also be conducted for purposes of identifying sources of particular biocontaminants.[40] Bioaerosol sampling may be "normal" or "aggressive". In the latter case, sampling follows a deliberate disturbance of a potential source such as porous ventilation system insulation.[47]

Media Selection

It is necessary to select an appropriate culture medium for the organisms under test. The use of malt-extract agar (MEA) and casein-soy peptone agar have been recommended by the ACGIH committee[40] for fungi and bacteria (including thermophilic actinomycetes), respectively. The recommended use of MEA is based on the comparative studies of Burge et al.[49] which demonstrated that MEA was superior to other mycological media in supporting the growth of common saprophytic fungi. It is also recommended because it is a diagnostic media for *Aspergillus* species and does not support the growth of bacteria. Though MEA is the most commonly used culture media for volumetric mold sampling, recent studies by Mididoddi et al.[50] at NIOSH laboratories indicate that streptomycin-amended rose-bengal agar (RBS) produced superior results to MEA both in total colony counts and the diversity of species cultured. Collection efficiency was particularly greater for *Cladosporium* and yeast, both of which have been shown to be associated with respiratory symptoms in the epidemiological study of Su et al.[51]

Though MEA and RBS provide good results relative to the recovery of a broad range of fungal species, they will not support the growth of *Stachybotrys atra*, a major toxigenic fungus that can grow indoors. Special media containing cellulose are required to culture and identify the presence of *S. atra* propagules in indoor environments.

In addition to media selection, it is important that culture media be relatively soft (1.5–2% agar) to increase collection efficiency. Willeke[38] has reported that increased media hardness (from 1.5 to 4% agar) can significantly reduce collection efficiencies.

Sampling Location/Numbers

Locations selected for bioaerosol sampling will depend primarily on sampling objectives. Samples should be collected in both complaint areas and non-complaint areas in assessing potential indoor air contamination. Outdoor samples should be collected to provide a reference for determining whether or not certain types of biocontaminants or species are being amplified in the indoor environment. Such samples should be collected away from obvious potential sources. Samples collected in complaint and non-complaint areas may provide the investigator with the opportunity to ascertain whether there are either quantitative or qualitative differences which may be important in diagnosing the nature of reported problems. Duplicate or triplicate samples are recommended for each culture medium used at each location.

Sampling Volume/Duration

The viable/culturable sampling devices commercially available differ in terms of their sampling or flow rate. As can be seen in Table 8.2, these range from 28.3 L/min for Andersen-type sieve impactors to 90 and 180 L/min for the SAS portable sieve impactor. The sampling rate of a device is an important determinant of sampling duration, which itself may be influenced by bioaerosol concentrations. Excessive collection of bioaerosol particles such as mold will result in multiple particle deposition at the same collection area, overgrowth of colonies, and an underestimation of airborne levels. It is often necessary to adjust the sampling duration and, therefore, sampling volume to minimize sample plate overloading. Sampling durations in the range of 0.5 to 5 minutes are commonly used with the N-6 sampler. Shorter sampling durations are employed when elevated concentrations are expected. Samplers (Andersen) using 100-mm petri dishes are suggested to be overloaded when fungi exceed 150 colonies/plate, with Rodac plates (SAS) overloaded above 40 colonies.[40]

As sampling volume and duration are decreased, variability between duplicate samples collected at the same time increases significantly. Stanevich and Petersen[39] have, for example, observed a sixfold increase in intrasampling variance with a sampling duration of 1 minute compared to 5 minutes using the N-6 sampler. In many cases, however, longer sampling durations such as 5 minutes will cause overloading of the sampling plate, making it more difficult to enumerate and identify colonies present, particularly when sampling for mold.

As previously mentioned, changing the sampling duration with the N-6 sampler produces anomalous results. Stanevich and Petersen,[39] observed that mold concentrations decreased as sampling duration increased from 1 to 3 and then 5 minutes (Table 8.3). This anomaly makes it difficult to recommend a standard sampling duration. Users would therefore be best served by selecting a standard sampling duration (e.g., 2 minutes) with which they feel comfortable and using it consistently so that measured values can be compared.

Sampler Disinfection

Because samplers can easily become contaminated during field use with the potential of cross-contamination of samples, the ACGIH committee has recommended that samplers be sterilized with a 70% ethanol solution (immersion for one minute) before use at each location.[44] In practice, this is generally not practical. As a consequence, it is more common to wipe down sampler surfaces with alcohol between samplings.[45]

Sampler Calibration

Since concentrations are calculated from sample volumes, it is essential that sampling flow rates are known with some degree of confidence. Flow rates may change with usage in some instruments. As a consequence, it is necessary to calibrate instrument flow rates frequently. Calibration is recommended prior to and after each field use for the Andersen N-6 and other viable/culturable samplers. This is a relatively conservative practice. It has been the author's experience that flow rates of N-6 samplers do not change even after years of use. This is due in great measure to the fact that the sieve plate acts as a limiting orifice maintaining a constant flow rate as long as the critical vacuum pressure is maintained and sieve holes are not obstructed. Swab wiping with alcohol before sampling appears to be effective in removing culture medium which deposits on the bottom of the sieve plate during sampling. Though it is not necessary (based on the author's experience) to calibrate sampling instruments as frequently as the ACGIH committee recommends, it is good practice to calibrate air flow on a periodic basis using a standardized protocol. The instrument should always be recalibrated if a different sampling pump is used.

Sampling devices for viable/culturable aerosols are not sold with air flow calibration devices. Because of the flow rates employed, they require the use of high air flow measuring devices for calibration. The ACGIH committee recommends that devices such as pitot tubes and vane anemometers be used to calibrate samplers with built-in air movers. Static measurements of sampling rates can also be made by exhausting a sampling bag of sufficient volume using the sampler and its air mover.[45]

It is common to calibrate bioaerosol samplers using a rotometer calibrated against a primary or secondary standard; these can include wet or dry test meters. Laboratory calibration can easily be conducted by attaching the activated sampling device to a wet-test meter. A special connector must be constructed to cover the inlet cone of the Andersen N-6 sampler. This can be easily constructed from a piece of flexible tubing and the dust cap that comes with the instrument. Centrifugal impactors cannot be mechanically calibrated and therefore users must depend on the manufacturer's calculated calibration.

Sample Handling and Analysis

After collection, samples can be incubated, counted, and individual organisms identified to specific taxa by the individual investigator or sent to an outside laboratory which provides such services. Proper shipping practices are required to insure both the viability and the integrity of samples.

Mold sample plates should be incubated at room temperature with some ultraviolet light (from fluorescent or Gro-Lux lights) to stimulate sporulation of dark-spored fungi. A minimum of 5 days incubation is desirable before counts are made and fungal types identified. Fungal colonies can often be identified to genus from their macroscopic morphology (color, colony form, and texture), by using low-power magnification (10–60×) or high-power magnification (400–1000×) of reproductive propagules. Species identification is often a difficult task and requires a level of expertise not available to the average investigator.

Colony counts obtained from sieve plate impactor samplers must be adjusted to account for multiple impactions using statistical charts specific for each sampler. Positive-hole correction methodology and charts have been published[52,53] for the Andersen sampler.

Sample handling for bacteria begins as soon as samples have been collected. Because delayed temperature control confers a selective advantage to those organisms able to grow at ambient temperatures, it is desirable that bacterial plates be transported to the laboratory within 24 hours in well-insulated packages. Bacterial plates must be incubated for one or more days to produce recognizable colonies. Culture temperatures depend on the organisms of interest. Human-source bacteria grow best at 35–37°C; bacteria commonly found in the environment, at temperatures of 25–30°C; and thermophilic organisms, at temperatures in excess of 50°C.

Common airborne bacterial genera such as *Bacillus*, *Micrococcus*, *Streptococcus*, and *Corynebacterium* can be identified using relatively simple procedures including colony morphology, gram-staining, and enzyme tests. Species identification requires a much higher level of proficiency and generally highly trained individuals to accomplish.

Sample Collection—Total Mold Spore/Particles

With the exception of media and culturing concerns, most of the sampling considerations described for viable/culturable bioaerosol sampling apply to total mold spore/particle sampling as well. Nevalainen et al.[36] and Willeke et al.[37] indicate that a sampling duration of 140 minutes would be required to achieve a statistically acceptable surface density of spores on the deposition area of the greased slide sampling surface. They, however, recommend a sampling duration of 14 minutes since 140 minutes is an unrealistic collection time and would result in overloading by dust particles. Due to dust-overloading problems, sampling durations of 5–10 minutes are commonly used with the Mini-Burkhard, one of the most commonly employed total mold spore/particle sampling devices.

Quantification

Samples do not require any immediate processing and can be stored for future microscopic evaluation and mold structure enumeration. Examination of slide samples at 1000× magnification is recommended.[40] Both spores and hyphal fragments are counted for a fraction of the collection surface (10 horizontal traverses). This count is then converted to a total count expressed as spores/m^3.

Enumeration of mold spores and hyphal fragments on sample slides requires the use of a microscope with a magnification capability of 1000 × (400 × minimum), a movable stage, and an ocular micrometer in the eye piece. It also requires technical skill to identify mold spores and yeast cells against an often intense and heterogeneous background of deposited particulate matter.

Calibration

Total spore samplers require frequent calibration since they are very prone to flow rate changes with time. Flow rates in the portable Burkhard sampler is voltage-regulated and needs to be readjusted periodically to bring it back to the target flow rate of approximately 10 L/min. Additionally, as the device is in most instances battery-operated, the flow rate can decrease markedly as the battery loses its charge (usually after the fourth consecutive sampling).

Data Interpretation

As with gaseous contaminants, the results of bioaerosol sampling must be amenable to some type of meaningful interpretation. Unfortunately, there is relatively little scientific information available to establish a dose-response relationship between exposures and adverse health effects. It is, as a consequence, very difficult to develop and establish guideline values or standards for bioaerosol exposures.

In their 1986 guidelines, the ACGIH committee[44] indicated that indoor/outdoor viable mold levels exceeding 0.33 were considered to be excessive in mechanically ventilated office buildings. They also suggested that the presence of any one mold species exceeding 500 CFU/m^3 in an office workspace was indicative of a building-related source and was in need of remedial action. Fungal genera recommended for enumeration and identification included *Acremonium, Alternaria, Aspergillus, Aureobasidium, Chaetomium, Cladosporium, Mucor, Penicillium, Stachybotrys*, and other predominant fungi.

The ACGIH committee also indicated that total bacterial levels in excess of 500 CFU/m^3 in workspaces of affected individuals but not in other indoor locations were indicative of poor ventilation, overcrowding, and in need of remedial action. In initial screening, *Bacillus* species and Gram negative bacteria were recommended to be enumerated and identified; in subsequent screening, *Staphylococcus aureus, S. epidermis*, other *Staphylococcus* spp., *Streptococcus salivaris*, other *Streptococcus* spp., *Corynebacterium* spp., *Acenetobacterium* spp., *Pseudomonas* spp., and *Micrococcus* spp. High levels of *Bacillus* spp. or Gram negative rods (>500 CFU/m^3) were suggested to be indicative of a building-related source.

The presence of thermophylic actinomycetes with indoor levels exceeding 500 CFU/m^3 was considered a risk factor for hypersensitivity lung disease. Thermophilic actinomycetes recommended for identification and enumeration included *Thermoactinomyces candidus, T. vulgaris, Micropolyspora faeni, Thermonospora* spp., and *Saccharomonospora* spp.

Though the ACGIH Bioaerosols Committee attempted to provide guidelines for the interpretation of bioaerosol sampling results in 1986, they have subsequently deleted any reference to such guidelines in their 1987[45] and 1989[40] updates. This more conservative posture reflected the lack of scientific data to support their initial guidelines and recognition of the difficulty in interpreting results inherent in bioaerosol sampling. Bioaerosol levels in both indoor and outdoor environments vary considerably over the course of time.[46] Variations in the range of an order of magnitude or more may occur as a consequence of changes in rates of generation and dissemination and as a result of random factors. Many mold species, for example, go through growth cycles where periods of vegetative growth may be associated with limited spore production and more stressful periods with reduced vegetative growth and high spore production. One-time sampling on a single day may, as a consequence, produce results which can be easily misinterpreted by less knowledgeable investigators.

In the absence of any officially recognized guidelines, investigators are left to their own devices in interpreting sampling results. In the case of mold, some investigators have used indoor viable/culturable levels of 1000 CFU/m^3 (or greater) as an indicator of excessive levels[54] and 500 CFU/m^3 (or greater) as an indicator of unsanitary conditions.[16] They may also use differential levels of species present as an indicator of a problem environment. *Penicillium*, for example, typically has higher prevalence rates indoors than outdoors. Some investigators consider the presence of toxigenic species such as *Stachybotrys atra* in bioaerosol samples as indicative of a potentially significant mold contamination problem. Interpretation of bioaerosol sampling results is enhanced when there is evidence of significant mold growth on building materials or furnishings and/or sensory perception of mold odor.

RECAPITULATION

The conduct of measurements for either chemical or biological contaminants is a common practice in problem-building investigations. Such measurements should be conducted in a thoughtful and systematic way to insure good results that will be useful in evaluating the nature of problems under investigation.

Elements of a standardized contaminant measurement protocol should include clearly defined sampling objectives and attention to a variety of sampling considerations such as sampling methods to be used, their advantages and limitations, the need for calibration and quality assurance, and factors such as sampling location, when to sample, number of samples, sampling duration, and administrative practices.

Various sampling methods available for conducting IAQ measurements of common chemical contaminants are discussed. These include direct-read real-time sampling methods such as gas sampling tubes and dynamic sampling instruments, active-integrated procedures such as gas bubblers and solid sorbent samplers, and passive-integrated sampling procedures such as diffusion samplers. These are described within the context of operating principles and advantages and disadvantages of their use for CO_2, CO, formaldehyde, VOCs, and particulate dusts.

Though many of the same considerations are applicable to all types of sampling, bioaerosol sampling has unique concerns. These reflect, in part, the methodology employed. There are significant uncertainties about the performance and reliability of samplers which are presently commercially available and widely used, and these uncertainties are unlikely to be resolved soon. As a consequence, investigators must use available methods, recognizing potential performance limitations and the significance of results obtained.

As with chemical contaminants, the conduct of bioaerosol measurements should be based on clearly defined objectives and a standardized measurement protocol to ensure relatively good results. Important sampling considerations include the type of sampling device used, media selection, sampling location and numbers, sampling volume and duration, sampler disinfection, equipment calibration, and sample handling after collection. Media selection, disinfection, and the nature of sample handling are unique to viable/culturable sampling techniques.

In addition to viable/culturable sampling procedures, biological contamination (particularly mold) can be assessed by collection of mold spores and vegetative fragments on greased slides with subsequent microscopic analysis. Viable/culturable and total spore/particle sampling procedures measure different aspects of biological contaminant concentrations. Their value is enhanced when they are used together.

The absence of any guideline values or general guidelines for the interpretation of bioaerosol sampling results poses significant challenges to IAQ investigators. In the absence of guidelines, investigators must use their best judgment based on a knowledge of potential hazards associated with certain types of biological contaminants and associated factors such as microbial growth and odor, which may indicate a problem.

REFERENCES

1. Thorsen, M.A. and L. Molhave. 1987. "Elements of a Standard Protocol for Measurements in the Indoor Atmospheric Environment." *Atmos. Environ.* 21:1411–1416.
2. McCarthy, J.F. et al. 1991. "Assessment of Indoor Air Quality." 82–108. In: Samet, J.M. and J.D. Spengler (Eds.). *Indoor Air Pollution: A Health Perspective.* Johns Hopkins University Press. Baltimore.
3. Balmat, J.L. 1986. "Accuracy of Formaldehyde Analysis Using the Draeger Tube." *Am. Ind. Hyg. Assoc. J.* 47:512.
4. Godish, T. 1985. "Residential Formaldehyde Sampling — Current and Recommended Practices." *Am. Ind. Hyg. Assoc. J.* 46:105–110.
5. Valbjorn, O. et al. 1990. "Climate and Air Quality Problems — Investigation and Remedy." SBI Report 212. Danish Building Research Institute.
6. Ancker, K. et al. 1989. "Evaluation of CO_2 Detector Tubes for Measuring Air Recirculation." *Environ. Int.* 15:605–608.
7. Norback, D. et al. 1992. "Field Evaluation of CO_2 Detector Tubes for Measuring Outdoor Air Supply Rate in the Indoor Environment." *Indoor Air* 2:58–64.
8. National Institute of Occupational Safety and Health. 1984. "Method 3500." *Manual of Analytical Methods.* 3rd. ed. Dept. of Health and Human Services (NIOSH) Pub. no. 84–100. Cincinnati, OH.
9. Levin, J.O. 1985. "Determination of Sub-part-per-million Levels of Formaldehyde in Air Using Active or Passive Sampling on 2,4 Dinitrophenylhydrazine-Coated Glass Fiber Filters and High Performance Liquid Chromatography." *Anal. Chem.* 57:1032–1035.
10. Levin, J.O. et al. 1986. "Passive Sampler for Formaldehyde in Air Using 2,4-Dinitrophenylhydrazine-Coated Glass Fiber Filters." *Environ. Sci. Technol.* 20:1273–1276.
11. Schirk, O. 1993. "Different Methods for the Measurement of Formaldehyde in Indoor Air." 357–362. In: *Proceedings of the Sixth International Conference on Indoor Air Quality and Climate.* Vol. 2. Helsinki.
12. Dingle, P. et al. 1993. "A Study of Formaldehyde in a New Office Building." 51–55. In: *Proceedings of the Sixth International Conference on Indoor Air Quality and Climate.* Vol. 2. Helsinki.
13. Osada, E. et al. 1993. "Development of a Passive Sampler of Formaldehyde and Its Application to Exposure Measurement." 317–322. In: *Proceedings of the Sixth International Conference on Indoor Air Quality and Climate.* Vol. 2. Helsinki.

14. American Society of Testing and Materials. 1993. "Standard Test for Measurement of Formaldehyde in Indoor Air (Passive Sampler Methodology)." ASTM D5014–89. 11.03:411–416. Philadelphia, PA.
15. Matthews, T.C. 1982. "Evaluation of a Modified CEA Instrument, Inc. Model 555 Analyzer for Monitoring Formaldehyde Vapor in Domestic Environments." *Am. Ind. Hyg. Assoc. J.* 43:547–552.
16. Gammage, R.B. et al. 1989. "Indoor Air Quality Investigations: A Practitioner's Guide." *Environ. Int.* 15:503–510.
17. Molhave, L. 1990. "Volatile Organic Compounds, Indoor Air Quality and Health." 15–33. In: *Proceedings of the Fifth International Conference on Indoor Air Quality and Climate*. Vol. 5. Toronto.
18. Wolkoff, P. 1990. "Some Guides for Measurement of Volatile Organic Compounds Indoors." *Environ. Tech.* 11:339–344.
19. Hodgson, A.T. and J.R. Girman. 1989. "Application of a Multisorbent Sampling Technique for Investigations of Volatile Organic Compounds in Buildings." 244–256. In: Nagda, N.L. and J.P. Harper (Eds.). *Design and Protocol for Monitoring Indoor Air Quality*. ASTM STP 1002. American Society for Testing and Materials. Philadelphia.
20. Walkinshaw, D.S. et al. 1987. "Exploratory Field Studies of Total Volatile Organic Compound Concentrations in Relation to Sources and Ventilation Rates." 139–149. In: *Proceedings of IAQ '87: Practical Control of Indoor Air Problems*. American Society of Heating, Refrigerating and Air-Conditioning Engineers. Atlanta.
21. Tsuchuya, Y. and M. Kanabes-Kaminska. 1990. "Volatile Organic Compounds in Canadian Indoor Air." 24–34. In: *Pilot Study on Indoor Air Quality, Report on a Meeting Held in Saint-Adele, Quebec, August 6–18*. NATO: Committee on the Challenges of Modern Society.
22. Wallace, L.A. 1991. "Volatile Organic Compounds." 253–272. In: Samet, J.M. and J.D. Spengler (Eds.). *Indoor Air Pollution: A Health Perspective*. Johns Hopkins University Press. Baltimore.
23. Pellizzari, E.D. 1977. "The Measurement of Carcinogenic Vapors in Ambient Atmospheres." U.S. Environmental Protection Agency. Research Triangle Park, NC.
24. Pellizzari, E.D. 1979. "Analysis of Organic Air Pollutants by Gas Chromatography and Mass Spectroscopy." U.S. Environmental Protection Agency. Research Triangle Park, NC.
25. Black, M.S. and C.W. Bayer. 1988. "Pollutant Measurement Methods Used in IAQ Evaluations of Three Office Buildings." 317–353. In: *Proceedings of IAQ '88: Engineering Solutions to Indoor Air Problems*. American Society of Heating, Refrigerating and Air-Conditioning Engineers. Atlanta.
26. Bayer, C.W. and M.S. Black. 1987. "Capillary Chromatographic Analysis of Volatile Organic Compounds in the Indoor Environment." *J. Chromatogr. Sci.* 25:60–64.
27. Brooks, B.O. and W.F. Davis. 1991. *Understanding Indoor Air Quality*. CRC Press. Boca Raton, FL.
28. Burge, H.A. 1990. "Bioaerosols: Prevalence and Health Effets in the Indoor Environment." *J. Aller. Clin. Immunol.* 86:687–701.

29. Chatigny, M.A. et al. 1989. "Sampling Airborne Microorganisms and Aeroallergens." 199–220. In: Hering, S.V. (Ed.). *Air Sampling Instruments for Evaluation of Atmospheric Contaminants.* 7th ed. American Conference of Governmental Industrial Hygienists. Cincinnati, OH.
30. Brachman, P.S. et al. 1964. "Standard Sampler for Assay of Airborne Mircroorganisms." *Science.* 144:1295–1297.
31. Clark, S. et al. 1981. "The Performance of the Biotest RCS Centrifugal Air Sampler." *J. Hosp. Infect.* 2:181–185.
32. Lach, V. 1985. "Performance of Surface Air System Air Samplers." *J. Hosp. Infect.* 6:102–107.
33. Macher, J.M. and M.W. First. 1983. "Reuter Centrifugal Air Sampler: Measurement of Effective Airflow Rate and Collection Efficiency." *Appl. Environ. Microbiol.* 45:1960–1962.
34. Verhoeff, A.P. et al. 1990. "Enumeration and Indentification of Airborne Viable Mould Propagules in Houses." *Allergy.* 45:275–284.
35. Samimi, B.S. et al. 1993. "Comparison of Three Methods of Bio-Aerosol Sampling: Andersen N–6, Biotest and Membrane Filter Under Field and Chamber Conditions." Presented at the American Industrial Hygiene Conference and Exposition '93, New Orleans.
36. Nevalainen, A. et al. 1992. "Performance of Bioaerosol Samplers: Collection Characteristics and Sampler Design Considerations." *Atmos. Environ.* 26A:531–540.
37. Willeke, K. et al. 1993. "Physical and Biological Sampling Efficiencies of Bioaerosol Samplers." 131–136. In: *Proceedings of the Sixth International Conference on Indoor Air Quality and Climate.* Vol. 4. Helsinki.
38. Willeke, K. et al. 1993. "Physical and Biological Sampling Efficiencies of Bioaerosol Samplers." Presented at the Sixth International Conference on Indoor Air Quality and Climate, Helsinki, July 4–8.
39. Stanevich, R. and M. Petersen. 1990. "Effect of Sampling Time on Airborne Fungal Collection." 91–95. In: *Proceedings of the Fifth International Conference on Indoor Air Quality and Climate.* Vol. 2. Toronto.
40. American Conference of American Governmental Industrial Hygienists — Bioaerosols Committee. 1989. "Guidelines for the Assessment of Bioaerosols in the Indoor Environment." ACGIH. Cincinnati, OH.
41. Burge, H.P. et al. 1977. "Comparative Recoveries of Airborne Fungus Spores by Viable and Nonviable Modes of Volumetric Collection." *Mycopathologia.* 61:27–33.
42. Godish, D. et al. 1993. "Mold Risk Factors in Latrobe Valley, Victoria, Australia Houses." 369–374. In: *Proceedings of the Sixth International Conference on Indoor Air Quality and Climate.* Vol. 4. Helsinki.
43. Godish, T. 1993. Unpublished results.
44. Morey, P.R. et al. 1986. "Airborne Viable Microorganisms in Office Environments: Sampling Protocol and Analytical Procedures." *Appl. Ind. Hyg.* 1:R19–R27.
45. Burge, H.A. et al. 1987. "Bioaerosols: Guidelines for Assessment and Sampling of Saprophytic Bioaerosols in the Indoor Environment." *Appl. Ind. Hyg.* 2:R10–R16.

46. Sugawara, F. and S. Yoshizawa. 1990. "On the Fluctuation of Airborne Microbiological Particle Concentration and Sampling Time." 97–102. In: *Proceedings of the Fifth International Conference on Indoor Air Quality and Climate.* Vol. 2. Toronto.
47. Morey, P.R. and C. Williams. 1990 "Porous Insulation in Buildings: A Potential Source of Microorganisms." 529–533. In: *Proceedings of the Fifth International Conference on Indoor Air Quality and Climate.* Vol. 4. Toronto.
48. Light, E.N. et al. 1989. "Abatement of an *Aspergillus niger* Contamination in a Library Building." 224–231. In: *Proceedings of IAQ '89: The Human Equation: Health and Comfort.* American Society of Heating, Refrigerating and Air-Conditioning Engineers. Atlanta.
49. Burge, H.P. et al. 1977. "Comparative Merits of Eight Popular Media in Aerometric Studies of Fungi." *J. Aller. Clin. Immunol.* 60:199–203.
50. Mididoddi, S. et al. 1993. "Comparison of Seven Media for Sampling Viable Airborne Fungi." Presented at the American Industrial Hygiene Conference and Exposition '93, New Orleans.
51. Su, H.J. et al. 1990. "Examination of Microbiological Concentrations and Association with Childhood Respiratory Health." 21–26. In: *Proceedings of the Fifth International Conference on Indoor Air Quality and Climate.* Vol. 2. Toronto.
52. Andersen, A.A. 1958. "New Sampler for the Collection, Sizing and Enumeration of Viable Airborne Particles." *J. Bacteriol.* 76:471–484.
53. Macher, J. 1989. "Positive-Hole Correction of Multiple Jet Impactors for Collecting Viable Microorganisms." *Am. Ind. Hyg. Assoc. J.* 50:561–568.
54. Morey, P.R. et al. 1984. "Environmental Studies in Moldy Office Buildings: Biological Agents, Sources and Preventive Measures." *Ann. ACGIH: Evaluating Office Environmental Problems.* 10:21–35.

9 SOURCE CONTROL

The mitigation of health/comfort complaints associated with work/building environments is, in many cases, a difficult problem. Successful mitigation, in theory, requires that the causes/sources be identified and appropriate measures implemented. However, as discussed in Chapter 7, identification of causal factors and sources is often elusive. This reflects, in good measure, our limited understanding of causal relationships between contaminant exposures (or elements of the work environment) and health effects reported by building occupants. It also reflects limitations of the various protocols used in conducting building investigations and skills of investigators. The fact that a variety of factors may be contributing to health complaints reported may further contribute to the difficulties inherent in mitigating an apparent indoor air quality (IAQ) problem.

In Chapters 2 through 6, a number of potential causal or contributing factors to work/building-related health complaints were discussed. Many of these may be present in individual building environments at the same time. It is not uncommon, for example, to conduct investigations observing that (1) bioeffluent levels are elevated due to inadequate ventilation; (2) clerical staff are exposed to carbonless copy paper, a variety of copying devices, and video-display terminals; and (3) a portion of the building population is allergic to common indoor antigens such as mold and/or dust mites. If the overall prevalence rate for general, mucous membrane and skin symptoms is due to a variety of individual causal factors, it will likely prove to be very difficult to mitigate "the problem".

Successful mitigation, of course, depends on how success is defined by those conducting the investigation. If one, for example, uses a complaint rate which does not exceed 20% of the building population as a measure of air quality acceptability, then successful mitigation by definition will mean attaining a building/work environment air quality with a complaint rate less than 20%. A complaint rate of less than 20% as a guideline for acceptability of air quality is consistent with ASHRAE recommendations[1] and Building Diagnostics investigative protocols.[2,3] Complaint rates of 10–20% are, of course, not insignificant when viewed from the perspective of those affected.

On a practical basis, successful mitigation for the average IAQ investigator means that the implementation of recommended measures has resulted in anecdotal testimony by building management that complaint levels have decreased and that no further requests for diagnostic services are requested. These, of course, do not confirm the success of the mitigation measures recommended or even that they have been carried out. Based on the author's experience, it is commonly the case that building management has, for a variety of reasons, not implemented mitigation measures recommended. This may, in part, be due to economic considerations. It also reflects the reality that many building managers view IAQ complaints in political rather than technical terms. The investigation itself may serve to mitigate the political dimensions of occupant complaints.

Mitigation of health complaints in a problem building is not a simple task. It has both significant technical and political-social constraints. Recognizing these constraints it is, nevertheless, the goal of every IAQ investigator to "solve the problem" and, despite differences in the conceptual and practical aspects of what that means, try to do so with the best science and technology available.

It is assumed in most cases that exposures which cause IAQ complaints are due to contaminants which become airborne after being emitted by a source. An exception to this assumption would be complaints of skin symptoms associated with exposure to contact irritants. This may be the case with handling carbonless copy paper, exposure to glass fibers, and potentially other work place materials.

With the exception of contact irritants, it is commonly assumed that causal contaminants are airborne, have been airborne, or have the potential to become airborne again and that contaminants are gaseous or particulate-phase materials. It is also commonly assumed that these substances can be controlled at their source before they contaminate building air or can be reduced in concentration once they become airborne by local or general ventilation or by air cleaning. Though source control is technically attractive because it has the potential for being relatively simple and effective, it has one very significant requirement, that the source/sources of the causal contaminant(s) have been identified, with some degree of assurance that they are indeed causing the problem. In NIOSH investigations,[4] the identification of causal contaminants apparently was only determined in about one third of cases reported.

Source control comprises a variety of principles and applications based on the nature of individual contaminants and contamination problems.[5] It includes (1) measures that prevent or exclude the use of various contaminant-emitting materials, furnishings, equipment, etc. in the building environment; (2) elements of building design and maintenance that prevent or minimize air contamination; (3) the treatment or modification of sources either directly or indirectly to reduce emissions; (4) physical removal of the source or source materials and replacement with materials or equipment with no or minimal emissions; (5) measures that prevent the amplification and entrainment of biological contaminants in indoor air; and (6) removal of particulate dusts from surfaces by cleaning.

Though source control can be described as a mitigation measure in a generic sense, it is in most instances case specific. Its application depends both on the contaminants and their sources.

CHEMICAL CONTAMINANTS

Prevention/Avoidance

It is, of course, more desirable to prevent a health-affecting indoor air contamination problem from occurring in the first place than to attempt to control it once it has manifested itself in the form of building-related health complaints. Prevention requires foresight and a knowledge of materials, furnishings, office equipment and supplies, and building design factors that may contribute to or minimize indoor contamination problems. Unfortunately, such foresight and knowledge among professionals who design buildings and their systems, select interior furnishings, office equipment and supplies, and operate and maintain building systems is very limited. At our present state of knowledge, it would, with a few exceptions, be difficult for IAQ professionals to make specific recommendations to architects, interior designers, and building managers as to what specific products to avoid to minimize air contamination problems and health complaints.

Emission/Source Characterization

A large variety of products have been evaluated for their vapor-phase emission characteristics in the past decade. These have included the first-of-its-kind study of Molhave on buildings and furnishings[6] and emission studies of carpets,[7,8] adhesives,[9] plastic floor and wall coverings,[10-15] and paints and finishes.[16-18] They have also included the multi-product studies of USEPA,[19-21] the building materials studies of Levin[22] in the U.S., Wolkoff in Denmark,[23] and Tirkkonen and colleagues in Finland.[24] These are in addition to the studies of formaldehyde-emitting products by Pickrell et al.[25] and Matthews et al.[26,27]

Despite the fact that significant source characterization efforts have been underway in Europe and the United States, and the availability of evidence from human and animal exposure studies that a variety of building materials/ furnishings emit VOCs which appear to cause sensory effects, efforts to identify problem products for purposes of recommending their avoidance in building design, construction, and furnishing to this date have had limited practical application in avoiding indoor contamination problems and occupant health complaints. This has been due in part to (1) the relatively limited database available on products emissions; (2) questions about whether chamber data can be reliably extended to larger, more complex indoor environments;

(3) questions about the significance of individual chemical species relative to total volatile organic compound (TVOCs) concentrations; and (4) the lack of toxicological/health effects information which would better facilitate the use of product emission characterization data. In addition to these technical problems, there is little knowledge among architects, interior designers, and building managers about the nature of IAQ problems and the need for them to actively become involved in designing and operating buildings for the purpose of attaining and maintaining good air quality.

Bioassays

Mouse bioassays and human exposure studies have been used to evaluate the toxicological significance of emissions from a variety of materials such as carpeting, floor tile, etc.[28-31] In the United States, the use of mouse bioassays and reported extreme results of some of these (test animals dying on exposure) has received considerable attention in the electronic and print media. It is not surprising then that the use of such bioassays, the methodologies employed, and the interpretation of results have been the subject of considerable controversy. Nevertheless, the use of bioassays has the prospect of being an important new tool in the evaluation of materials for their potential in contributing to both building contamination and occupant complaints.

"Bad" Products

It would be desirable for simplicity's sake to be able to test materials and to rate them relative to their potential IAQ hazards. Given the uncertainties involved, it would be very difficult to compile a list of "good" and "bad" products. Because of such uncertainties, it is improbable that any authority (at least in the United States), be it government or otherwise, would "go out on a limb", so to speak, and recommend that specific "bad" products be avoided. A precedent of reticence in making recommendations relative to formaldehyde-emitting products in the United States has already occurred. Despite strong epidemiological evidence to implicate formaldehyde as a cause of building-related symptoms among occupants of houses and other buildings[32-36] as well as asthma in children,[37] the USEPA has avoided any recommendation in its publications or policies that building products which are known to emit significant quantities of formaldehyde and produce health-affecting indoor levels be avoided. In addition, the ASHRAE Standards Project Committee 62-1989, in response to an alleged threat of litigation by the formaldehyde industry, deleted its rather liberal indoor air quality guideline of 0.10 ppm[38] from its 1989 standards.[1]

Prospective USEPA Policies

Tucker,[39] former head of USEPA's indoor air research program, has proposed policy initiatives which would, if implemented, establish a significant database for building designers to draw from and provide manufacturers with an incentive to improve products voluntarily. Proposed requirements on manufacturers and suppliers of building products which may cause indoor air contamination are summarized in Table 9.1. Based on data obtained from manufacturers and suppliers, USEPA would model indoor concentrations and employee inhalation exposures as a function of time for the building design conditions and product applications being considered.

Tucker,[39] considering the complexities involved in the acceptability of both product emissions and indoor concentrations, has attempted to classify low-emitting products (and, therefore, presumably acceptable materials/products) based on what he describes as engineering judgment. This classification (Table 9.2) is based on the studies of Molhave[40] which suggest that the threshold for mucous membrane irritation for TVOCs is in the range of 0.16–5 mg/m^3. The maximum acceptable indoor TVOC concentration from any source would be 0.5mg/m^3.

The policy initiatives described above[39] have not been formerly proposed under USEPA's rule-making authority. If USEPA should attempt to implement such policies, it would involve an arduous process of official proposal, commentary from affected parties, and then subsequent rules promulgation. It would likely be opposed by many small product manufacturers on the basis that the costs of such testing would be onerous and threatening of the proprietary nature of the product. Nevertheless, as a practical matter, it makes sense to require product manufacturers or suppliers to characterize emissions from their products, to provide a database for product selection by building and interior designers, and to identify potential problem contaminants.

Washington State Initiative

Although USEPA policy is at best prospective with an uncertain future, the government of the state of Washington has embarked on a program to improve air quality in new or renovated state office buildings. In their "Healthy Buildings" program, manufacturers of materials, furnishings, and finishes are required to provide emission testing information with their bids to ensure compliance with IAQ specifications and emission profile data detailing how product emissions will perform over time.[41,42]

The designer/builder of a state office building must develop and implement an indoor source control plan and assure that the minimum emission rate standards set for interior construction materials, finishes, and furnishings are met. Emission rate standards apply to formaldehyde, TVOCs, 4-PC, particles, regulated pollutants, and pollutants for which a threshold limit value (TLV) has been specified. For products emitting formaldehyde, the emission rate (mg HCHO/m$^2 \cdot$ h) is to be sufficiently low that indoor concentrations do not exceed 0.05 ppm (61 µg/m^3); for TVOCs, product emission rates are not to exceed 0.5 mg/m^3; for 4-PC, 1 ppb; and for particles, 50 µg/m^3. These are to be met at the anticipated loading conditions (m^2/m^3) within 30 days of installation. In addition,

Table 9.1 Proposed manufacturer/supplier requirements of products which may cause indoor air contamination.

1. Conduct emission rate testing of
   - Coatings such as paints, varnishes, waxes
   - Vinyl and fabric floor and wall covering
   - Adhesives
   - Furniture and furnishings with pressed wood or fabrics
   - Ductwork materials
   - Office equipment and supplies
   - Building maintenance materials
2. Provide material safety data sheets for chemicals used in product manufacture
3. Provide testing data that documents emission factors
   - For the five major organic compounds emitted
   - Any specified product compound that is toxic or irritating at an air concentration of 5 mg/m^3 or less
   - At three ages of the product
   - Of office machines
4. Provide documentation of
   - Chamber testing conditions
   - Product storage and handling procedures

From Tucker, W.G., 1990. *Proceedings Fifth International Conference on Indoor Air Quality and Climate.* Vol. 3. Toronto.

Table 9.2 Classification of low-emitting materials and products, TVOCs, and O_3.

Material/product	Maximum emission rate[a]
Flooring materials	0.6 $mg/h/m^2$
Floor coatings	0.6 $mg/h/m^2$[b]
Wall coverings	0.4 $mg/h/m^2$
Wall coatings	0.4 $mg/h/m^2$[b]
Movable partitions	0.4 $mg/h/m^2$
Office furniture	2.5 mg/h per workstation
Office machines (central)	0.25 mg/h per m^3 office space
O_3 emissions	0.01 mg/h per m^3 office space
Office machines (personal)	2.5 mg/h per workstation
O_3 emissions	0.1 mg/h per workstation

[a] Assumptions: ventilation rate = 0.5 ACH outdoor air; indoor air well mixed; maximum prudent increment of TVOCs from any single source is 0.5 mg/m^3; maximum prudent increment of O_3 is 0.02 mg/m^3 (0.01 ppm); and volume of concern for dispersion is 10 m^3.

[b] Within several hours.

From Tucker, W.G., 1990. *Proceedings Fifth International Conference on Indoor Air Quality and Climate.* Vol. 3. Toronto.

any substance regulated as a primary or secondary ambient air pollutant must meet an emission standard that will not exceed USEPA ambient air quality standards and one tenth the TLV of other substances of concern.

Another significant specification is the identification, quantification, and disclosure of any substances listed in the International Agency for Research on Cancer as category 1, 2, 2A, and 2B carcinogens, on the cancer list of the National Toxicology Program, and reproductive toxins listed in the latest edition of the catalog of Teratogenic Agents. Other specifications include the use of the least feasible amount of wet materials (e.g., paints, caulks, adhesives, sealants, glazes, etc.) and the installation of dry materials (carpeting, furnishings, etc.) only after wet materials have been allowed to "dry". All dry products are to be "aired-out" or pre-conditioned prior to installation. The rationale for the installation of "wet" materials first, followed by drying, is to minimize sink effects which can significantly increase the time period that elevated VOC levels may occur in building spaces.[43]

The State of California's Indoor Air Quality Program[44] has recently drafted voluntary guidelines for the reduction of exposure to VOCs from newly constructed or remodeled office buildings. These guidelines include recommendations for design, construction, commissioning, and operation during initial building occupancy. Source control recommendations include (1) selecting building materials to avoid unnecessarily strong VOC emitters, (2) isolating construction zones in partially or fully occupied buildings, (3) scheduling construction or installing furnishings to minimize build-up of elevated contaminant levels before occupancy, (4) installing adsorptive materials after applying "wet" materials, and (5) using low-emitting maintenance and housekeeping materials. The California guidelines also suggest that building bakeouts (see section, "Building Bake-Out", in this chapter) be considered prior to occupancy or reoccupancy.

Private Initiatives

Levin[45,46] has been at the forefront of developing and recommending control measures which minimize emissions and reduce exposures to contaminants from materials used in building construction and furnishings. His private sector architectural approach focuses on careful planning; specifications; the selection, modification, and treatment of products; special installation procedures; and ventilation system operation.

Conceding that there is insufficient knowledge regarding the health and irritation effects of exposures to very low concentrations of most indoor air contaminants and the difficulty of interpreting the effects of exposures to low levels of a large number of different VOCs, Levin,[45] nevertheless, has proposed that a careful review of available data concerning emissions and health effects and judicious use of ventilation can effectively reduce occupant exposures to irritants/toxins from building materials, products, and furnishings.

Levin[45] places major emphasis on the evaluation of building materials as sources of indoor contaminants. He describes elements of the materials selection process as identifying and screening target products, emission testing, evaluations, and recommendations.

Since hundreds of products are used in most building projects, target products are identified on the basis of their potential for exposing occupants to harmful contaminants. Such exposures are suggested to be a function of product emission characteristics and the quantity and nature of materials used. In identifying target products, consideration is given to the overall building design, the anticipated use of the space, the material and products to be selected, and the quantities and applications contemplated for each major product.

In selecting target products, considerable emphasis is given to those materials which have very large surface areas such as textiles, fabrics, and insulation materials. Screening of these products is conducted by evaluation of their quantity and distribution in the building, their chemical composition, the stability of contaminants of concern, and the toxic/irritancy potential of major chemical substances. Materials such as floor coverings, ceiling tiles, office workstation surfaces (desk tops), and interior workstation partitions are considered to be potentially significant contaminant sources because of their high surface area. Using floor areas as a reference, the relative surface area for different floor coverings may vary from some fraction to 100%; ceiling tiles that serve as a base for return air plenums, 200%; workstation furniture, 15–35%; and interior workstation partitions, 200–300%.

Potential emissions from target products are assessed from published general and specific information on building products and materials, information provided from building interior designers, and from material safety data sheets (MSDSs) provided by the manufacturer or supplier. Chemical stability is determined from reviewing vapor pressure and molecular weight data for chemicals of concern. The toxicity/irritation potential of product emissions is evaluated by using standard reference sources such as NIOSH's Registry of Toxic Effects of Chemical Substances[47] and Sax's Dangerous Properties of Industrial Materials.[48]

Emission testing of selected materials is deemed essential for the determination of chemical content, emission rates, and changes in emissions due to environmental conditions. The burden of emission testing is expected to be borne by manufacturers/suppliers or others since it would be impractical for the building designer to do so.

A variety of factors are reviewed and evaluated. These include emission test results and recommendations for materials selection, modifications, or handling to control indoor contamination. It also involves negotiation with product manufacturers, suppliers, and installers to modify products, their installation, and use. Levin[45] suggests that the most effective way of obtaining "clean products" is to place as much responsibility as possible on manufacturers and suppliers to control emissions and provide data. Certain materials are singled out as more important sources than others. These include carpets, vinyl flooring products, adhesives, caulks, sealants, paints, insulation, and office workstations. Model "guide" specifications for carpeting which can be adapted for other building products are given in Table 9.3.

In addition to specifications for low emission products, Levin[45] recommends the use of temporary special ventilation during and immediately following carpet installation, adhesive application, installation of vinyl composition tile and other floor coverings, caulks, sealants, paints, and workstation panels. Temporary exhaust ventilation through doors, operable windows, stair towers,

Table 9.3 Model specifications for carpeting to be used in new building construction or renovation.

1. Carpets shall be designed, manufactured, handled, installed and maintained in a manner which will produce the least harmful effects on occupants of the building.
2. The manufacturer shall avoid unnecessary use of chemicals that are toxic or irritating to humans in the manufacture, treatment, or handling of the product.
3. The manufacturer shall implement measures to reduce, as much as possible, installed product emissions that are toxic or irritating to humans.
4. The manufacturer shall provide a specification for the installation of the product with the least quantity of adhesive necessary to satisfactorily maintain the performance of the product.
5. Manufacturers shall submit for review by the owners and their agents the following:
    - A list of chemicals used in the manufacture of the product including a breakdown of the contents by weight, volume, or both.
    - A description of any procedures used by the manufacturer to minimize emissions of VOCs from the product.
    - A description of all testing performed by the manufacturer, its agents, contractors, or anyone that provides information on the chemical composition of the finished product; emission rates of VOCs; listing of all chemicals found in emission testing; description of test methods used, history, and conditioning of samples prior to testing, etc.

From Levin, H., 1989. *Problem Buildings: Building-Associated Illness and the Sick Building Syndrome. Occupational Medicine: State of the Art Reviews.* Philadelphia. Hanley and Belfus, Inc. With permission.

and emergency exits can reduce the potential for the HVAC system to act as a sink for VOCs. Installed materials are to be protected with sealed plastic coverings to the extent feasible during the application of VOC-containing finishing products such as adhesives and paints. Protection is also to be provided to fiber-lined HVAC ducts and return-air plenums to avoid contamination of system components.

TVOCs and Source Control

The use of source control principles in designing and constructing "Healthy Buildings" has focused, for the most part, on limiting VOC emissions from building products and furnishings. This focus is based on the assumption that reduction in TVOCs will result in fewer occupant complaints of poor IAQ in new or newly renovated buildings. It is based in good measure on the human exposure studies of Molhave and his co-workers who observed significant irritant and apparent neurotoxic effects of multiple VOC exposures in the range of 3–25 mg/m^3 TVOC.[40]

In the absence of other studies which would link building-related health complaints to specific chemical exposures, it is not unreasonable to attempt to reduce TVOCs from what would be relatively elevated levels in new construction or renovations to avoid exposures which may result in sick building-type complaints. At the same time, it does require "a leap of faith" to believe that TVOC reductions will, in and of themselves, be effective in preventing or minimizing IAQ problems in many cases.

Results of epidemiological studies have been mixed relative to symptom prevalence rates and TVOC concentrations. The building materials study of Oie et al.[49] also suggests that TVOC emissions may not be a good indicator of the potential effect of product emissions on air quality. They failed to observe any relationship between TVOC product emissions and human sensory responses (in decipols). The mouse bioassay studies of Anderson[31] of carpet emissions also suggest that TVOC emissions may not be a good indicator of the effects of carpeting on indoor air quality. Despite studies which show rapid decreases in TVOC emissions from carpeting with time[8,9] and efforts by carpet and rug manufacturers to reduce TVOC emissions from their products, Anderson's studies[31] suggest that some component of carpeting emissions produces significant irritant and neurotoxic effects in animal exposures. However, as indicated in Chapter 4, USEPA has reported that it has failed to replicate Anderson's carpet studies.[50]

While source control of TVOC emissions is intuitively desirable and is based on sound scientific theory supported by well-executed research studies, it is plausible, and in fact likely, that specific contaminants or contaminant mixtures may be responsible for sick-building complaints as well, contaminants whose levels are mostly unaffected by efforts to reduce product TVOC emissions. These would, for example, include dust or other particulate-phase contaminants.

Product Labeling

Given the inherent problems associated with identifying and regulating "bad" products, it may be desirable to use a labeling system for building products which would give manufacturers an incentive to produce low-emitting and safer products. Such a labeling concept has been used voluntarily in the United States by carpet manufacturers relative to TVOC emissions. This effort has been very controversial since the so-called "green label" implied that a carpet product was safe without any toxicological verification, and reports of controversial bioassay studies which indicated that low TVOC emission carpeting produced both irritation and neurotoxic effects in test animals.[31,50]

The concept of product labeling has given serious attention by the Danish Ministry of Housing. Danish scientists[51] have made considerable progress in developing the scientific background for a system of labeling building products according to their impact on IAQ. The system developed to date combines chemical emission testing over time, modeling, and health evaluation. The principle objective is to determine the time in months required to reach an acceptable indoor air concentration for odor and/or mucous membrane irritation. This time value is then used to rank the various products evaluated for their potential impact on health and IAQ.

The Danish program for developing "healthy building materials" begins with a qualitative headspace analysis of emitted VOCs followed by quantitative measurements of single volatile compounds under controlled laboratory chamber conditions. Emitted compounds are selected for further evaluation based on their odor detection thresholds, thresholds for mucous membrane irritation and TLVs. Dominant VOCs having the lowest odor thresholds and irritative VOCs both singly and additively are used to determine the time $t(C_m)$ to reach acceptable values (C_m).

A model is used to determine $t(C_m)$ under standard conditions if the emission profile of a product is above one or more acceptable concentrations. Model assumptions include (1) standard physical and environmental room conditions, (2) odors and irritants are important, (3) all materials are allowed an equal emission per square meter, (4) the maximum contribution from building materials based on odor and mucous membrane irritation thresholds is set at 50%, (5) emission factors are assumed to be the same in test chambers and in the standard room provided the emission process is independent of exchange rate, and (6) models are valid for long periods (months) as well.

Dominant and persisting VOCs having low C_m values based on odor and irritation are selected for the determination of $t(C_m)$. Ranking is established using the criterion that materials reaching their C_m based on odor or irritation for all potential VOCs within the shortest time based on relevant threshold values are the best materials. Fifty percent of the odor dilution threshold was used by Wolkoff and Nielsen[51] for C_m values for odor acceptability. A C_m value 5% of the odor detection threshold was recommended to account for individual differences and hyperadditive effects in perception. This would provide a safety factor of ten. Acceptable C_m values for irritation effects are determined as 50% of total summed irritation thresholds (IT) which are determined from mouse bioassays ($IT = RD_{50} \times 0.03$) and TLVs. In the latter case, TLVs are divided by 40, which accounts for 24 hours of exposure per day (as compared to 8 hours/day 5 days per week) and a safety factor of ten.

Wolkoff and Nielsen[51] evaluated samples from three carpets. They calculated that $t(C_m)$ based on odor acceptability of 4-phenylcychohexene (4-PC) would have been approximately 98, 61, and 94 months. For other compounds such as 2-ethylhexanol, less than 1 month; for benzyl alcohol for one carpet type, approximately 3 months and less than 1 month for another; for acetic acid in one carpet type, approximately 6 months.

Design Considerations

The avoidance of "bad products" is best accomplished during the design stage of new building construction or building renovation.[45] Avoiding potential problem products is but one of many design considerations which can potentially prevent IAQ complaint problems. Other design factors include site planning/design, a variety of overall architectural considerations, and ventilation/climate control.[52] Outdoor sources can be evaluated prior to site acquisition and project planning to avoid ambient pollution problems which may exist nearby. Optimally, a new building would be located on a site distant from ambient sources with parking lots located, and vehicle traffic directed, to minimize pollutant concentrations at the building edge, especially at outdoor air intakes and their openings.[52]

Other design features recommended by Levin[52] include (1) placing motor vehicle access to garages, loading docks, and pedestrian drop-off points away from air intakes and building entries and locating openings through which contaminants might enter distant and downwind from pollutant sources outside the building and building exhausts; (2) providing operable windows for emergency ventilation and psychological benefits to occupants; and (3) locating pollutant-generating activities such as printing, food preparation, tobacco smoking, etc. in areas where air flow is controlled to avoid transferring contaminants to adjacent spaces or to recirculating air. The selection of materials that minimize the adsorption of dirt and moisture (by eliminating or reducing the use of fibrous insulation inside duct work), providing improved drainage from drip pans of fan coil units, and using steam rather than cool mist humidification are recommended to reduce potential microbial amplification within HVAC systems.[52]

Implementation of Design Criteria

A variety of attempts have been made to implement innovative design criteria for the purpose of achieving good-superior air quality and comfort in new office buildings. Turner et al.[53] have reported their experiences in designing and constructing a 25,000 ft^2 (2322 m^2) corporate headquarters building. They attempted a degree of source control in addition to features designed to ensure the adequacy and effectiveness of general ventilation and air filtration. Interior finish materials with high volatility solvents for rapid emission were chosen as well as low-emission building furnishings. As an example of the latter, workstation dividers were finished with sheet rock as opposed to conventional office divider partitions. This was expected to eliminate long-term emission of VOCs from fabric finishes and internal sound-absorbing materials.

A number of material specifications were included in the design of another "Healthy Building".[54] These included the use of (1) paints which putatively did not emit VOCs, water-based clear finishes on natural wood, and wood-staining off-site; (2) low VOC, short-piled commercial carpet; (3) low-release, no odor adhesives and non-formaldehyde-bonded plywoods; (4) double-walled air handling units with no exposed duct liner, filter housing, and VAV boxes; and (5) monolithic flooring in areas expected to get wet.

Architects designing the main San Francisco library building[55] included a number of specifications to minimize the contamination of indoor air by a variety of products. Specifications included (1) the prohibition of the use of particleboard in construction and furniture, (2) the use of mineral floor surfaces for entry level floors to reduce deposit of particulate matter on carpeting, (3) selection of carpeting with no detectable levels of 4-PC and carpet airing before installation, (4) the use of "low-emitting" solvent-free carpet adhesive, (5) the use of low VOC-emitting paints which contain no preservatives or antifreeze, and (6) furniture with low VOC-emitting adhesives and finishes and "airing out" prior to delivery.

The Danish Technological Institute (DTI) has collaborated with architects and the building committee of a large insurance company in the construction of a new 65,000 m^2 headquarters building.[56] A variety of design features which focused on source control were recommended by DTI staff to minimize IAQ problems on occupancy. These are summarized in Table 9.4. Levels of a number of target contaminants such as dust, formaldehyde, methylchloroform, toluene, etc. were reported to be low in the occupied building, and a questionnaire survey of building occupants was reported to show "great satisfaction" with IAQ among building occupants.

Source Removal/Modification/Treatment

Avoidance is a control measure that attempts to prevent contamination problems from occurring in new buildings or buildings undergoing renovation. In many cases, problem buildings will have been constructed at some time in the past with air quality only a recent issue or concern. Mitigation measures can be recommended to remediate an air quality problem if causal factors and sources have been identified. Applicable control measures may include source removal/replacement, treatment, or modification in some way.

Source Removal

In theory, the most effective mitigation measure would be to remove the source and replace it with less-polluting material or equipment. In the former case, materials which emit significant levels of formaldehyde (such as pressed wood furnishings) can be replaced with materials that do not emit formaldehyde or emit at very low levels. This would include replacing pressed wood desks with steel ones, particleboard or MDF workstation surfaces with steel or very low-emission wood materials such as softwood plywood or oriented-strand board. The major limitation on using source removal and replacement in such cases is cost. In many office environments, the removal and replacement of furniture and workstations could cost tens of thousands of dollars (U.S.). This would also be true if carpeting were replaced with vinyl floor covering. Removal/replacement in such circumstances would be considered by building management to be excessively costly.

In the case of office equipment, removal of wet-process photocopiers which liberate substantial quantities of VOCs and replacement with electrostatic devices would be a desirable mitigation measure. Significant considerations in implementing removal/replacement in such cases would be size and number of wet-process photocopiers present and whether they were owned or rented.

Replacement could be used, in theory, to reduce health complaints associated with carbonless copy paper (CCP). Its effectiveness would depend on

Table 9.4 Source control design features.

- Using pre-fabricated units factory-baked to reduce emissions from adhesives and paints.
- On-site emission reduction from surface treatment agents such as adhesives and joint fillers by forced ventilation, room or building bake-out.
- Conducting finishing processes in specially adapted, ventilated rooms.
- Limiting the use of exposed fibrous products unless stability and durability of binders is documented.
- Using movable walls based on skeleton of steel laths covered with factory-painted or baked-out fiber-gypsum plates.
- Placing shelves on partition walls in pre-made inserts to avoid dust emission.
- Using mineral wool ceiling materials sealed by surface treatments documented to be emission free.
- Filling floors with cement-based product and covering with linoleum.
- Using factory pre-painted doors.
- Using furniture with low emission characteristics.

From Frederiksen, E., 1990. *Proceedings Fifth International Conference Indoor Air Quality and Climate.* Vol. 3. Toronto.

the identification of CCP which does not cause work-related health problems or the use of other types of copy paper (e.g., carbon paper) for the tasks involved in work being conducted. The handling of CCP could not be avoided entirely since it would be difficult to control exposure to CCP which originates from outside the building as a result of normal business transactions.

Only a few cases have been reported in the literature where source removal/replacement has been used to mitigate IAQ problems. Hijazi et al.[57] identified plastic panels used as office dividers as a significant source of a potent lachrymator, dimethyl acetamide, in an investigation of a Connecticut office building. Panel removal was observed to be accompanied by a reduction, but not complete elimination, of occupant complaints. In another study,[58] an office building in Oregon was subject to an odor problem associated with polyvinyl chloride-backed modular carpeting. Significant quantities of a 2-ethyl-1-hexanol and heptanol were emitted as a result of the hydrolytic breakdown of phthalate plasticizers. The problem was mitigated by removing the carpeting, its adhesive, as well as allowing a 9-month period for both plasticizers and alcohols sorbed in floor concrete to be released.

Norback and Torgen[59] attempted to mitigate illness complaints in two Swedish schools by removing wall-to-wall carpets. Overall symptom frequencies decreased significantly after carpet removal, as compared to a reference group. Symptom frequency nevertheless remained relatively high, notably symptoms of the respiratory airways.

Source Treatment

Source removal/replacement may in a number of cases be deemed to be too costly and impractical. Its perceived impracticality may be related to a number of factors. In addition to cost, these may include unacceptable interruption of work activities, uncertainties related to suitable replacement materials, many and diffuse sources, and uncertainties involved in removing sources based on the recommendations of a single individual who believes but does not "actually" know that they are the cause of the problem. In such instances, it may be desirable to treat or somehow modify the source.

The efficacy of source control has been reported for only limited uses of pressed-wood products emitting formaldehyde in residential applications. These treatments include the use of coatings over raw particleboard used as subflooring. Effectiveness of such treatment measures have varied from 43% for two applications of polyurethane, 53% for alkyd resin and 70% for nitrocellulose-based varnishes, and 17–87% for specially formulated formaldehyde-scavenging coatings which are no longer available on the U.S. market.[60]

Several European investigations[61,62] have evaluated the efficacy of surface treatments and barriers to formaldehyde emissions from pressed-wood products under laboratory chamber conditions. Effectiveness varied depending on materials used. Efficacy under real-world conditions was not evaluated.

Based on these studies, it appears that source treatment (or use of surface barriers) has the potential for effectively reducing formaldehyde emissions both in laboratory chamber studies and, in some cases, under real-world building conditions. Such measures have not been evaluated for more complex formaldehyde-emitting sources such as workstations, desks, tables, counter tops, storage cabinets, etc. In addition to raw surfaces of particleboard and MDF, some of these furnishings contain numerous joints that may not be accessible to treatment. Wood furniture is in many cases finished with acid-cured coatings which are themselves significant sources of formaldehyde. As such, there is a considerable degree of uncertainty as to how effective source treatment measures would be in mitigating a formaldehyde problem in a building.

Building Bake-Out

A unique application of source treatment is a procedure called "building bake-out". Its purpose is to accelerate the emission of VOCs from the many and diffuse sources found in newly constructed or renovated buildings. Bake-outs are based on the theory that the normal reduction of VOC emissions from building products with time can be accelerated by elevating building temperatures for several days. Elevated temperatures increase vapor pressures of residual product solvents which are expected to significantly increase emission rates and thus reduce their future emission potential.

The effectiveness of a building bake-out appears to be determined by several factors.[63] These include elevated building air temperatures in the range of 30–35°C, bake-out duration, and ventilation conditions both during and after the bake-out period. It is desirable to optimize all three factors to achieve the maximum degree of VOC reduction possible in a particular building environment.

A number of studies have been conducted to evaluate the effectiveness of building bake-out as a VOC mitigation measure. An evaluation of these studies and their results indicates that conducting building bake-outs is a difficult task and that their effectiveness in actual application may fall below expectations.

Complicating the task of conducting a bake-out is the difficulty of bringing the building into the desired temperature range and maintaining optimal bake-out temperatures for a sufficient duration.[64,65] Constraints include the inability of some HVAC systems to attain desired temperatures without the use of supplemental heating systems, concerns that elevated temperatures may result in damage to materials and even the building structure, and the problem of providing sufficient ventilation to flush emitted VOCs (so that they are not readsorbed by building materials) from the building environment.[64] The use of outdoor air for ventilation often makes it very difficult to achieve and maintain optimal bake-out temperatures. Investigators have in some cases chosen to limit ventilation during the bake-out period to achieve desired building temperatures, reasoning that the latter is of greater importance in reducing VOC levels in the building. Because of the thermal mass involved, it typically requires 2[66] to 3[67] days to reach desired bake-out temperatures. As a consequence, a bake-out should last at least 48 hours.

Achievement of desired bake-out temperatures and further maintaining them for at least another 48 hours is constrained by the availability of the facility for a sufficient duration to optimize bake-out conditions. The time period available to adequately conduct a bake-out is often limited because of on-going construction activities and the desire of building managers to occupy spaces as soon as possible. In multiple-story building projects, construction activity varies from floor to floor with some floors completed and occupied, while others are in various stages of construction. Bake-outs may, therefore, be conducted on a floor-by-floor basis. Construction activities on other floors often significantly complicate the process. Such problems make it very difficult to standardize the various factors necessary to conduct and evaluate the effectiveness of the bake-out procedure.

Several investigators have attempted to determine the efficacy of bake-outs in reducing both VOCs and formaldehyde levels in a variety of buildings. Girman et al.[64] conducted a bake-out in a newly renovated San Francisco office building. Bake-out conditions included building temperatures between 32 and 39°C maintained for a 24-hour period at an outdoor air exchange rate of 1.59 ACH. A modest decrease of 29% in VOC levels was observed. Girman et al.[65] reported a reduction of 15–20 target VOCs in the range of 20–30% with TVOC concentration reductions of 13 to 95% in a bake-out evaluation of five office buildings. Bake-out conditions included temperatures in the range of 29–39°C, durations of 1 to 4 days, and outdoor air exchange rates of 0.5–1.5 ACH. In general, TVOC reductions were associated with increased bake-out intensity (temperature and duration) and initial TVOC concentration. The most effective bake-out occurred in the building with the highest initial TVOC levels (13 mg/m^3, 3.5 ppm).

In another study, Lindstrom et al.[66] baked-out 4 floors of a 21-story office building which was in various stages of construction. Bake-out conditions included a temperature range of 30–34°C for a period of 2–4 days at an outside air exchange rate of 0.60 ACH. TVOC source strengths decreased by 67% on the two floors which had the highest initial TVOC concentrations and 38% on those with lower initial concentrations. This study was characterized by a long period of sustained temperatures (circa 60 hours) and 60 hours of maximum outdoor air exchange rates before post-bake-out sampling. These conditions were suggested to be responsible for the relatively successful reduction in TVOC source strengths of 38–76% from pre-bake-out levels.

Hicks et al.[68] conducted a bake-out on a new five-story wing of a hospital building. Bake-out temperatures varied from 32–43°C (with a peak of 34–43°C) over a period of 72 hours at an air exchange rate of approximately 0.20 ACH. Results were mixed, due in considerable measure to the confounding occurrence of outdoor VOC levels that were considerably higher in the pre-bake-out measurement period. Post-bake-out TVOCs (compared with pre-bake-out values) were higher on three floors and lower on two floors. Consistent declines were only observed for dioxane and formaldehyde. The significant decline in formaldehyde levels was anomalous since no changes in formaldehyde levels were reported by Girman et al.[64,65] and Lindstrom et al.[66] and because of the reservoir effect of urea-formaldehyde-based products, significant reductions in formaldehyde levels due to bake-out would not be expected.[56]

Offermann et al.[69] conducted a bake-out in a recently constructed, furnished and occupied two-story California office building. Though a building temperature of 34°C was attained, elevated temperatures could not be maintained over the 3-day bake-out period. A significant decline in VOC levels was observed in the week following the bake-out with emissions of some of the more volatile compounds such as 2-proponone, 2-propanol, and methyl chloroform decreasing by more than a factor of 20. By the second week, however, VOC emissions returned to pre-bake-out levels.

Bayer[70] conducted bake-outs of office partitions and particleboard at controlled chamber conditions of 50°C and an air exchange rate of 0.5 ACH for a duration of 3 to 4 days. No reduction in TVOC emissions from these materials was observed.

Results of bake-out studies have been quite mixed and, as a consequence, it is difficult to make any definitive conclusions about the effectiveness of this procedure. The field studies of Hicks et al.[68] and Offermann et al.[69] on single buildings and the laboratory chamber studies of Bayer[70] indicate that bake-out is not effective in reducing VOC emissions from buildings and some building materials and furnishings. On the other hand, the studies of Girman et al.[64,65,67] conducted in six buildings and that of Lindstrom et al.[66] on a single building indicate that bake-outs are effective albeit their effectiveness varies widely. Bake-out studies appear to be most effective[64-67] when initial TVOCs are high and when they are more intense (temperature and duration). Field studies[68,69] indicating that bake-outs were not effective in reducing VOC levels were based on only two buildings and the study of Offermann et al.[69] which had significant problems in maintaining bake-out conditions.

When results of bake-out studies are viewed collectively, it is evident that serious questions remain as to its applicability and efficacy in reducing VOC levels in newly constructed or renovated buildings. In addition, there is no systematic evidence available that demonstrates that it is effective in reducing symptom prevalence rates in buildings. That is not surprising since the bake-out procedure is intended to be used prior to building occupancy. There are, however, several anecdotal reports of decreased odor and occupant dissatisfaction in several occupied buildings where bake-outs have been conducted.[63,64] Despite the uncertainties indicated above, the Indoor Air Quality Program of the California Department of Health recommends that bake-outs be considered a control option for VOCs in newly constructed or renovated buildings.[45]

Source Modification

Source treatment, in a way, can be considered a form of source modification and the reverse is true as well. Here, however, source modification is discussed in the context of measures employed to improve products before they are introduced into buildings or after they are already in place.

Considerable efforts have been expended by wood products manufacturers to reduce formaldehyde emissions from materials manufactured with urea-formaldehyde-based adhesive resins. These improvements have included (1) changes in mole ratios of formaldehyde to urea in resin formulations; (2) changes in material processing factors such as wood chip moisture levels, press temperatures, etc.; (3) the addition of formaldehyde scavengers to resin formulations; and (4) post-pressing fumigation with ammonia.[71-75] Given sufficient regulatory incentive, it is likely that emissions of formaldehyde from such products can be reduced even further.

Due to emission concerns associated with new carpeting, manufacturers have made major efforts to reduce initial TVOC emissions from their products. Similar efforts have been made by manufacturers of carpeting adhesives. Though such efforts have been successful in reducing TVOC levels, studies using mouse bioassays have questioned the safety of even low emission carpet products.[31]

Carbonless copy paper appears also to have undergone modifications as a result of health concerns. Manufacturers of CCP discontinued the use of paratoluene sulfonate of Michler's hydrol because it apparently caused skin irritation.[76] Based on emission measurements made by the author, it appears that some CCP manufacturers have modified their formulations to reduce/ eliminate formaldehyde emissions and exposures. As described in Chapter 4, formaldehyde in CCP may have been a contributing factor to symptoms reportedly associated with handling CCP.

Other contaminant sources can be designed and manufactured so that they produce no or minimal emissions. Braun-Hansen and Anderson,[77] for example, have reported significant differences in O_3 production associated with electrostatic copiers. Increased O_3 production was observed to be associated with increases in voltage and alternating current (AC). Those machines using AC produced O_3 at twice the level of those using positive direct current (DC), with negative DC producing ten times as much O_3 as positive DC. Ozone emissions were affected by the use and effectiveness of filters incorporated into electrostatic copiers for O_3 control. It is evident, therefore, that O_3 emissions can be significantly reduced from electrostatic copy machines (and such devices as laser printers) by designing them to produce low O_3 emissions. Significant reduction in O_3 emissions can also be achieved by proper servicing of existing equipment. Allen et al.[78] observed that O_3 emissions were reduced to minimal levels after copy machines were serviced.

BIOLOGICAL CONTAMINANTS

Source control will, in most cases, be the method of choice in mitigating building-related health problems associated with biological contaminants. This is true for cases of Legionnaires' disease, hypersensitivity pneumonitis/humidifier fever, allergic symptoms due to exposure to dust mite or mold antigens, and possibly to other microbially produced contaminants such as fungal volatiles, mycotoxins, and endotoxins.

Legionnaires' Disease

Legionnaires' disease (LD) has been associated with exposures to aerosols from contaminated cooling towers and evaporative condensers and hot water storage tanks and recirculation lines where elevated temperatures favor the growth of the causative organism *Legionella pneumophila.*

Biocidal Treatments

Biocidal treatments are commonly used in an attempt to control the growth of *L. pneumophila* in cooling tower waters. These have included quaternary ammonium compounds, 1-bromo-3-chloro-5, 5-dimethylhydantoin, *bis* (tri-*n*-butyltin) oxide, *n*-alkyl-1,3-propanediamine, methylene-*bis* (thiocyanate), dithiocarbamates, and chlorine.[79-82]

Though a variety of biocides appear to be effective in controlling *L. pneumophila* under laboratory conditions, they may not be as effective under cooling tower operating conditions. Witherell et al.[83] observed no significant differences between treated and untreated waters relative to the presence and prevalence of *L. pneumophila* in sampling waters of 130 operating cooling towers in Vermont (52% of which received some type of biocidal treatment). Ironically, increased levels of *L. pneumophila* were observed among units treated weekly with quaternary ammonium compounds. Such treatments may have eliminated competing organisms. Studies on a variety of commercially available biocides intended for controlling *L. pneumophila* conducted by Braun[84] showed that dithiocarbamates, bis (tri-*n*-butytin) oxide, and *n*-alkyl dimethylbenzol ammonium chlorides were not effective when applied to waters of cooling towers and evaporative condensers.

Broadbent and Bentham,[85] conducting studies on cooling towers in Australia, observed that biocidal treatments with slug doses of fentichlor (chlorinated phenolic thioether) at relatively high concentrations (e.g., 200 ppm) periodically (usually weekly) or BCD (1 bromo-3-chloro-5,5-dimethylhydantoin) applied continuously to maintain a minimum residual were very effective in reducing *L. pneumophila* to nondetectable levels in previously heavily contaminated waters. Counts were below limits of detection in the first week after application and remained so for 2 weeks after treatment was stopped. A third biocide, BNPD (bromo-nitro-propane-diol), was effective in less than 50% of tower waters treated.

Two factors appear to reduce the efficacy of biocidal treatments of cooling tower waters. These include the ability of *L. pneumophila* to undergo intracellular multiplication when ingested by amoebae and a ciliate, *Tetrahymena* spp.[86,87] The presence of *L. pneumophila* within these organisms may provide protection from the action of biocides. Additionally, field studies have shown that *L. pneumophila* can remain at high concentrations in the sludge which normally forms at the base of cooling towers. Sludge concentrations of *L. pneumophila* may be on the order of 10^7–10^8 organisms/gram and are reported to be much more stable than concentrations found in recirculating waters.[88] The presence of sludges is likely to decrease the effectiveness of most biocidal treatments.

The use of chlorine or chlorine-releasing compounds has been suggested to be the most effective biocidal treatment under field-use conditions with a chlorine residual of 2 ppm under normal cooling tower operating conditions.[89] These results are not consistent with studies which indicate that *L. pneumophila* is relatively resistant to chlorine.[90] *L. pneumophila*, as a consequence, is commonly found in potable water where chlorine residuals of 2 ppm are maintained as standard practice. Because of the relative resistance of *L. pneumophila* to chlorine, high chlorine concentrations (hyperchlorination) are often applied to cooling tower waters as a shock treatment.[91]

Disinfecting cooling tower waters by hyperchlorination to control LD has been recommended by the U.S. Centers for Disease Control.[92] The procedure includes establishing and maintaining a neutral pH with the careful addition of acid, the addition of a hypochlorite solution sufficient to maintain a free chlorine residual of 100–250 ppm (maintaining this residual for 48 hours), reflushing and refilling the system, and, if necessary, maintaining a free chlorine residual of 2–5 ppm until the source of contamination has been completely removed. The application of such disinfection measures as well as the removal and replacement of HVAC system filters in a new San Francisco office building where an epidemic outbreak of LD was reported appeared to result in no new cases over a 12-month period.[92]

The use of ultraviolet light irradiation has been suggested as a potentially effective disinfectant treatment of cooling tower waters. Yamayoshi and Tatsumi[93] observed that 10 hours of irradiation from submerged UV lights (dose of 76.8 joules) decreased *L. pneumophila* concentrations by over 90% to 1–2 CFU/ml under field conditions.

Source Modifications

The potential for LD outbreaks may be reduced by using mist eliminators that limit drift aerosol loss from cooling towers. Miller[94] reported that new and well-maintained systems reduced drift loss to 0.05 to 0.2% of circulating water flow rates compared to older towers which had drift losses of 5–6 times that amount. Newer designs developed in the early 1970s reportedly have only 0.004% of the recirculating water lost as mist.[95] Guidelines for hospitals in the United Kingdom recommend replacing existing evaporative cooling towers with dry towers when the former reach the end of their useful life and avoiding wet towers in new applications.[96] Guidelines for the operation and maintenance of cooling towers have been published by the World Health Organization.[97]

Potable Water Systems

Legionnaires' disease may result from exposures associated with potable water systems. The disease organism survives chlorination and proliferates in the favorable environmental conditions of hot water systems in hospitals, hotels, nursing homes, and likely residences as well. Such systems have been implicated as sources of *L. pneumophila* which has caused LD. Most recent attention has focused on the control of *L. pneumophila* in the water distribution systems of hospitals.

The effectiveness of a control measure for any contaminant depends in good measure on our understanding of the factors which contribute to the contamination problem. Risk factors observed by Vickers et al.[98] included elevated hot water temperature, vertical configuration of hot water tanks, older tanks, and elevated magnesium and calcium water concentrations. Hot water tanks with set point temperatures above 60°C were more likely to be free of *L. pneumophila*; lower temperatures were much more commonly associated with *L. pneumophila* contamination.

The significance of older water tanks and levels of calcium and magnesium appears to be related to scaling or accumulation of mineral deposits. *L. pneumophila* has been shown to concentrate in the scale and sediment of water distribution systems. These deposits apparently serve as a source of nutrients and provide physical shelter.[99]

The application of periodic maintenance procedures by hospitals surveyed by Vickers et al.[98] had no apparent effect on the presence of *L. pneumophila.* Preventative maintenance typically consisted of cleaning or flushing hot water tanks on a weekly to annual basis. In such cases, residual chlorine was maintained at a standard level of 1–2 ppm, a level considered to be inadequate for *L. pneumophila* control.

Effectiveness of Control Measures

Several studies have been conducted to evaluate the efficacy of eradicating *L. pneumophila* from contaminated water distribution systems in buildings. Muraca et al.,[100] studying the outbreak of Legionnaires' disease among workers in a plastics plant, reported that a water distribution system which provided quench water for an injection mold process was contaminated with *L. pneumophila.* Biocidal treatment with a formulation containing 4.3% of 5-chloro-2-methyl-4-isothiazolin-3-one and 1.35% 2-methyl-4-isothiazolin-3-one directed at the cooling water system only decreased positive sites for *L. pneumophila* by 10%. This was followed by an attempt to remove scale from the system by a biofilm dispersant circulated for 24 hours. Positive sites for *L. pneumophila* dropped to 50% after physically cleaning and disinfecting all holding tanks and flushing with fresh water. A more drastic and stringent decontamination regime was then initiated. This included purging the system with an acid scale remover (pH 1.0–2.0) for 6 hours, flushing with fresh water, circulating a caustic neutralizer (pH 12.0–13.0) for 4 hours, flushing with fresh water, and then shock-treating with 100 ppm chlorine for 8 hours. *L. pneumophila*-positive sites were reduced to 7%. Continuous application of the biocide and biofilm dispersant resulted in no detectable *L. pneumophila* in the circulating water system within 2 weeks of the additional biocidal treatment.

Muraca et al.[101] conducted studies of the efficacy of four different *L. pneumophila* eradication measures in a water distribution system under a variety of test conditions. All four procedures (Figure 9.1) were observed to be effective in eradicating *L. pneumophila* from the model plumbing system. This included the use of chlorination (continuous chlorination at 4–6 ppm), heat (50–60°C), ozonation (1–2 ppm), and UV light disinfection (continuous irradiation at 30,000 $\mu Ws/cm^2$). The application of chlorine, O_3, and UV light resulted in 5–6 orders of magnitude decrease in *L. pneumophila* concentrations in a 6-hour period with shorter exposures producing less effective control. Heat disinfection resulted in complete eradication of *L. pneumophila* from the test system.

The use of hyperchlorination[102,103] at rates of 4-6 ppm and heat treatment[104,105] at 50–60°C have reportedly been effective in eradicating *L. pneumophila* in *in situ* systems. The effectiveness of ozonation was inconclusive in one study.[106]

Muraca et al.[107] reviewed the various advantages and disadvantages (including cost) of *L. pneumophila* control systems in institutional environments.

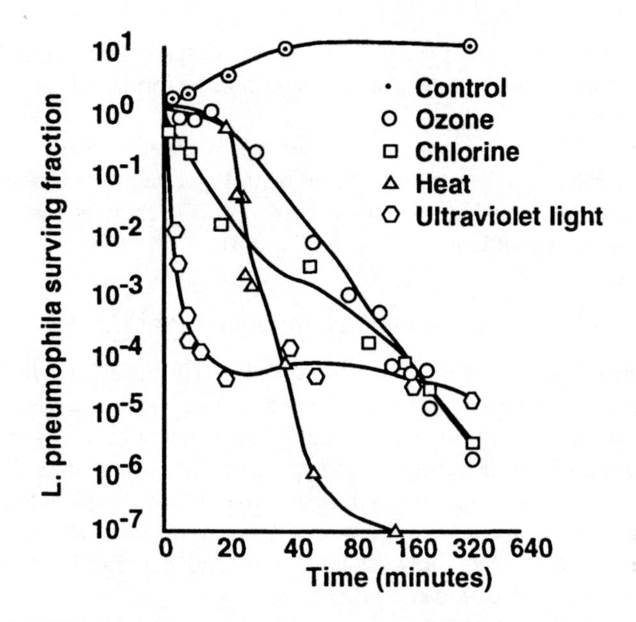

Figure 9.1 Effect of treatment measures on *L. pneumophila* populations in a model plumbing system. (From Muraca, P. et al., 1987. *Appl. Environ. Microbiol.* 53:447–453. With permission.)

Thermal eradication, the so-called "heat and flush" method, was reported to be the most widely used *L. pneumophila* eradication method. The basic technique requires that hot water tank temperatures be elevated to 70°C followed by flushing all faucets and showerheads with hot water. This procedure only provides short-term control. By maintaining hot water tank temperatures at 60°C, the "heat and flush" method is required only once every 2 to 3 years. Costs are relatively low, and no specialized equipment is required. The method is, however, tedious and time-consuming and has the potential for scalding hot water users.

A variation on the heat treatment theme is to use instantaneous steam heating systems. These operate by flash heating water to 88°C, then blending with cool water to achieve the desired temperature. The major advantage of this approach is that hot water storage tanks that serve as a colonization habitat for *L. pneumophila* are not needed. Disadvantages include cost ($15–30,000 U.S.) and the fact that treatment is only limited to the hot water system. Steam heating systems are more efficient, require less space, and need no specially trained staff to operate them.

Hyperchlorination has been successfully used to control *L. pneumophila* in a variety of systems and is widely used as a consequence. Its primary advantage is that it can provide a residual concentration throughout the entire system rather than being limited to a specific area or site. It does have a number of disadvantages which include cost ($30–45,000 U.S. in the first year), difficulty in maintaining stable residual chlorine, the need for qualified personnel to conduct monitoring programs and calibrate and operate residual chlorine analyzers, the potential for system corrosion, and the formation of carcinogenic trihalomethanes.

Once an institution embarks on the disinfection of water distribution systems, it requires a long-term commitment. Program effectiveness depends on both equipment and techniques used and on the continued attention and cooperation of institutional staff.

Hypersensitivity Pneumonitis/Humidifier Fever

Outbreaks of hypersensitivity pneumonitis (HP) in offices, institutional, and industrial buildings have been related to biological aerosols of various forms (thermophilic actinomycetes, mold, Flavobacterium and its endotoxins, thermotolerant bacteria, serotypes of *L. pneumophila*); humidifier fever (HF), with amoebae and endotoxins of Gram negative bacteria. Such outbreaks have been associated with HVAC system condensate drip pans, duct work, humidifiers, air washers, open spray cooling units, and microbially contaminated materials associated with water intrusions.

Remediation Measures

The successful remediation of a problem environment where an outbreak of HP or HF has occurred depends on correct diagnosis and the identification of the principal sources/causes of contamination. There are few guideposts for building investigators in such circumstances. Control measures have, for the most part, been implemented on a "trial and error" or "best judgment" basis without reference to any systematic studies. The actual effectiveness of mitigation measures employed is, for the most part, unknown. The non-reoccurrence of HP episodes after the implementation of certain measures is suggestive of their effectiveness but is not conclusive. Of course, reoccurrence of HP indicates the ineffectiveness of a control measure or at least its unsuccessful application.

Weiss and Soleymani,[108] in one HP outbreak, removed a water spray unit from a HVAC system and replaced it with an air cooled system; symptoms did not reoccur. Banazak et al.[109] removed water systems and steam-cleaned the HVAC system. Aranow et al.[110] replaced the HVAC system as well as all furnishings in the occupied space. Scully et al.[111] attempted to control an HP problem by cleaning duct work. Bernstein et al.[112] attempted to control an HP outbreak apparently caused by mold by cleaning components of the HVAC system with ammonia and chlorine and providing regular maintenance. Despite significant reductions in mold levels, HP symptoms continued to occur. Ganier et al.[113] tried to decontaminate a humidifier without success and, as a consequence, removed it from the HVAC system. Cockroft et al.[114] attempted to mitigate a HF problem by draining and cleaning a contaminated humidifier every 2 weeks. The process was ineffective in preventing the buildup of antigenic materials. The use of three biocidal treatments including a bismethylene chlorophenol compound, a mixture of pixcloxydine, octyl phenoxy polyethanol, and benzalkonium chloride, and Metronidazole (an antiamoebic drug) were ineffective in controlling the amoebic organisms suspected to be the cause of the problem.

Morey et al.[115] conducted detailed investigations in five office buildings where a number of HP cases had been associated with different contamination problems. Remedial recommendations were specific to each case. Microbial contamination of the HVAC system was considered responsible in one building. Recommendations included the cleaning of all nondisposable building contents including books, desks, carpets, drapes, etc.; HEPA (high-efficiency particulate absolute) vacuum cleaning of HVAC system ductwork and water spray direct expansion surfaces; and discarding and replacing materials not amenable to cleaning. The problem in a second building was apparently associated with a flooded cafeteria. Remediation measures included (1) making plumbing changes; (2) removing and disposing of damaged carpeting and ceiling tiles; (3) cleaning all upholstered furniture, wall partitions, and office materials needing to be reused with a HEPA-type vacuum cleaner; and (4) disinfecting the floor with bleach. Despite the implementation of these measures, HP symptoms reoccurred when office partitions were subsequently disturbed. Remedial measures recommended in a third building included (1) providing adequate drainage of cooling coil condensate waters; (2) using deep, sealed, water-filled traps; and (3) cleaning and disinfecting cooling coils and drain pans of fan coil units. The installation of adequately sized access doors to the cooling deck of air handling units (AHUs) was recommended in a fourth building to facilitate a preventative maintenance program of slime and stagnant water removal. Remedial recommendations for a fifth building which had a history of roof leaks and excessive relative humidities included preventing moisture incursions into occupied spaces from drain pan overflows, replacing filters from fan coil and AHUs routinely and frequently, and operating the main AHUs during all occupied times.

Burge[116,117] has suggested that cleaning and maintaining water reservoirs in situations where epidemics of HP or HF are occurring is effective in preventing further symptoms. This includes removal of mineral scale and disinfecting the system with sodium hypochlorite. Water reservoirs must be free of scale at all times and cleaned with biocides when they are not being operated. The use of biocides in humidification or air washing units is not recommended since they can themselves cause air contamination problems.

Avoidance

In an HP outbreak reported by Hodgson,[118] investigators did not feel confident that remediation would solve the problem. For safety reasons, they recommended that the building be abandoned. This is an extreme example of what could be described as avoidance.

The prevention of HP outbreaks in offices, institutional, and industrial buildings requires the implementation of a variety of proactive measures which have the potential for decreasing the likelihood of microbial growth and subsequent contamination of building air handling systems and materials. Recommendations for the avoidance/prevention of HP outbreaks in large mechanically ventilated buildings have been made by Morey et al.[115,119] These include (1) the use of steam rather than recirculating cool water for humidification, (2) use of once-through potable water flow if cool water humidifiers are to be used, (3) periodic inspection and preventative maintenance of water-generating HVAC systems (with particular emphasis on condensate water drip pans), (4) maintenance of building relative humidities below 70%, and (5) maintaining of pumping systems and roofing materials in good condition to reduce the likelihood of water incursions in buildings.

Asthma/Allergic Rhinitis

Although there have been relatively few reported prevalence studies of asthma and allergic rhinitis in office and institutional buildings, a variety of studies have suggested that exposure to microbial organisms and antigens, dust mite antigens, and macromolecular organic dust (MOD) may contribute to allergy or, more rarely, asthmatic-type symptoms.

Microorganisms

It is probable that in HP outbreaks some portion of the exposed population will experience allergy or asthmatic-type symptoms in building environments which have significant microbial contamination. Avoidance and remediation measures recommended by Morey et al.[115,119] and Burge[116,117] would be expected to be of considerable benefit in mitigating exposures to allergic and asthmatic individuals as well.

Because they are more generalized in nature, let us discuss two concerns associated with potential microbial contamination problems and their remediation not previously discussed in the context of HP. These are the use of porous insulation (fiberglass duct liner) and filters used in HVAC systems.

Morey and Williams[120] reported that fibrous glass duct liner materials used in HVAC system components serve as amplification sites for microorganisms which can be released in large quantities when disturbed during routine maintenance activities. Because they trap dust that serve as a microbial medium and cannot be easily cleaned, their use is not recommended, particularly in areas of high humidity (cooling coil compartments) or where air washers are a part of the HVAC system.

Media filters used in HVAC systems have also been reported to amplify microbial populations on accumulated dust[121,122] which penetrate the filter medium. Elixmann et al.[123] report that filters can serve as a source of antigenic material which can pass into building air undetected by ordinary bioaerosol sampling procedures. These study results suggest that frequent changing of HVAC system filters is desirable to reduce potential occupant exposures.

Several case studies of remediation efforts in buildings heavily contaminated by mold have been reported. These case investigations are of considerable interest because of the extreme nature of the mold contamination problems investigators encountered and the intensive remediation efforts that were required to make buildings safe for reoccupancy.

Light et al.[124] reported on an abatement effort in a heavily mold-contaminated library section of a large educational facility. In-house maintenance staff initially implemented a variety of remediation measures associated with the water incursion and mold contamination problem. These included repairing the roof and implementing the following cleaning actions: (1) removing mold stains on walls with bleach cleaner, (2) discarding moldy books, (3) shampooing carpet with a disinfectant, (4) damp-wiping shelves and walls, and (5) installing a dehumidifier. This initial remediation effort failed to mitigate allergy-type symptoms in building staff. A reinspection of the building environment indicated that the initial remediation was only partially effective. A more comprehensive abatement program was then implemented using a commercial cleaning contractor. This more rigorous abatement program included systematic cleaning, disinfection and removal of mold sources over a 7-day period. The HVAC system was turned off, the contaminated area was isolated by plastic barriers and maintained under neutral or negative pressure, and access to the work area was restricted. Over 12,000 books were examined with approximately 1% discarded due to visible mold growth. Others were HEPA-vacuumed. All carpeting was removed and discarded. Intensive cleaning of all surfaces was conducted by means of HEPA vacuuming, wet wiping, and using detergent/disinfectant compounds. Mold concentrations were reduced to background levels, and *Aspergillus niger*, the predominant mold species found in initial sampling of the contaminated building, was completely eliminated.

These intensive remediation efforts were modeled on abatement procedures used for asbestos removal in buildings. They included containment of the abatement area, disposal of all contaminated materials in heavy duty plastic bags, rigorous cleaning, monitoring both inside and outside of the abatement area, and achieving clearance criteria.

Biocontaminant abatement based on engineering and work practices used for asbestos abatement have also been reported for a mold-infested building where the toxigenic fungus *Aspergillus versicolor* was prevalent and in a second building where toxigenic fungi *A. versicolor* and *Stachybotrys atra* dominated. The Hazard Communication Standard was used to inform occupants of potential exposure risks from airborne allergens and mycotoxins.[125] Similar abatement practices were reported for another building where large quantities of *S. atra* were found in an office building after persistent flooding.[126] A special medical surveillance program was also implemented.

The case studies reported above are trend setting in defining good practice in buildings heavily contaminated with mold. They have employed abatement measures which treat potential exposures as being similar to hazardous materials such as asbestos. They are likely to become the standard for such abatements, particularly when toxigenic fungi have been identified as being present in significant concentrations.

Dust Mites

As indicated in Chapter 6, several studies have implicated dust mites as potentially contributing to SBS symptoms. In studies conducted by the British Building Research Establishment,[127,128] intensive dust cleaning measures were observed to simultaneously decrease dust mite populations on some surfaces and the prevalence of a number of SBS symptoms. Similar results[129] were observed when building surfaces were treated with liquid nitrogen which had been previously reported to be an effective mite controlling agent.[130] The case study of Lundblad[131] also indicated that the use of a miticide on office carpeting was effective in mitigating symptoms in two office employees.

SURFACE DUSTS

A number of studies have implicated surface dusts as a risk factor for SBS symptoms. Potential causal factors include dust mite antigens,[127] macromolecular organic dust,[132,133] and fibrous glass.[134] Surface dust levels and potential human exposures can be reduced by a variety of control measures including source removal, sink removal, and surface cleaning.

Source Removal

As with other contaminants, it would be desirable to identify the sources of surface-deposited dust and remove them from the building environment. Since only limited characterization of surface dust has been conducted, it would, in many cases, be difficult to specifically identify components which may be responsible for causing SBS symptoms. Such dusts are likely to be very heterogeneous, indicating diverse origins. Nevertheless, it would be reasonable to anticipate that sources of health-affecting particulate-phase contaminants will be identified in future studies. These may include fibrous glass materials used in ceiling tile and duct liner materials. If such a causal connection were to be made, then a case could be made for removing source materials and replacing them with other materials which would not pose similar or new problems.

Reservoir Removal

A number of studies have implicated carpeting as a reservoir for dust mite antigens and MOD which may cause SBS-type symptoms. The prevalence of SBS symptoms in such cases could, in theory, be reduced by removing the reservoir, wall-to-wall carpeting. Carpet removal has been reported to have resolved complaints of extreme fatigue, malaise, and slight fever in occupants of a problem building with poorly cleaned wall-to-wall carpeting.[135] Studies which have shown significant associations between MOD and SBS symptoms have also observed significantly higher MOD concentrations in carpeting. As such, carpeting appears to serve as a sink for MOD, suggesting that carpet removal would be an appropriate mitigation measure.

Surface Cleaning

Studies conducted at the British Building Research Establishment[127-129] have shown that SBS symptom prevalence rates can be significantly reduced by intensive cleaning of building surfaces including steam cleaning of carpeting. Leinster et al.[127] conducted studies of the effects of a variety of environmental changes in an office building on SBS symptom prevalence rates and acceptability of air quality and environmental conditions. These included two intensive cleaning regimes wherein fabric surfaces such as carpets and chairs were shampooed, hard surfaces wet wiped, and files vacuumed with high-efficiency filters. The two regimes differed in that one used cool shampoo and the other, steam cleaning. Significant reductions in SBS symptoms and decreased environmental discomfort were reported for occupants in building areas that had been intensively surface-cleaned using steam cleaning of carpet and fabric furnishings. In a second study, significant reductions in SBS symptoms were only observed when files were also HEPA-vacuumed.[129] Because intensive surface cleaning and the application of liquid nitrogen to fabric surfaces were both observed to reduce SBS symptom prevalence and dust mite populations, Leinster et al.,[127] Raw et al.,[128] and Roys et al.[129] suggested that reduced exposure to dust mite antigens may have been responsible for observed decreases in SBS prevalence rates. Evidence to implicate dust mites is at best indirect. The cleaning regimes may have also affected the nature and quantity of MOD and other particulate-phase surface deposits.

Duct Cleaning

In the United States, duct cleaning is widely advertised by commercial interests as a means of improving air quality. Though HVAC system ducts in some cases can be shown to be significantly contaminated with dust deposits, there is no evidence to support claims that such dust accumulations are associated with SBS symptoms and that duct cleaning will decrease symptom prevalence rates. The latter is supported by the studies of Roys et al.[129] who observed that HVAC system cleaning did not decrease SBS symptom prevalence rates.

HVAC system ducts may have the potential for serving as an amplification medium for microorganisms, particularly under optimal environmental conditions (e.g., high humidity or moisture levels). Studies which address this concern need to be conducted.

Dust accumulations in HVAC system ducts may be a "symptom" rather than a cause of poor IAQ. The accumulation of dust in ducts is likely to be associated with significant surface deposition in occupied spaces as well. Both are indicative of the presence of major dust-generating sources.

Duct cleaning firms typically spray biocides on interior surfaces after dust has been removed by vacuuming or mechanical means. The use of biocides such as glutaraldehyde on surfaces which come in direct contact with building supply air does not appear to be prudent.

RECAPITULATION

Source control is a desirable approach to both preventing and mitigating IAQ problems as it is often technically more difficult to control contaminants once they are airborne than to prevent their generation or release in the first place. Source control, however, can only be effectively applied where causal factors that contribute to building-related health problems are known and amenable to control by limiting emissions. Unfortunately, causal factors and sources are not often definitively identified in a large percentage of building investigations.

Source control of chemical contaminants can be achieved at two levels: avoidance of known significant sources in designing and equipping buildings and implementation of remedial measures once an IAQ manifests itself. The avoidance/prevention approach requires characterization of emissions from sources so that choices can be made relative to product selection. The database for such emission characterization is presently limited, and our understanding of the toxicological significance of emissions from most products is relatively primitive. Nevertheless, efforts are being made in the U.S. and other countries (particularly the Nordic countries) to characterize product emissions in order to select low emission products for use in newly constructed and renovated buildings. Several public and private initiatives are presently going forward in the United States and the Nordic countries to construct "healthy buildings" through architectural specifications and labeling which attempt to limit emissions of contaminants such as formaldehyde, TVOCs, and 4-PC.

After an IAQ problem manifests itself, it may be mitigated by the removal and replacement of offending sources or treating or modifying them in some way. Few studies of the efficacy of source removal have been conducted so that in many cases recommendations for source removal and replacement with other less-offending products has to be done on an intuitive basis. It is often difficult to identify replacement products which themselves do not contribute to an IAQ problem. Source removal and replacement may not in many cases be practical because of the uncertainties and the often significant cost involved.

Only a relatively few source treatment measures have been evaluated for purposes of reducing source emissions. Most of these have focused on formaldehyde emissions from urea-formaldehyde-bonded wood products. The source treatment method of the greatest potential significance in mitigating potential building IAQ and health complaints is described as building bake-out. It has been evaluated by a number of investigators under field and laboratory conditions. Results have been mixed and there remains some uncertainty about its applicability and effectiveness.

Source control is the primary mitigation method employed for IAQ problems associated with biological contaminants, exposures which cause Legionnaires' disease, hypersensitivity pneumonitis, humidifier fever, asthma, chronic allergic rhinitis, or sick building symptoms. Contaminants of concern include microorganisms and their antigens, dust mite antigens, and macromolecular organic dust.

With the exception of Legionnaires' disease, only relatively limited attention has been given to systematic evaluations of control measures applied to the remediation of contamination/health complaints associated with contaminants of biological origin. Recommended mitigation measures are based on case studies which have employed a trial and error approach.

Source control measures can be applied to the control of surface dusts which have been implicated as a risk factor for SBS symptoms. Appropriate control measures include identification and removal of sources, removal of reservoirs, and surface cleaning. Intensive surface cleaning using steam cleaning of fabrics, wet wiping of hard surfaces, and high-efficiency vacuuming of office files has been shown to significantly reduce symptom prevalence. Duct cleaning, though widely promoted as a solution to indoor air quality problems, has received little research attention. Limited studies indicate that duct cleaning does not reduce SBS symptoms and the use of biocidal treatments with such cleaning is not considered to be prudent.

REFERENCES

1. ASHRAE Standard 62–1989. 1989. "Ventilation for Acceptable Indoor Air Quality." American Society of Heating, Refrigerating and Air-Conditioning Engineers. Atlanta.
2. Woods, J.E. et al. 1989. "Indoor Air Quality Diagnostics: Qualitative and Quantitative Procedures to Improve Environmental Conditions." 80–98. In: Nagda, N.L. and J.P. Harper (Eds.). *Design and Protocol for Monitoring Indoor Air Quality*. ASTM STP 1002. American Society for Testing and Materials. Philadelphia.
3. Lane, C.A. et al. 1989. "Indoor Air Quality Diagnostic Procedures for Sick and Healthy Buildings." 237–248. In: *Proceedings of IAQ '89: The Human Equation: Health and Comfort*. American Society of Heating, Refrigerating and Air-Conditioning Engineers. Atlanta.

4. Seitz, T.A. 1990. "NIOSH Indoor Air Quality Investigations—1971 through 1988." 163–171. In: Weekes, D.M. and R.B. Gammage (Eds.). *Proceedings of the Indoor Air Quality International Symposium: The Practitioner's Approach to Indoor Air Quality Investigations*. American Industrial Hygiene Association. Akron, OH.
5. Godish, T. 1989. *Indoor Air Pollution Control*. Lewis Publishers, Chelsea, MI.
6. Molhave, L. 1982. "Indoor Air Pollution Due to Organic Gases and Vapors of Solvents in Building Materials." *Environ. Int.* 8:117–127.
7. Girman, J.R. et al. 1986. "Volatile Organic Emissions from Adhesives with Indoor Applications." *Environ. Int.* 12:317–321.
8. Bayer, C.W. and C.D. Pananicolopoulos. 1990. "Exposure Assessments to Volatile Organic Compound Emissions from Textile Products." 725–730. In: *Proceedings of the Fifth International Conference on Indoor Air Quality and Climate*. Vol. 3. Toronto.
9. Black, M. 1990. "Environmental Chamber Technology for the Study of Volatile Organic Compound Emissions from Manufactured Products." 713–718. In: *Proceedings of the Fifth International Conference on Indoor Air Quality and Climate*. Vol. 3. Toronto.
10. Wallace, L.A. et al. 1987. "Emissions of Volatile Organic Compounds from Building Materials and Consumer Products." *Atmos. Environ.* 21:385–393.
11. Sheldon, L.S. et al. 1988. "Indoor Air Quality in Public Buildings." EPA/600/6–88/09a.
12. Sheldon, L.S. et al. 1988. "Indoor Air Quality in Public Buildings." EPA/600/6–88/09b.
13. Christiansson, J. et al. 1993. "Emission of VOCs from PVC-Floorings—Models for Predicting the Time-Dependent Emission Rates and Resulting Concentrations in the Indoor Air." 389–394. In: *Proceedings of the Sixth International Conference on Indoor Air Quality and Climate*. Vol. 2. Helsinki.
14. Bremer, J. et al. 1993. "Measurement and Characterization of Emissions from PVC Materials for Indoor Use." 419–424. In: *Proceedings of the Sixth International Conference on Indoor Air Quality and Climate*. Vol. 2. Helsinki.
15. Gustafsson, H. and B. Jonsson. 1993. "Trade Standards for Testing Chemical Emission from Building Materials; Part 1: Measurements of Flooring Materials." 437–442. In: *Proceedings of the Sixth International Conference on Indoor Air Quality and Climate*. Vol. 2. Helsinki.
16. Jensen, B. et al. 1993. "Characterization of Linoleum; Part 1: Measurement of Volatile Organic Compounds by Use of the Field and Laboratory Emission Cell, 'FLEC'." 443–448. In: *Proceedings of the Sixth International Conference on Indoor Air Quality and Climate*. Vol. 2. Helsinki.
17. Kirchner, S. et al. 1993. "Characterization of Volatile Organic Compounds Emission from Floor Coverings." 455–460. In: *Proceedings of the Sixth International Conference on Indoor Air Quality and Climate*. Vol. 2. Helsinki.
18. De Bortoli, M. et al. 1993. "Emission of Formaldehyde, Vinyl Chloride, VOCs and Plasticizers from Different Wall-Coating Materials." 413–418. In: *Proceedings of the Sixth International Conference on Indoor Air Quality and Climate*. Vol. 2. Helsinki.
19. Tichenor, B.A. and Z. Guo. 1991. "The Effect of Ventilation on Emission Rates of Wood Finishing Materials." *Environ. Int.* 17:317–323.

20. Clausen, P.A. et al. 1991. "Long-term Emission of Volatile Organic Compounds from Water Borne Paints—Methods of Comparison." *Indoor Air*. 4:562–576.
21. Gehrig, R. et al. 1993. "VOC Emissions from Wall Paints—A Test Chamber Study." 431–436. In: *Proceedings of the Sixth International Conference on Indoor Air Quality and Climate*. Vol. 2. Helsinki.
22. Levin, H. 1987. "The Evaluation of Building Materials and Furnishings for a New Office Building." 88–103. In: *Proceedings of IAQ '87: Practical Control of Indoor Air Problems*. American Society of Heating, Refrigerating and Air-Conditioning Engineers. Atlanta.
23. Wolkoff, P. 1990. "Proposal of Methods for Developing Healthy Building materials: Laboratory and Field Experiments." *Environ. Technol.* 11:327–338.
24. Tirkkonen, T. et al. 1993. "Volatile Organic Compound (VOC) Emission from Some Building and Furnishing Materials." 477–482. In: *Proceedings of the Sixth International Conference on Indoor Air Quality and Climate*. Vol. 2. Helsinki.
25. Pickrell, J.A. et al. 1983. "Formaldehyde Release Coefficients from Selected Consumer Products: Influence of Chamber Loading, Multiple Products, Relative Humidity and Temperature." *Environ. Sci. Technol.* 17:753–757.
26. Matthews, T.G. et al. 1983. "Formaldehyde Emissions from Combustion Sources and Solid Resin-Containing Products: Potential Impact on Indoor Formaldehyde Concentrations and Possible Corrective Measures." In: *Proceedings of ASHRAE Symposium on Management of Atmospheres in Tightly Enclosed Spaces*. Santa Barbara, CA.
27. Matthews, T.G. et al. 1985. "Modeling and Testing of Formaldehyde Emission Characteristics of Pressed Wood Products." *Technical Report XVIII*. U.S. Consumer Product Safety Commission.
28. Johnsen, C.R. et al. 1991. "A Study of Human Reactions to Emissions from Building Materials in Climate Chambers. Part 1. Clinical Data, Performance and Comfort." *Indoor Air*. 4:377–388.
29. Wolkoff, P. et al. 1991. "A Study of Human Reactions to Emissions from Building Materials in Climate Chambers. Part II. VOC Measurements, Mouse Bioassay, and Decipol Evaluation in the 1–2 mg/m^3 TVOC Range." *Indoor Air*. 4:389–403.
30. Anderson, R.C. 1991. "Measuring Respiratory Irritancy of Emissions." 19–23. In: *Post-Conference Proceedings of IAQ '91: Healthy Buildings*. American Society of Heating, Refrigerating and Air-Conditioning Engineers. Atlanta.
31. Anderson, R.C. 1993. "Toxic Emissions from Carpets." 651–656. In: *Proceedings of the Sixth International Conference on Indoor Air Quality and Climate*. Vol. 1. Helsinki.
32. Broder, I. et al. 1988. "Comparison of Health and Home Characteristics of Houses Among Control Homes and Homes Insulated with Urea-formaldehyde Foam. II. Initial Health and House Variables and Exposure Response Relationships." *Environ. Res.* 45:156–178.
33. Ritchie, I.M. and R.H. Lehnen. 1987. "Formaldehyde-Related Health Complaints in Residents Living in Mobile and Conventional Houses." *Am. J. Pub. Health*. 77:323–328.
34. Olsen, J.H. and M. Dossing. 1982. "Formaldehyde-Induced Symptoms in Day Care Centers." *Am. Ind. Hyg. Assoc. J.* 43:366–370.

35. Godish, T. et al. 1990. "Residential Formaldehyde—Increased Exposure Levels Aggravate Adverse Health Effects." *J. Environ. Health.* 47: 300–305.
36. Liu, K.S. et al. 1987. "Irritant Effects of Formaldehyde in Mobile Homes." 610–614. In: *Proceedings of the Fourth International Conference on Indoor Air Quality and Climate.* Vol. 2. West Berlin.
37. Kryzanowski, M. et al. 1990. "Chronic Respiratory Effects of Indoor Formaldehyde Exposure." *Environ. Res.* 52:117–125.
38. ASHRAE Standard 62–1981. 1981. "Ventilation for Acceptable Indoor Air Quality." American Society of Heating, Refrigerating and Air-Conditioning Engineers. Atlanta.
39. Tucker, W.G. 1990. "Building with Low-Emitting Materials and Products: Where do We Stand?" 251–256. In: *Proceedings of the Fifth International Conference on Indoor Air Quality and Climate.* Vol. 3. Toronto.
40. Molhave, L. 1990. "The Sick Building Syndrome (SBS) Caused by Exposures to Volatile Organic Compounds (VOCs)." 1–18. In: Weekes, D.M. and R.B. Gammage (Eds.). *Proceedings of the Indoor Air Quality International Symposium: The Practitioner's Approach to Indoor Air Quality Investigations.* American Industrial Hygiene Association. Akron, OH.
41. State of Washington. 1990. *East Campus Plus Program, Indoor Air Quality Specifications.* Olympia, WA.
42. Black, M. et al. 1993. "Material Selection for Controlling IAQ in New Construction." 611–616. In: *Proceedings of the Sixth International Conference on Indoor Air Quality and Climate.* Vol. 2. Helsinki.
43. Tichenor, B.A. et al. 1991. "The Interaction of Vapor Phase Organic Compounds with Indoor Sinks." *Indoor Air.* 1:23–35.
44. Hayward, S.B. and J.J. Wesolowski. 1993. "Guidelines for Reduction of Exposure to Volatile Organic Compounds (VOC) in Newly Constructed or Remodeled Office Buildings." 605–610. In: *Proceedings of the Sixth International Conference on Indoor Air Quality and Climate.* Vol. 2. Helsinki.
45. Levin, H. 1989. "Building Materials and Indoor Air Quality." 667–693. In: Cone, J.E. and M.J. Hodgson (Eds.). *Problem Buildings: Building-Associated Illness and the Sick Building Syndrome. Occupational Medicine: State of the Art Reviews.* Hanley and Belfus, Inc. Philadelphia.
46. Levin, H. 1990. Indoor Air Quality Update. Cutter Information Corporation, February.
47. National Institute of Occupational Safety and Health. 1987. *Registry of Toxic Effects of Chemical Substances, 1985–86.* Vols. 1–5. NIOSH. Cincinnati, OH.
48. Sax, N.I. 1989. *Dangerous Properties of Industrial Materials.* 7th ed. Van Norstand Reinhold. New York.
49. Oie, L. et al. 1993. "Selection of Building Materials for Good Indoor Air Quality." 629–634. In: *Proceedings of the Sixth International Conference on Indoor Air Quality and Climate.* Vol. 2. Helsinki.
50. Tucker, W.G. 1993. "Bioresponse Testing of Sources of Indoor Air Contaminants." 621–627. In: *Proceedings of the Sixth International Conference on Indoor Air Quality and Climate.* Vol. 1. Helsinki.
51. Wolkoff, P. and P.A. Nielsen. 1993. "Indoor Climate Labeling of Building Materials. Chemical Emission Testing, Modeling and Indoor Odor Thresholds." National Institute of Occupational Health. Copenhagen, Denmark.

52. Levin, H. 1990. "Critical Building Design Factors for Indoor Air Quality and Climate: Current Status and Predicted Trends." 301–306. In: *Proceedings of the Fifth International Conference on Indoor Air Quality and Climate*. Vol. 3. Toronto.
53. Turner, W.A. et al. 1990. "Design of a Healthy Modern Office Building." 269–274. In: *Proceeding of the Fifth International Conference on Indoor Air Quality and Climate*. Vol. 3. Toronto.
54. Turner, W.A. et al. 1993. "Design of a Healthy Modern Office Building and Occupied HVAC/IAQ 'Training Laboratory'." 711–716. In: *Proceedings of the Sixth International Conference on Indoor Air Quality and Climate*. Vol. 2. Helsinki.
55. Bernheim, A. 1993. "Healthy Building: San Francisco Main Library." 623–628. In: *Proceedings of the Sixth International Conference on Indoor Air Quality and Climate*. Vol. 2. Helsinki.
56. Frederiksen, E. 1990. "Practical Attempts to Predict Indoor Air Quality." 239–244. In: *Proceedings of the Fifth International Conference on Indoor Air Quality and Climate*. Vol. 3. Toronto.
57. Hijazi, N.R. et al. 1983. "Indoor Organic Contaminants in Energy Efficient Buildings." In: *Proceedings of Measurement and Monitoring of Noncriteria (Toxic) Contaminants in Air*. Air Pollution Control Association. Pittsburgh.
58. McLaughlin, P. and R.A. Raigner. 1990. "Higher Alcohols as Indoor Air Pollutants: Source, Cause, Mitigation." 587–592. In: *Proceedings of the Fifth International Conference on Indoor Air Quality and Climate*. Vol. 3. Toronto.
59. Norback, D. and M. Torgen. 1987. "A Longitudinal Study of Symptoms Associated with Wall-to-Wall Carpets and Electrostatic Charge in Swedish Schools." 572–576. In: *Proceedings of the Fourth International Conference on Indoor Air Quality and Climate*. Vol. 2. West Berlin.
60. Godish, T. et al. 1989. "Control of Residential Formaldehyde Levels by Source Treatment." *Environ. Int.* 15:609–613.
61. Sundin, B. 1978. "Formaldehyde Emission from Particleboard and Other Building Materials: A Study from the Scandinavian Countries." 251–273. In: Maloney, T.M. (Ed.). *Proceedings of the Twelfth Annual WSU Particleboard Symposium*. Washington State University. Pullman, WA.
62. Haneto, P. 1986. "Effect of Diffusion Barriers on Formaldehyde Emissions from Particleboard." 202–208. In: Meyer, B. et al. (Eds.). *Formaldehyde Release from Wood Products*. American Chemical Society Symposium Series, 316.
63. Girman, J.R. 1989. "Volatile Organic Compounds and Building Bake-out." 695–712. In: Cone, J.E. and M.J. Hodgson (Eds.). *Problem Buildings: Building- Associated Illness and the Sick Building Syndrome. Occupational Medicine: State of the Art Reviews*. Hanley and Belfus, Inc. Philadelphia.
64. Girman, J. et al. 1987. "Bake-out of an Office Building." 22–26. In: *Proceedings of the Fourth International Conference on Indoor Air Quality and Climate*. Vol. 1. West Berlin.
65. Girman, J.R. et al. 1990. "Building Bake-out Studies." 349–354. In: *Proceedings of the Fifth International Conference on Indoor Air Quality and Climate*. Vol. 3. Toronto.
66. Lindstrom, A. et al. 1991. "Bake-out of a Portion of a New Highrise Office Building." Paper #91–62.3. Presented at the 84th Annual Meeting Air and Waste Management Association, Vancouver, BC.

67. Girman, J.R. et al. 1989. "Bake-out of a New Office Building to Reduce Volatile Organic Concentrations." Paper #89–80.8. Presented at the 82nd Annual Meeting of the Air and Waste Management Association, Anaheim, CA.
68. Hicks, J. et al. 1990. "Building Bake-out During Commissioning: Effects on VOC Concentrations." 413–418. In: *Proceedings of the Fifth International Conference on Indoor Air Quality and Climate*. Vol. 3. West Berlin.
69. Offermann, F.J. et al. 1993. "Indoor Contaminant Emission Rates Before and After a Building Bake-out." 687–692. In: *Proceedings of the Sixth International Conference on Indoor Air Quality and Climate*. Vol. 6. Helsinki.
70. Bayer, C.W. 1991. "The Effect of 'Building Bake-out' Conditions on Volatile Organic Compounds." 101–114. In: Kay, J.G. et al. (Eds.). *Indoor Air Pollution — Radon, Bioaerosols and VOCs*. Lewis Publishers, Chelsea, MI.
71. Sundin, B. 1985. "The Formaldehyde Situation in Europe." 255–275. In: Maloney, T.M. (Ed.). *Proceedings of the Nineteenth International WSU Particleboard/Composite Materials Symposium*. Washington State University. Pullman, WA.
72. Versar, Inc. 1986. "Formaldehyde Exposure in Residential Settings: Sources, Levels, and Effectiveness of Control Options." EPA Contract No. 68–02–3968.
73. Myers, G.E. 1984. "How Mole Ratios of UF Resin Effect Formaldehyde Emissions and Other Properties: A Literature Critique." *For. Prod. J.* 34: 35–41.
74. Myers, G.E. 1985. "Effect of Separate Additions to Furnish or Veneer on Formaldehyde Emission and Other Properties: A Literature Review (1960–1984)." *For. Prod. J.* 35:57–62.
75. Simon, S.I. 1980. "Using the Verkor FD-EX Chamber to Prevent Formaldehyde Emission from Boards Manufactured with Urea-formaldehyde Glues." 163–169. In: Maloney, T.M. (Ed.). *Proceedings of the Fourteenth Annual WSU Particleboard Symposium*. Washington State University. Pullman, WA.
76. Murray, R. 1991. "Health Aspects of Carbonless Copy Paper." *Contact Derm.* 24:321–333.
77. Braun-Hansen, T. and B. Anderson. 1986. "Ozone and Other Air Pollutants from Photocopying Machines." *Am. Ind. Hyg. Assoc. J.* 47: 659–665.
78. Allen, R.J. et al. 1978. "Characterization of Potential Indoor Sources of Ozone." *Am. Ind. Hyg. Assoc. J.* 39:456–460.
79. Grace, R.D. et al. 1981. "Susceptibility of *L. pneumophila* to Three Cooling Tower Microbiocides." *Appl. Environ. Microbiol.* 41:233–236.
80. Skaily, P. et al. 1980. "Laboratory Studies of Disinfectants Against *L. pneumophila*." *Appl. Environ. Microbiol.* 40:48–57.
81. Soracco, R.J. and D.H. Pope. 1983. "Bacteriostatic and Bacteriocidal Modes of Action of Bis (Tributyltin) on *L. pneumophila*." *Appl. Environ. Microbiol.* 45:48–57.
82. Sorocco, R.J. et al. 1983. "Susceptibility of Algae and *L. pneumophila* to Cooling Tower Biocides." *Appl. Environ. Microbiol.* 45:1254–1260.
83. Witherell, L.E. et al. 1986. "*Legionella* in Cooling Towers." *J. Environ. Health.* 49:134–139.
84. Braun, E.B. 1982. Ph.D. Dissertation. Rensselear Polytechnic Institute, Troy, NY.

85. Broadbent, C. and R. Bentham. 1993. "Field Evaluation of Control Strategies for *Legionella* in Cooling Water Systems." 345–350. In: *Proceedings of the Sixth International Conference on Indoor Air Quality and Climate*. Vol. 4. Helsinki.
86. Barbaree, J.M. et al. 1986. "Isolation of Protozoa from Water Associated with a Legionellosis Outbreak and Demonstration of Intracellular Multiplication of *L. pneumophila*." *Appl. Environ. Microbiol.* 50:422–425.
87. Anand, C.M. et al. 1983. "Interaction of *L. pneumophila* and a Free Living Amoeba." *J. Hygiene*. 91:167–178.
88. Howland, E.B. and D.H. Pope. 1983. "Distribution and Seasonality of *L. pneumophila* in Cooling Towers." *Curr. Microbiol.* 9:319–324.
89. Fliermans, C.B. et al. 1982. "Treatment of Cooling Systems Containing High Levels of *L. pneumophila*." *Water Res.* 16:903–909.
90. Muraca, P.W. et al. 1990. "Disinfection of Water Distribution Systems for Legionella: A Review of Application Procedures and Methodologies." *Infect. Contr. Hosp. Epidemiol.* 11:79–88.
91. American Society of Heating, Refrigerating and Air-Conditioning Engineers. 1984. *Systems*. American Society of Heating, Refrigerating and Air-Conditioning Engineers. Atlanta.
92. Conwill, D.E. et al. 1982. "Legionellosis. The 1980 San Francisco Outbreak." *Am. Rev. Respir. Dis.* 126:666–669.
93. Yamayoshi, T. and N. Tatsumi. 1993. "Ultraviolet Inactivation of *Legionella* in Running Water with Submerged Light Bulbs." 413–416. In: *Proceedings of the Sixth International Conference on Indoor Air Quality and Climate*. Vol. 4. Helsinki.
94. Miller, R.P. 1979. "Cooling Towers and Evaporative Condensors." *Ann. Int. Med.* 90:667–670.
95. Winstrom, G.K. and J.C. Ovard. 1973. "Cooling Tower Drift: Its Measurement, Control and Environmental Effects." Cooling Tower Institute Annual Meeting, Houston.
96. HMSO. 1988. *The Control of Legionellae in Health Care Premises: A Code of Practice*. Department Health and Social Security and the Welsh Office. London.
97. World Health Organization. 1986. *Environmental Aspects of the Control of Legionellosis*. Environmental Health Report 14. World Health Organization. Copenhagen.
98. Vickers, R.M. et al. 1987. "Determinants of *Legionella pneumophila* Contamination of Water Distribution Systems: 15-Hospital Prospective Study." *Infect. Contr.* 8:357–363.
99. Stout, J.E. et al. 1985. "Ecology of *Legionella pneumophila* Within Water Distribution Systems." *Appl. Environ. Microbiol.* 49:221–228.
100. Muraca, P.W. et al. 1988. "Legionnaires' Disease in the Work Environment: Implications for Environmental Health." *Am. Ind. Hyg. Assoc. J.* 49:584–590.
101. Muraca, P. et al. 1987. "Comparative Assessment of Chlorine, Heat, Ozone, UV Light for Killing *Legionella pneumophila* Within a Model Plumbing System." *Appl. Environ. Microbiol.* 53:447–453.
102. Baird, I.M. et al. 1984. "Control of Endemic Nosocomial Legionellosis by Hyperchlorination of Potable Water." 333–339. In: *Proceedings of the Second International Symposium of the American Society of Microbiologists: Legionella*. Washington, D.C.

103. Shands, K. et al. 1985. "Potable Water as a Source of Legionnaires' Disease." *J. Am. Med. Assoc.* 253:1412–1416.
104. Best, M. et al. 1983. "Legionellaceae in the Hospital Water Supply: Epidemiological Link with Disease and Evaluation of a Method of Control of Nosocomial Legionnaires' Disease and Pittsburgh Pneumonia." *Lancet.* ii:307–310.
105. Plouffe, J.F. et al. 1983. "Relationship Between Colonization of Hospital Buildings with *Legionella pneumophila* and Hot Water Temperatures." *Appl. Environ. Microbiol.* 46:769–770.
106. Edelsteen, P.H. et al. 1982. "Efficacy of Ozone in Eradication of *Legionella pneumophila* from Hospital Plumbing Fixtures." *Appl. Environ. Microbiol.* 44:1330–1334.
107. Muraca, P.W. et al. 1990. "Disinfection of Water Distribution Systems for *Legionella*: A Review of Application Procedures and Methodologies." *Infect. Contr. Hosp. Epidemiol.* 11:79–88.
108. Weiss, N.S. and Y. Soleymani. 1971. "Hypersensitivity Lung Disease Caused by the Contamination of an Air-Conditioning System." *Ann. Allergy.* 29:154–156.
109. Banazak, E.F. et al. 1970. "Hypersensitivity Pneumonitis Due to Contamination of an Air Conditioner." *New Eng. J. Med.* 283:271–276.
110. Aranow, P. et al. 1978. "Early Detection of Hypersensitivity Pneumonitis in Office Workers." *Am. J. Med.* 64:236–242.
111. Scully, R.E. et al. 1979. "Case Records of the Massachusetts General Hospital — Case 47–1979." *New Eng. J. Med.* 301:1168–1174.
112. Bernstein, M. et al. 1983. "Exposures to Respirable, Airborne *Penicillium* from a Contaminated Ventilation System: Clinical, Environmental, and Epidemiological Aspects." *Am. Ind. Hyg. Assoc. J.* 44:161–169.
113. Ganier, M. et al. 1980. "Humidifier Lung: An Outbreak in Office Workers." *Chest.* 77:183–187.
114. Cockcroft, A. et al. 1981. "An Investigation of Operating Theatre Staff Exposed to Humidifier Fever Antigens." *Br. J. Ind. Med.* 38:144–152.
115. Morey, P.R. et al. 1984. "Environmental Studies in Moldy Office Buildings: Biological Agents, Sources, and Preventative Measures." *Ann. ACGIH: Evaluating Office Environmental Problems.* 10:21–35.
116. Burge, H.A. 1987. "Approaches to the Control of Indoor Microbial Contamination." 33–37. In: *Proceedings of IAQ '87: Practical Control of Indoor Air Problems.* American Society of Heating, Refrigerating and Air-Conditioning Engineers. Atlanta.
117. Burge, H.A. 1990. "Bioaerosols: Prevalence and Health Effects in the Indoor Environment." *J. Aller. Clin. Immunol.* 86:687–701.
118. Hodgson, M.J. 1987. "An Outbreak of Recurrent Acute and Chronic Hypersensitivity Pneumonitis in Office Workers." *Am. J. Epidemiol.* 125:631–638.
119. Morey, P. et al. 1986. "Environmental Studies in Moldy Office Buildings. Sources and Preventative Measures." *ASHRAE Trans.* 92. Pt. 1.
120. Morey, P.R. and C. Williams. 1990. "Porous Insulation in Buildings: A Potential Source of Microorganisms." 529–533. In: *Proceedings of the Fifth International Conference on Indoor Air Quality and Climate.* Vol. 4. Toronto.
121. Schata, M. et al. 1987. "Allergies to Molds Caused by Fungal Spores in Air-Conditioning Plants." 777–780. In: *Proceedings of the Fourth International Conference on Indoor Air Quality and Climate.* Vol. 2. West Berlin.

122. Elixmann, J.H. et al. 1990. "Fungi in Filters of Air-Conditioning Systems Cause the Building-Related Illness." 193–196. In: *Proceedings of the Fifth International Conference on Indoor Air Quality and Climate.* Vol. 1. Toronto.
123. Elixmann, J.H. et al. 1989. "Can Airborne Fungal Allergens Pass Through an Air-Conditioning System?" *Environ. Int.* 15:193–196.
124. Light, E.N. et al. 1989. "Abatement of an *Aspergillus niger* Contamination in a Library." 224–231. In: *Proceedings of IAQ '89: The Human Equation: Health and Comfort.* American Society of Heating, Refrigerating and Air-Conditioning Engineers. Atlanta.
125. Morey, P.R. 1993. "Use of Hazard Communication Standard and General Duty Clause During Remediation of Fungal Contamination." 391–395. In: *Proceedings of the Sixth International Conference on Indoor Air Quality and Climate.* Vol. 4. Helsinki.
126. Johanning, E. et al. 1993. "Remedial Techniques and Medical Surveillance Program for the Handling of Toxigenic *Stachybotrys atra.* 311–316. In: *Proceedings of the Sixth International Conference on Indoor Air Quality and Climate.* Vol. 4. Helsinki.
127. Leinster, P. et al. 1990. "A Modular Approach to the Investigation of Sick Building Syndrome." 287–292. In: *Proceedings of the Fifth International Conference on Indoor Air Quality and Climate.* Vol. 1. Toronto.
128. Raw, G.J. et al. 1991. "A New Approach to the Investigation of Sick Building Syndrome." 339–343. In: *Proceedings of the CIBSE National Conference.* London.
129. Raw, G.J. et al. 1993. "Sick Building Syndrome—Cleanliness is Next to Healthiness." *Indoor Air* 3:237–245.
130. Coloff, M.J. 1986. "Use of Liquid Nitrogen in the Control of House Dust Mite Populations." *Clin. Allergy.* 16:41–47.
131. Lundblad, F.P. 1991. "House Dust Mite Allergy in an Office Building." *Appl. Occup. Environ. Hyg.* 6:94–96.
132. Gravesen, S. et al. 1990. "The Role of Potential Immunogenic Components of Dust (MOD) in the Sick Building Syndrome." 9–13. In: *Proceedings of the Fifth International Conference on Indoor Air Quality and Climate.* Vol. 1. Toronto.
133. Skov, P. et al. 1990. "Influence of Indoor Climate on the Sick Building Syndrome in an Office Environment." *Scand. J. Work. Environ. Health.* 16:363–371.
134. Hedge, A. et al. 1993. "Effects of Man-Made Mineral Fibers in Settled Dust on Sick Building Syndrome in Air-Conditioned Offices." 291–296. In: *Proceedings of the Sixth International Conference on Indoor Air Quality and Climate.* Vol. 1. Helsinki.
135. Nexo, E. et al. 1983. "Extreme Fatigue and Malaise Syndrome Caused by Badly Cleaned Wall-to-Wall Carpets?" *Ecol. Dis.* 2:415-418.

10 CONTAMINANT CONTROL

Though source control is, in theory, the best way of preventing and mitigating indoor air quality (IAQ) problems, it is not practical in many cases. This is particularly true when building investigators have failed to identify causal factors in building-related health complaints. In such instances, investigators have little choice but to recommend a contaminant control measure such as general dilution ventilation in the hope that it will mitigate reported IAQ problems. Based on reports of NIOSH investigations, this may be true in over 60% of cases investigated.[1]

Source control, of course, is also impractical when humans themselves produce effluents which may be the cause of odor and comfort complaints. Bioeffluents have been historically controlled by general dilution ventilation or in some cases, by sorbent filtration media.

Contaminant control has been widely used to reduce particulate matter levels in HVAC (heating, ventilating, and air-conditioning) system airstreams and in building spaces. The use of filtration to remove particles entrained in outdoor air supplies as well as those generated from many diffuse sources in a building is in many cases both practical and relatively effective.

Contaminant control measures are often used because they are considered to be both practical and technically appropriate to the task. Typical contaminant control measures include general dilution ventilation, proper operation of HVAC systems, local exhaust ventilation, and air cleaning.

GENERAL DILUTION VENTILATION

General or dilution ventilation is the most popular contaminant control measure used in large office, institutional, and commercial buildings. It is employed generically to provide occupants with a relatively comfortable and odor-free building environment and as a specific measure to remediate IAQ problems whose cause is unknown. General ventilation can be provided by either "natural" or mechanical means. The former has the longer history; ventilation is provided by simply opening windows when occupants sense discomfort. The use of natural ventilation is still common in many older office and institutional buildings in the United States and in countries where moderate summertime climatic conditions have not been a driving force in the design and construction of year-round climate-controlled buildings. In the United States and other "developed" countries, there has been an increasing trend toward building designs that feature year-round climate control and retrofit installation of climate-control systems in previously naturally ventilated buildings.

These trends reflect the view of building designers that climate control systems which provide heating, cooling, and ventilation are best able to assure thermal comfort and more effectively manage building energy use. Most of these buildings are designed with unopenable windows so as to better control airflows and heating and cooling demands.

Buildings designed with mechanical heating, ventilation, and air-conditioning (HVAC) systems with unopenable windows must be operated to provide an adequate amount of outdoor air to minimize both human odor and the general feelings of discomfort or stuffiness that appears to be associated with poorly ventilated spaces. Unfortunately, HVAC systems in many cases have not been properly designed or operated to provide for the ventilation needs of occupied spaces.

General Dilution Theory

The use of mechanical ventilation to maintain relatively comfortable and odor-free indoor spaces is based on principles of general dilution theory.[2] By doubling the volume of air available for dilution under static or constant conditions of contaminant generation, a 50% reduction in contaminant concentration can be expected. If the air volume is doubled again, the concentration will be reduced to 25% of its original value and so on. The converse is also true. By reducing the air volume available for dilution, contaminant concentrations would be expected to correspondingly increase. If similar dilution phenomena were to occur in buildings, one would expect a significant decrease in contaminant concentrations with increased outdoor ventilation rates and a significant increase in contaminant levels with reduced or low air exchange rates associated with buildings with insufficient design outdoor air capacity, reduced outdoor air flows associated with energy management policies, and poorly operated and maintained HVAC system equipment.

Applicability of General Dilution Theory

Is general dilution theory applicable to the problems of contaminant control in mechanically ventilated buildings in general and for mitigating IAQ problems in buildings? The answer unfortunately is not a simple one. One would expect that general dilution associated with increased outdoor air flow rates would be relatively effective where contaminants would be produced episodically (e.g., tobacco smoking, emissions from intermittent photocopier use, etc.) or where source emissions are constant or fairly so (e.g., human bioeffluents). Indeed the dominant historical use of general dilution ventilation has been to control bioeffluent levels in occupied spaces and ventilation guidelines[3,4] have been based on human bioeffluents such as CO_2.

Exceptions to General Dilution Theory

Dilution ventilation may not in many cases be as effective as general dilution theory would predict. In those instances where contaminants such as formaldehyde and volatile organic compounds (VOCs) are released from sources by diffusion, emissions will vary in response to changes in environmental conditions such as temperature, relative humidity, and ventilation rates themselves. General dilution theory does not quite apply in such cases. Increased ventilation rates, for example, produce vapor pressure changes around sources which cause emission rates to increase. This phenomenon has been reported for formaldehyde[5] and a number of VOCs.[6,7] With VOCs, a sixfold increase in ventilation rate will result in a twofold increase in source strength. Such emission rate increases offset in part the effect of increased air exchange on contaminant concentrations. This effect has been observed to compromise the effectiveness of mechanical ventilation systems installed to control formaldehyde levels in residential structures.[8-10] Similar responses would be expected with other gas-phase contaminants emitted continuously from sources by diffusion.

Systematic Ventilation Studies

Health and Comfort Concerns

Ventilation as a potential risk for SBS symptoms and occupant dissatisfaction with air quality was discussed in detail in Chapter 3. Though much of that discussion is relevant to the use of ventilation for contaminant control/problem building remediation, it will not be repeated here. Those studies for the most part reported mixed results. It appears that ventilation rate increases from 5–20 CFM (2.5–10 L/s) per person are associated with decreased SBS symptom prevalence rates, whereas ventilation rate increases above 20 CFM (10 L/s) per person are not. This indicates that under very low ventilation conditions, increased ventilation may be effective in reducing SBS symptom prevalence while under other conditions it may provide no benefit. As such, ventilation cannot be viewed as a generic solution to IAQ problems.

The limited effectiveness of general dilution ventilation in reducing the prevalence of SBS symptoms may in part be due to the nature of toxic exposures that cause irritation symptoms. In the studies of Alarie[11] and Anderson[12] irritant effects of formaldehyde and other VOCs in mouse bioassays have been demonstrated to be log linear (see Figure 5.5, Chapter 5). What this means in a practical sense is that a 90% reduction in exposure concentration is required to reduce irritant response by 50%. The log-linear dose-response relationship for irritant chemicals indicates that order of magnitude increases in outdoor air exchange rates would be required to have a significant reduction in irritant symptom response. Achievement of such air exchange conditions in most buildings is impractical if not impossible.

Contaminant Concentrations

Studies designed to evaluate the effect of increased outdoor ventilation rates have also reported mixed results. In a study of a five-story office building, Palonen and Seppanen[13] observed no apparent relationship between ventilation rates and measured indoor concentrations of carbon dioxide (CO_2), formaldehyde, VOCs, particles, bacteria, mold, and radon. In the studies of Nagda et al.,[14] there were no significant effects of outdoor ventilation rates on concentrations of formaldehyde, nicotine, and total volatile organic compounds (TVOCs) with marginally significant increases of CO_2 and carbon monoxide (CO) under lower ventilation rate conditions. Similarly, in studies conducted by Collett et al.,[15] no significant reductions in CO_2, formaldehyde, respirable particles (RSP), nicotine, and microbial concentrations were observed under nominally higher outdoor ventilation rate conditions. The apparent ineffectiveness of nominally higher ventilation rates in reducing contaminant concentrations in the studies of Nagda et al.[14] and Collett et al.[15] were in good measure due to their inability to control actual air exchange conditions because of the confounding effect of infiltration. Though target outdoor air ventilation rates in the former were 10 L/s and 2.5 L/s per person, actual differences in air exchange were only in the range of 20–38%. Despite efforts by Collett et al.[15] to configure HVAC systems to provide outdoor ventilation rates of 10 and 5 L/s per person, actual differences in air exchange were slight.

In contrast to the studies reported above, Farant et al.[16] observed significant associations between outdoor ventilation rates (based on floor surface area, $L/s/m^2$) and the natural log of the indoor/outdoor ratios of CO_2, TVOCs, formaldehyde, and the oxides of nitrogen. In a second office building study, Baldwin and Farant[17] reported very strong correlations ($r = -0.86$) between outdoor ventilation rates and natural log values of the indoor/outdoor TVOC ratio. A twofold increase in ventilation rate was observed to result in an approximate 75% reduction in indoor TVOC levels. Significant correlations ($r = -0.50$) have also been reported between ventilation rate conditions and indoor RSP levels by Norback et al.[18] in a study of 14 Swedish primary schools.

Significant differences between planned and observed outdoor air ventilation have also been reported (Table 10.1) in the studies of Menzies et al.[19] This inability of investigators to control building ventilation rates based on HVAC system adjustments may have significant implications for those who attempt to use increased outdoor ventilation rates by increasing volumetric airflows through

Table 10.1 Differences between planned and observed office building ventilation rates (L/s/person).

Week	Planned	Observed
1	24.0	21.6
2	9.6	15.1
3	4.8	11.0
4	24.0	14.8
5	9.6	11.1
6	4.8	9.9

From Menzies, R.I. et al., 1990. *Proceedings Fifth International Conference on Indoor Air Quality and Climate.* Toronto. Vol. 1.

outdoor air intakes for the purpose of reducing contaminant levels and mitigating SBS health complaints. It appears that in attempting to increase outdoor ventilation rates, one is in a way "flying blind", that is, one hopes that damper adjustments indeed result in a significant increase in building air exchange rates.

Controlling Human Bioeffluents

The reduction of contaminants generated by building occupants has historically been the primary reason for ventilating buildings. These contaminants described as human bioeffluents produce unpleasant odors which are perceived by individuals entering a building or building spaces. They are also believed to be responsible for sensory perceptions of "stuffiness", "stale air", and discomfort associated with poorly ventilated buildings. The studies of Fanger[20] have confirmed that a portion of the dissatisfaction with air quality determined from panel ratings is due to human bioeffluents, albeit only a relatively small portion of the total dissatisfaction reported.[21]

Contaminants emitted by humans into building spaces have received only limited chemical characterization.[22] As a consequence, bioeffluent levels are described in terms of CO_2, the bioeffluent produced in the greatest concentration and the most easily measured. Carbon dioxide levels are used as a measure of the acceptability of indoor air relative to ventilation adequacy.[3,4]

The relationship between CO_2 generation rates, occupant density, and outdoor ventilation rates has been well-established and serves as the basis for ASHRAE Standards and World Health Organization (WHO) guidelines for acceptable IAQ. Because bioeffluent levels determined from CO_2 measurements have a relatively constant generation rate for a given occupant density, increased actual outdoor ventilation rates have a predictable effect on bioeffluent levels in buildings. Effects of high and low outdoor ventilation rates as well as occupant density on CO_2 levels in a San Francisco office building[23] can be seen in Figures 5.1a and 5.1b (Chapter 5).

Ventilation Standards/Guidelines

A consensus has formed over the years that adequate ventilation is essential for the maintenance of a comfortable and relatively human-odor-free building environment. As a consequence, standards and guidelines have been developed to assist both designers and building operators in providing adequate outdoor ventilation air to meet the needs of building occupants. The lead professional organization for developing and recommending ventilation standards and guidelines for acceptable IAQ in North America has been the American Society of Heating, Refrigerating and Air-Conditioning Engineers (ASHRAE). In Europe, such guidelines have been typically developed under the auspices of governmental agencies, regional collaboration, and the WHO.

Different approaches have been used or suggested for developing ventilation standards/guidelines for the purpose of addressing odor, comfort, and health concerns. These are described as ventilation rate (VR), indoor air quality (IAQ), and perceived air quality (PAQ) procedures.

Ventilation Rate Procedure

The VR procedure has a long history of use. It specifies a rate of outdoor ventilation that can be reasonably expected to provide acceptable air quality related to human odor and comfort. It is based wholly on contaminants generated by humans themselves. Ventilation guidelines reflect the percentage of outdoor air that will maintain a building steady-state CO_2 concentration which does not exceed a maximum acceptable value. They are usually specified in terms of cubic feet per minute (CFM) per person or liters per second (L/s) per person. Under the design occupancy, the provision of the specified minimum ventilation rate would be expected to maintain CO_2 levels below guideline values. As such, the VR procedure prescribes the minimum amount of outdoor ventilation air required to meet air quality goals.

The minimum outdoor ventilation rate for office buildings in both ASHRAE 62-73[24] and 62-1981[3] was 5 CFM (2.4 L/s)/person. At maximum design building occupancy, this ventilation rate would cause CO_2 levels to rise to 2500 ppm. This ventilation guideline reflected a significant concern for energy conservation and a reduction in the importance of human comfort and health. The explosion in sick building complaints in the late 1970s and throughout the 1980s and attempts to mitigate them by general ventilation led many in the IAQ community to conclude that prescribed ventilation rates were too low and were in fact a major contributing factor to the sick building "epidemic". As a consequence, ASHRAE guidelines were revised to provide a minimum ventilation rate of 20 CFM (10 L/s)/person in the typical mechanically ventilated office space with other ventilation rates for more specialized circumstances such as smoking lounges, 60 CFM (30 L/s)/person.[4] In the former case, the CO_2 concentration would not be expected to exceed 800 ppm. This is similar to the WHO guideline.[25] The ASHRAE IAQ guideline for CO_2 is, however, 1000 ppm.

Though originally developed to address IAQ and comfort concerns associated with human bioeffluents, guidelines based on the VR procedure have unfortunately become the generic measure of IAQ acceptability irrespective of what the contaminants and contaminant sources are. Indeed, the document developed for ASHRAE Standard 62-1989[4] states that prescribed ventilation rates will result in acceptable levels of CO_2, RSP, odors, and other contaminants. No evidence, however, was presented to show that there was any relationship between indoor CO_2 levels and other contaminants or between CO_2 levels and sick building complaints. ASHRAE guidelines as a measure of IAQ acceptability other than for human bioeffluents have no scientific basis. Their use as a measure of acceptability of air quality in many building investigations has therefore been inappropriate.

The scientific shortcomings of ASHRAE Standard 62-1989 based on the VR procedure have been addressed by Grimsrud and Teichman.[26] Grimsrud,[27] a member of the ASHRAE committee which developed the consensus ventilation guidelines by means of the VR procedure, suggests that ASHRAE Standard 62-1989 should be seen as a transition document to be updated when new scientific data on health effects and emission rates of indoor contaminants become available.

Indoor Air Quality Procedure

The IAQ procedure was developed as an alternative option to prescriptive ventilation rates in ASHRAE Standard 62-1981.[3] In this approach, a building designer may use any amount of outdoor air as long as it can be shown that specific contaminants will be maintained below guideline values. The IAQ procedure was intended to encourage innovative energy conserving solutions to the problem of building ventilation.[28]

Grimsrud[27] has suggested that an ideal ventilation standard would be based on performance criteria like those of the IAQ procedure in ASHRAE 62-1981 and 62-1989.[3,4] Performance standards would be developed from a knowledge of the known health effects of all the pollutants in a building. Acceptable levels could then be set after the relative risks of exposures to relevant pollutants had been determined. Such standards would have a prescriptive path to give building designers a well-understood and specific procedure to satisfy the performance standard. This would include upper bounds on source strength and lower bounds on ventilation rates required. These bounds would be determined from IAQ models which would be anticipated to accurately simulate contaminant emissions and transport, ventilation parameters, and occupant exposures.

Though ventilation standards based on the IAQ procedure first proposed by ASHRAE in 1981 and included in its 1989 revisions would in theory be an ideal approach to the problem of providing adequate ventilation for health and comfort needs, the IAQ procedure is not likely to ever be seriously considered by many building designers. It is a complicated procedure and it would be unrealistic to expect the average building designer to have the requisite knowledge to use it effectively.

Because of anticipated litigation by affected industries, ASHRAE deleted formaldehyde from its IAQ guidelines in its 1989 revisions. It is, of course, ironic that ASHRAE should do so because formaldehyde is one contaminant for which the health effects and emissions database is particularly strong. Since ASHRAE "caved" in to industry pressure and potential legal challenges on formaldehyde, can one realistically expect that IAQ guidelines could be adopted for other contaminants where associations with illness symptoms are likely to be considerably weaker? The ASHRAE precedent on formaldehyde is indeed an unfortunate one.

A variation of the IAQ procedure is being used by the state of Washington's Department of General Administration, East Campus Plus Program which requires that products used in the construction of new state office buildings meet emission limits.[29] Contaminant emissions are limited per product or activity to maximum values of 50 $\mu g/m^3$ TVOC; 50 $\mu g/m^3$ RSP; and 60 $\mu g/m^3$ formaldehyde.

The achievement of IAQ standard(s) or guideline(s) for a particular contaminant or group of contaminants would likely be based on a combination of source control and the use of general ventilation. Source control for purposes of new building design and construction would best be achieved by the use of product standards.

Perceived Air Quality Procedure

Fanger,[30] recognizing the shortcomings of existing ventilation standards, has proposed that future standards be based on providing sufficient outdoor air to reflect the pollution load of the building and the resultant PAQ. The pollution load is determined from the annoying/irritating effects of building contaminants generated by humans, tobacco smoking, and emissions from building materials on the olfactory and common chemical sense. The two olfactory organs combine to determine whether the air is perceived fresh and pleasant or various degrees of stale, stuffy, or irritating.

Perceived air quality is expressed in units called decipols, where one decipol is the PAQ in a space with a pollution strength of one olf ventilated by 10 L/s of clean air. An olf is defined as pollution from a standard person. Other pollution sources are characterized relative to their effects on PAQ using the standard person as a reference. The olf rating for a source indicates the number of standard persons required to make the air as annoying as the pollution source. The effect of ventilation on the perceived dissatisfaction rate based on L/s/olf can be seen in Figure 10.1. The curve is derived from panel ratings of the acceptability of bioeffluents from more than 1000 sedentary men and women. The corresponding relationship between PAQ in decipols and the percentage of dissatisfied individuals is illustrated in Figure 10.2.

Fanger[30] described three levels of PAQ based on the percentage of dissatisfied individuals and the corresponding required ventilation rates. Ventilation

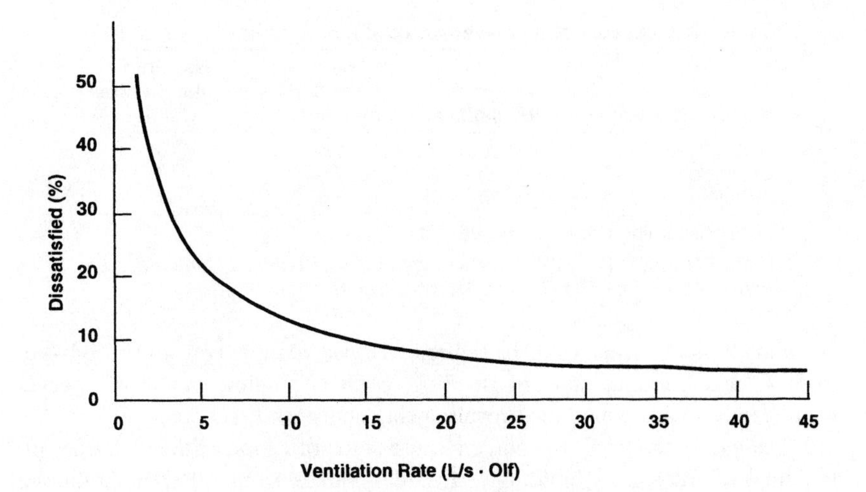

Figure 10.1 Relationship between ventilation rate and dissatisfaction with air quality. (From Fanger, P.O., 1990. *Proceedings 5th International Conference on Indoor Air Quality and Climate.* Toronto. Vol. 5.)

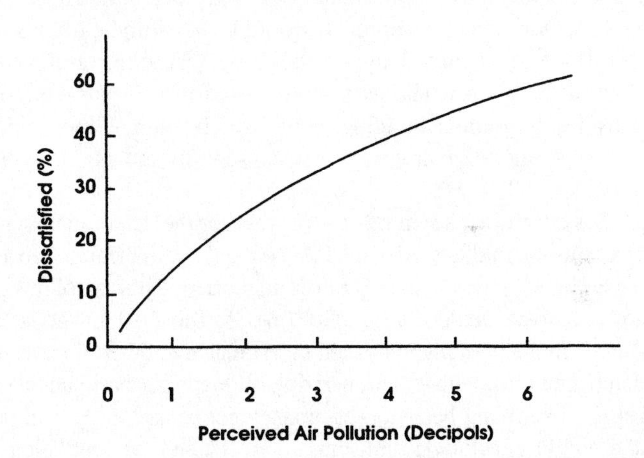

Figure 10.2 Relationship between perceived air quality (decipols) and percentage (%) dissatisfied judges. (From Fanger, P.O., 1990. *Proceedings 5th International Conference on Indoor Air Quality and Climate.* Toronto. Vol. 5.)

rates for different levels of PAQ are indicated in Table 10.2. These ventilation rates do not take into account the potential effect of human adaptation to contaminants in indoor spaces. Adaptation to indoor air contaminated by human bioeffluents, tobacco smoke, and contaminants from building materials has been reported to occur within 16 minutes of exposure.[31] Significant adaptation has been observed for human bioeffluents, moderate adaptation to tobacco smoke, with very little adaptation to contaminants from building sources. Gunnarsen and Broals[31] indicate that if adaptation were taken into consideration, the source

Table 10.2 Levels of perceived air quality.

Indoor air quality	% Dissatisfied	PAQ (decipols)	Required ventilation rates L/s/olf
High	10	0.6	16
Standard	20	1.4	7
Minimum	30	2.5	4

Note: Assuming clean air and VE = 1.

From Fanger, P.O., 1990. *Proceedings Fifth International Conference on Indoor Air Quality and Climate.* Toronto. Vol. 5.

strength for bioeffluents would be negligible, tobacco smoking would be reduced to 50%, and building materials to 70%. Such adaptation would, of course, significantly reduce ventilation requirements under the PAQ procedure.

The use of the PAQ procedure requires quantification of the olf values of pollution sources in a building space. Olf values from different pollution sources including people (1 olf/individual), smoking (6 olfs/individual smoker) and building materials are added together. Fanger[30] suggests that the source strength be determined by adding the olf values per square meter (m^2) of all materials present. Olf values unfortunately are only available for a relatively few materials. A more feasible approach would be to estimate the olf load per square meter floor area caused by the building, carpeting and ventilation system. A low olf building would have an olf load (olf/m^2 floor area) of 0.05–0.1. Olf loads for individual buildings vary widely both within and among buildings. Olf loadings for a variety of offices, schools and assembly halls are reported in Table 10.3.

Fanger[30] has developed a step-by-step process for the determination of ventilation requirements in buildings using the PAQ procedure. In the first step a desired IAQ in the ventilated space (Table 10.2) is identified. This is followed by an estimation of perceived outdoor air quality (PAQ), Table 10.4. The strength of pollution sources in the building must then be estimated to determine the olf load. These are determined from tables summarizing olf loads per occupant at different activity levels and smoking behavior, occupancy/per square meter and pollution caused by the building including furnishings, carpets, and the ventilation system. The total load is determined by adding the various olf loads described.

Ventilation required to handle the total olf load and obtain the PAQ can be calculated using the following equation:

$$Q_c = 10\frac{G}{C_i - C_o} \times \frac{1}{E_v} \tag{10.1}$$

where Q_c = ventilation rate (L/s)
G = total pollution load (olfs)
C_i = perceived air quality desired (decipols)
C_o = perceived air quality (decipols)
E_v = ventilation effectiveness

Table 10.3 Olf loads in buildings caused by furnishings, carpets and ventilation systems.

	Olf load (olf/m^2 floor)	
Existing buildings	**Mean**	**Range**
Offices (N = 24)	0.3	0.02–0.95
Schools (N = 6)	0.3	0.12–0.54
Assembly halls (N = 5)	0.5	0.13–1.32

From Fanger, P.O., 1990. *Proceedings Fifth International Conference Indoor Air Quality and Climate.* Toronto. Vol. 5.

Table 10.4 Typical perceived air quality in outdoor environments.

Location	Decipol
Rural—mountains, at sea	0
Urban, high AQ	<0.1
Urban, medium AQ	0.2
Urban, low AQ	>0.5

From Fanger, P.O., 1990. *Proceedings Fifth International Conference Indoor Air Quality and Climate.* Toronto. Vol. 5.

Ventilation effectiveness is defined as the relationship between the pollution concentration in the return air (C_e) and in the breathing zone (C_r), Equation 10.2. Ventilation effectiveness depends on air distribution, and location of pollution sources. When complete mixing occurs, $E_v = 1$; if air quality in the breathing zone is better than in the exhaust, E_v = greater than one; if air quality at the breathing zone is lower than exhaust air, E_v = less than one.

$$E_v = \frac{C_e}{C_r} \tag{10.2}$$

The PAQ technique is designed to provide sufficient outdoor ventilation air to maintain human comfort. Fanger[30] proposes using an approach similar to that described for the IAQ procedure to avoid health problems.

Reprise

Fanger's PAQ procedure attempts to address the shortcomings of the VR procedure by accounting for all the sources of pollution which contribute to occupant dissatisfaction with IAQ. Because it is more scientifically sound than the VR procedure, it represents a considerable advancement in providing ventilation that has a reasonable expectation of improving occupant satisfaction with air quality. Because its focus is primarily on comfort rather than health concerns, it represents an intermediate approach between using ventilation to control bioeffluents (VR procedure) and to minimize health effects (IAQ procedure). However, because the PAQ procedure takes the irritant effects of indoor contaminants into consideration, it is likely to provide significant health benefits as well.

Despite the fact that the PAQ procedure has fewer of the scientific shortcomings of the VR procedure, it is probably too complicated to be used by building designers and facility managers. The major difficulty associated with the PAQ procedure is estimating olf values associated with building materials and furnishings prior to building construction. Sufficient ventilation capacity must therefore be installed into buildings to mitigate worst case olf-loading conditions.

Despite its shortcomings, the VR procedure is likely to continue to be the ventilation approach of choice among designers of new buildings in North America and in other parts of the world. It has the advantage of simplicity and a long history of use. A minimum ventilation rate is specified by a consensus process and the design engineer simply has to select the equipment to meet the numbers specified.

Ventilation Effectiveness

The effectiveness of general dilution ventilation for controlling contaminants in building spaces may be diminished further by the poor mixing of contaminants and ventilation air. As a consequence, air quality may vary considerably within putatively well-ventilated building spaces. Of greatest concern is the quality of air in the breathing zone of room occupants.

Inadequate mixing of contaminants with both outdoor and recirculated air supplied to a space may occur as a consequence of short-circuiting.[32,33] In short-circuiting, a portion of the supply air bypasses the occupied zone and quickly leaves the space through return air outlets which are often located nearby. The degree of short-circuiting depends on the proximity of supply inlets and return outlets (typically located in the ceiling), the supply air temperature, and the nominal ventilation rate.[32,33] In addition, Farant et al.[34] have reported that the amount of outdoor air received by a person in an office building depends on the type of supply air diffuser and the type and size of return air outlets.

The design and layout of office spaces may also affect ventilation effectiveness (VE). Ventilation effectiveness was observed to be significantly better in closed office spaces as compared to open-concept office spaces. In the former, the proximity of supply and return air inlets and outlets respectively, and the supply air temperature were observed to be major determinants of VE.[34] Warmer supply air temperatures decreased VE. The size (area) of return air outlets was observed to have a significant effect on VE. When the area of return air outlets was reduced from 1240 to 310 cm^2, VE increased by a factor of 2.4. Considerable variation in the amount of outdoor air supplied to a space was also observed. However, VE was not affected by the volumetric flow rate of outdoor air.

In the office studies of Farant et al.,[34] ventilation was most effective in reducing contaminant levels in closed offices with dual slot-type diffusers and small egg-crate return air inlets followed by partitioned peripheral offices with two-way slot-type diffusers whose airflow was directed toward the windows. Lowest VE values were observed in the inner zones of partitioned (open plan) offices served by equidistant four-way conical air diffusers within 3 meters of large air returns. Ventilation effectiveness in different office space areas can be seen in Figure 10.3.

When supply air fails to mix with the air of occupied spaces or workstations, it can be said to be stratified, that is, forms two zones.[28,35] The phenomena of short-circuiting and stratification can be seen in Figure 10.4. Ventilation efficiency is, of course, significantly affected by the degree of stratification which is primarily determined by the supply air temperature and the recirculation rate.

Fisk et al.[36] observed in several San Francisco office buildings minimal short-circuiting in rooms when supply air temperatures were lower than indoor temperatures and a moderate degree of short-circuiting in rooms with heated supply air. Persily and Dols[37] observed that VE measurements were consistent with good distribution of outdoor air and mixing in occupied spaces of the John Madison Building of the Library of Congress. These studies as well as those of Farant et al.[34] indicate that VE is quite variable both among buildings and within building spaces.

Farant et al.[34] suggested that the poor VE they observed was consistent with the fact that most IAQ complaints originated from occupants in open-concept workspaces. In a study designed to evaluate the effect of varying levels of outdoor ventilation on SBS symptoms, Menzies et al.[38] observed that individuals in open-concept workspaces had significantly higher SBS prevalence rates. In the USEPA headquarters building study,[39] however, no significant spatial variation in health effects or evidence of "hot spots" for SBS symptoms were observed. Results were very mixed in another study[40] which focused on symptom assessment on the same day of questionnaire administration. Females, for example, working in open areas reported fewer eye and ergonomic symptoms whereas males showed a positive association between nasal symptoms and hours spent at a VDT. Mid-height partitions showed a positive association with air quality perception for males and a negative association for females.

Flush-Out Ventilation

Despite the limitations of general dilution ventilation for contaminant control in buildings, it may be used to reduce initially high TVOC concentrations in new or recently remodeled buildings during periods of carpet installa-

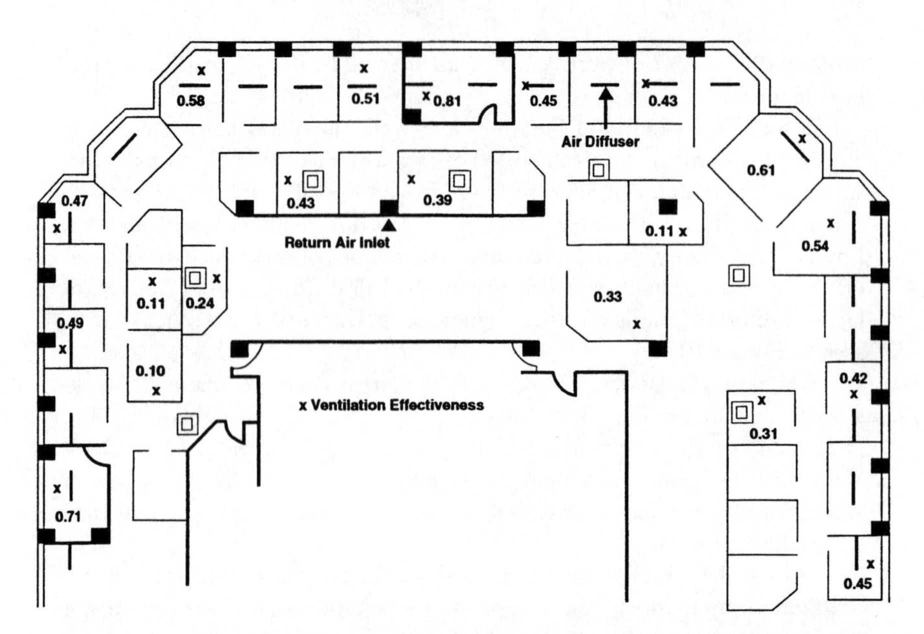

Figure 10.3 Ventilation effectiveness in different spaces of an office building. (From Farant, J.P. et al., 1991. *Proceedings IAQ '91: Healthy Buildings.* American Society of Heating, Refrigerating, and Air-Conditioning Engineers. Atlanta. With permission.)

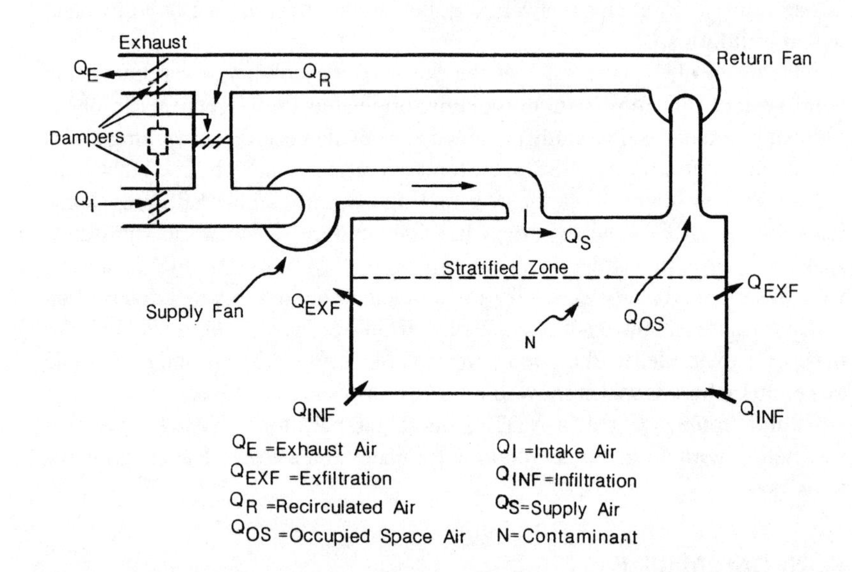

Figure 10.4 Diagrammatic representation of stratification phenomenon in mechanically ventilated building space. (From Janssen, J., 1984. *Ann. ACGIH: Evaluating Office Environmental Problems.* 10:59–67. With permission.)

tion and application of surface coatings.[41] In addition, the use of maximum outdoor ventilation rates appears to have the potential for accelerating the natural decrease of VOC emissions with time.

The principle of applying maximum ventilation rates for extended periods in newly constructed or remodeled buildings is described as flush-out ventilation. The use of maximum outdoor ventilation rates for a 6-month period was first reported in the Scandinavian countries.[42] More recently, this principle has been adopted by the state of Washington in their new office buildings program.[29] A flush-out period using 100% outside air for 90 days has been recommended. Flush-out begins prior to furniture installation to minimize potential sink effects. The use of flush-out ventilation for VOC control in newly constructed or remodeled office buildings has also been proposed in California.[43] Flush-out employing 100% outdoor air continuously would begin during construction and continue for a minimum of 4 weeks after occupancy. After this initial period, the hours of operation and the fraction of outdoor air would be reduced over a 6-month period.

The use of flush-out ventilation is based on both laboratory[6,44,45] and field studies[42,46,47] that have demonstrated that VOC and formaldehyde levels decrease significantly with time. Laboratory studies with formaldehyde[48-50] and VOCs[6] have demonstrated that high ventilation rates increase contaminant emission rates thus accelerating their normal decay with time.

Ventilation Innovations

A variety of efforts have been made to improve the effectiveness of conventional ventilation systems for contaminant control and to develop alternative systems which may be more effective than general dilution ventilation in reducing contaminant levels in occupied spaces. In the former case, demand-controlled systems increase and decrease outdoor air volumetric flow rates in response to changes in contaminant concentrations. In the latter case, displacement ventilation is designed to more effectively move contaminants generated in occupied spaces to exhaust outlets in the ceiling. Both innovations have the potential for significantly improving air quality in building spaces.

Demand-Controlled Ventilation

Demand-controlled ventilation (DCV) systems differ from conventional systems in that outdoor air flow rates are varied in response to contaminant concentrations measured by sensors located in the return air stream of a room or building. These systems have the advantage of supplying air in response to ventilation needs based on contaminant concentrations.[51-56] As such, they can be used to decrease outdoor air flows and conserve energy when selected contaminants are below acceptable levels.

Major problems associated with DCV include determining what contaminants are the best indicators of acceptable air quality and the availability of sensors that can perform the requisite task at an affordable cost. To date, sensors have been developed and are available for CO_2, humidity, and mixed gases.[54-56] Though CO_2 is a relatively poor indicator of IAQ as related to SBS-type symptoms, such sensors, nevertheless, are increasingly becoming popular for use in DCV. This reflects their relatively low cost, reliability, and simplicity of use and the historical role of the VR procedure as being a determinant of acceptable IAQ. Carbon dioxide sensors would, of course, be very effective in controlling bioeffluent levels in buildings. The effect of CO_2 levels on outdoor air flow rates in an auditorium can be seen in Figure 10.5.[53]

Infrared sensors are used to measure CO_2 levels. Photoacoustic sensors are also available. Air flow rates to spaces can also be controlled by "presence" sensors which are based on passive infrared absorption. Room air flows are reduced to a basic flow rate and increased to full flow when a room is occupied.[53]

Mixed-gas sensors are available commercially in Europe. Their chemical sensitivity is typically based on a group of compounds such as oxidized gases, non-oxidized gases, etc. Mixed-gas sensors are reported to be very sensitive to tobacco smoke and therefore have been successfully used in bars and restaurants. Raatschen[54] has reported, however, that mixed-gas sensors available on the European market have problems with stability and reliability.

Wenger et al.[55] evaluated the use of tin-oxide sensors (listed by the manufacture for the detection of ethanol, hydrogen sulfide, ammonia, cooking gases, organic vapors, combustible gases, air quality, and CO_2), 11 optical filters on a photoacoustic instrument, and measurements of temperature and relatively humidity. Using pattern recognition techniques, they were able to select those sensors, photoacoustic filters, and climate variables that demonstrated the best association with decipol values as determined by a trained panel. Though relatively good results were obtained, evaluations were conducted on only a limited variety of spaces. As such, the authors cautioned that observed results may not be applicable to all occupancy conditions, construction types, system performance levels, and climatic conditions. They did, however, suggest that a similar sensor array and pattern recognition technique developed over a broader data base could probably be used to reliably predict decipol levels over a broader range of spaces. These could be used as diagnostic tools, for on-line monitoring, and, of course, for determining ventilation needs.

Meckler[56] reports the use of CO_2 and mixed-gas sensors in a proprietary variable air volume/bypass filtration system (VAV/BFS). The mixed-gas or IAQ sensor is integrated into the VAV/BFS where it measures the concentration of VOCs in one or more zones or in a common return air duct and independently resets the supply air temperature leaving the air handling unit (AHU) to increase or decrease airflow through a filter/cleaner assembly in the bypass duct to maintain putatively satisfactory VOC levels. The sensor can be

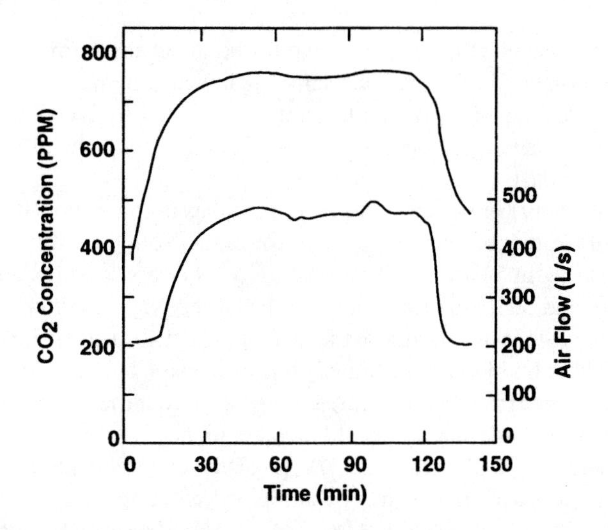

Figure 10.5 Demand-controlled increases in outdoor air flow rates in response to increased indoor CO_2 levels. (From Strindehog, O., 1991. *Proceedings IAQ '91: Healthy Buildings.* American Society of Heating, Refrigerating and Air-Conditioning Engineers. Atlanta. With permission.)

located in either the return-air duct or in an occupied space. Neither the nature of the mixed-gas sensor nor its reliability were reported.

Displacement Ventilation

In general ventilation, contaminant levels are reduced by mixing ventilation air with room air. Maximum contaminant reduction occurs under conditions of complete or perfect mixing which in many cases does not occur.

Displacement ventilation (DV) has been developed to more effectively ventilate spaces in industrial and office environments.[57-60] It is based on the principle of moving contaminated air upward to exhaust grilles in the ceiling in a piston-like fashion. It requires the upward flow of air from floor level (or near floor level). The objective of DV is to send excess heat and contaminants toward the ceiling to significantly improve the quality of air in the occupied zone.[60]

In DV, relatively low air flow rates (less than 10 L/s) are initiated at floor level at a temperature of approximately 18°C. Air is warmed by sources in the space and/or by the use of heaters. As air flows upward by convection, it creates by piston-action stratified zones of temperature and contaminants with the highest temperatures and contaminant levels near the ceiling. Displacement ventilation appears to work best when the convective flow rates and supply air flow rates are equal.[60] The "shift zone" is defined as the height where the contaminant or tracer gas concentration is one third of the concentration in the exhaust or extract air. Concentrations in the occupied zone would be less than or equal to one third of the concentration in exhaust air. In practical terms, this would mean that in DV air quality would be about three times better than general dilution ventilation with the same airflow rate. In maintaining good air quality, it is desirable to maintain the shift zone above the occupied zone. The relationship of tracer gas concentration with height and the height of the shift zone (Figure 10.6).

Displacement ventilation is a concept that has been developed, evaluated, and applied for indoor contaminant control in the Nordic countries[57-60] where it commands about 50% and 25% respectively, of new ventilation applications in industrial and office environments.[59] It has received little commercial attention elsewhere. This may be due to the fact that it may be better suited to Nordic countries where buildings are relatively tightly constructed, well-insulated, and heat loads are generally small compared to other countries where ventilation requirements are different and heat loads are high.[61]

The potential applicability of DV to office environments in Japan was evaluated by Koganei et al.[61] under a simulated office space of approximately 3 meters height. A room with heat loads of 30–170 W/m^2 was observed to have been divided into two zones: a lower unmixed "clean" zone with temperature stratification connected by heat source plumes to an upper mixed (dirty) zone with a relatively uniform temperature distribution, Figure 10.7. Contaminants added to the upper zone mixed with the turbulent air flow of that zone while contaminants emitted to the lower zone did not mix but ascended until entrained by heat source plumes. The height of the "clean zone" was approximately 1.6 meters. Maximum particle and CO concentrations were 20 and 5 times higher, respectively, in the upper "dirty" zone. Paradoxically, no vertical gradients of CO_2 emitted by smokers were observed.

Koganei et al.[62] conducted modeling studies of vertical DV. Their model assumed that (1) piston flow occurred in the "clean" zone and uniform mixing in the "dirty" zone, (2) no recirculation between zones except by heat source plumes, and (3) that the "clean" zone height could be expressed as a function of heat loads and supply air conditions. Though the model performed relatively well in predicting contaminant removal effectiveness, experimental data showed that the lower zone did not have pure piston flow since low levels of contaminants were measured despite a clean air supply to the test room. They also observed considerable variability in their contaminant data that they suggested may have been due to different diffusion and recirculation rates associated with different contaminant densities and varying temperatures of sources.

Unlike the studies of Koganei et al.,[61] Kim and Homma[63] observed significant effects of upward ventilation air flows coupled with free convection from human subjects on CO_2 levels, particularly when compared to CO_2 levels associated with conventional downward flow ventilation. Carbon dioxide concentrations in the former case increased with height with a maximum about 1 meter over the heads of seated subjects; in the latter, little variation with height was observed indicating that air in the chamber was relatively well mixed. Upward ventilation coupled with free convection was observed to have a

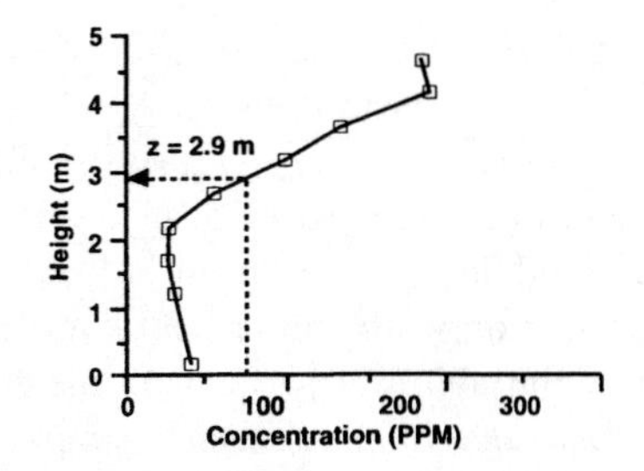

Figure 10.6 Shift zone associated with displacement ventilation. (From Laurikainen, J., 1991. *Proceedings IAQ '91: Healthy Buildings.* American Society of Heating, Refrigerating and Air-Conditioning Engineers. Atlanta. With permission.)

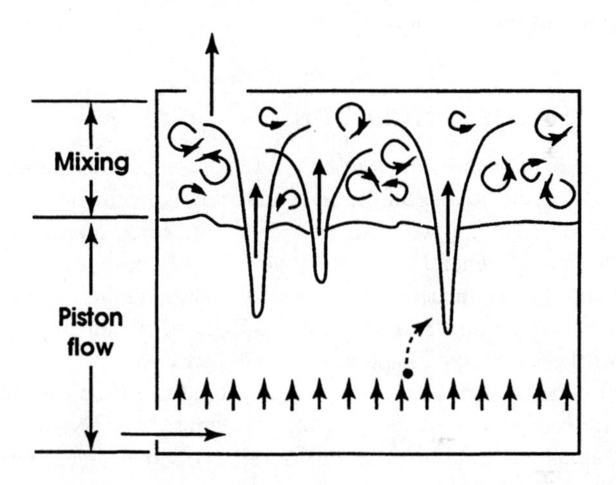

Figure 10.7 Diagrammatic representation of the effect of displacement ventilation on contaminants generated in a space. (From Koganei, M. et al., 1991. *Proceedings of the Sixth International Conference on Indoor Air Quality and Climate.* Vol. 5. Helsinki.)

significant effect on VE, extracting 1.50 times more CO_2 than conventional downward flow ventilation.

Breum[64] studied the effectiveness of DV in a printing plant. The estimated air exchange efficiency was 54% compared to the 100% for ideal displacement flow and 50% for complete mixing. Though there was a tendency toward displacement flow, it was very modest. In a study in an industrial environment (electroplating plant), Breum et al.[65] observed that DV was 1.6 to 6.6 times more effective in supplying "fresh" air to occupied zones.

Fisk et al.[36] in the United States reported preliminary results of what they described as "task ventilation", a method of ventilation which allows occupants to adjust local air supply parameters such as supply flow rate, direction, or temperature. They suggested that task ventilation can result in improved air quality because supply air can be delivered more directly to the region around the occupant. In some situations, task ventilation may produce displacement air flow patterns because slightly cool, denser air is supplied at floor or desk level and is typically removed at or near the ceiling.

Of the two task ventilation systems described by Fisk et al.,[36] the "floor supply" system was more likely to produce displacement-type ventilation. In this system, air supply modules were installed in a raised panel floor. Each module contained a fan that drew air from a subfloor plenum and discharged it into the room or space through four plastic 0.13-meter square grilles. Both air flow rate (40–90 L/s) and direction of air supply were controlled by occupants. Measurements of the air exchange effectiveness in 20 tests of these systems with and without recirculation did not indicate significant enhancement of ventilation in the breathing zone of occupants. A significant improvement of ventilation effectiveness indicative of displacement air flow patterns was observed in only one test.

LOCAL EXHAUST VENTILATION

Local exhaust ventilation can be used in special circumstances to reduce contaminant levels in buildings. It is particularly well-suited to situations where the source is known, emission levels are high, and for a variety of reasons the use of general ventilation would not be acceptable. In industrial environments, local exhaust ventilation is the control method of choice because it can remove contaminants from the workspace and presumably, the breathing zone of workers effectively and at relatively low cost.

Local exhaust ventilation is commonly used in office and other public-access buildings to control odors from lavatories. It is also used to control combustion and cooking by-products as well as odors from cafeteria kitchens. It is, of course, used to either passively (natural draft) or actively (fan-powered) exhaust combustion gases from space and water-heating combustion systems that serve most buildings. Local exhaust ventilation is also used to remove the high emissions of ammonia associated with blueprint machines commonly used in the engineering departments of counties and municipalities and planning departments of architectural and engineering firms, as well as other large institutions.

Increasingly, local exhaust ventilation is being used to remove tobacco-generated contaminants from designated smoking areas such as lounges and conference rooms. It makes sense to concentrate this activity in a confined area and to exhaust the contaminants to the outdoors before they contaminate air which would be recirculated throughout the building or zone served by a particular AHU.

Local exhaust ventilation may also be amenable to other applications as well. A variety of studies have implicated office equipment as sources of contaminants such as O_3,[66-68] TVOCs[69-73] and RSPs,[69] and several studies have implicated emissions from office equipment as contributing to health complaints.[69,74-76] Although it would be more desirable to control emissions from such sources as electrostatic photocopiers, wet-process photocopiers, laser printers, and even spirit duplicators by source control measures, particularly in future product development, local exhaust ventilation may be desirable for those situations where office equipment is used intensively, and as is often the case when such equipment is located in the most poorly ventilated spaces in a building. It may be desirable to locate office equipment in a "designated office equipment" room much like smoking areas, with these rooms being both locally exhausted and on a separate AHU. In the case of wet-process photocopiers, VOC emissions are so high that they may cause significant zonal, floor, or building-wide contamination and as a consequence, they would most appropriately be used in spaces provided with local exhaust ventilation.[70-72]

ENTRAINMENT/RE-ENTRY/CROSS-CONTAMINATION

As indicated in Chapter 3, indoor air contamination problems associated with contaminant entrainment, re-entry and cross-contamination are for the most part relatively easily mitigated. This is due to the fact that they are more easily diagnosed than most other IAQ problems.

Each of these IAQ problems is caused in one way or another by some deficiency in the design and operation of building ventilation systems. As such, their mitigation depends on eliminating those deficiencies in system design and operation which have created the problem.

In the case of entrainment of contaminants associated with outdoor sources, mitigation measures may include the relocation of outdoor air intakes relative to loading docks, parking areas or other sources of contaminants[77] or exhausting air from lower-level parking garages.

In the case of re-entry, mitigation measures may include the relocation of intakes relative to exhaust systems or exhaust systems relative to intakes. This would involve considerations of both proximity and direction of contaminant travel associated with outdoor air flows. The mitigation of re-entry problems may require changes in exhaust systems to increase both exhaust stack height and velocity[78] and to correct ventilation system imbalances.[79] In the last case, significant re-entry can occur by infiltration when the building is "starved for air", that is when exhaust exceeds outdoor air make-up needs. In balanced ventilation, typically only about 1% of exhausted gases returns to the building. In buildings with severe imbalance (excess of exhaust over intake) the re-entry of exhaust gases may be as high as 10–15%.[80] Such problems can be mitigated or avoided by providing sufficient outdoor air to compensate for air exhausted by local ventilation systems.

Pressure imbalances between different building areas served by separate AHUs are the principle cause of cross-contamination, a problem not uncommon in many multi-purpose buildings. In cross-contamination, contaminants generated in a production area, a printing facility, storage area, etc., may move to office areas where they are out of place and less tolerated. Mitigation in such instances requires that the air pressure in both areas be balanced relative to the other or that the area to be protected from contaminants be under positive pressure relative to contaminant-generating areas.

Problems of entrainment, re-entry, and cross-contamination are usually seen as being resolvable by the application of "common sense" engineering solutions. As such, there are only a few case reports on how effective such mitigation measures actually were. Bahnfelt and Govan[80] investigated a contamination problem in an office building which resulted from the horizontal discharge of grease fumes from a hamburger restaurant into an alley with subsequent entry into the outdoor intake of an office building. In response to complaints about odor and increased absenteeism among office employees, the horizontal mushroom-type exhaust unit was replaced by a centrifugal blower and a vertical discharge duct. This mitigation action was deemed moderately successful in that absenteeism declined to "normal" levels but some re-entry of restaurant vapors/odors continued to occur. Further mitigation efforts were made by extending the exhaust duct to the south side of the building with the exhaust plume being directed in a southwesterly direction.

Ludwig et al.[81] conducted tracer gas studies of the dilution of effluents from exhaust stacks of chemical fume hoods from a university laboratory complex. They observed that increased stack heights significantly increased dilution. The variable effects of turbulence were also decreased for greater stack heights (from 3.4 to 7.9 meters). Despite high dilution rates achieved, some chemical odors continued to be detected on the rooftop and the outdoor intake of the building. These were reported to be intermittent in nature and not expected to be health-affecting. Nevertheless, the fact that building occupants occasionally sensed odors was a continuing cause for concern.

AIR CLEANING

Although widely used in HVAC systems for the limited control of particulate dusts in buildings air streams, air cleaning has been for the most part rarely used as a mitigation measure for solving problem building complaints. This has been due in good measure to the fact that air cleaning has been historically focused on dust control whereas IAQ problems have been considered to be due to gas/vapor-phase contaminants. Because ventilation systems are in place in most cases, the use of dilution ventilation to resolve building air contamination problems and associated occupant complaints is the first and usually only choice of building investigators. Air cleaning for gaseous contaminants, on the other hand, would require the installation of expensive new filtration systems with which investigators have little familiarity and, as a consequence, less confidence that they in fact can resolve "the problem".

The potential role of particulate-phase materials such as dust in problem building complaints was discussed in detail in Chapter 5. Because airborne and surface dusts appear to be related to SBS symptom prevalence rates, the application of air cleaning for dust and other particulate-phase materials would appear to be a potentially effective mitigation measure in some problem buildings. The potential benefits of air cleaning may be limited since the strongest associations with SBS complaints appear to be associated with surface rather than airborne dusts.[82-84] It has been reported, for example, that office workers are exposed to dust levels which are three to five times higher than airborne levels.[85] This may be due to the fact that such individuals create their own "dust cloud" during the course of their work by stirring up settled dust.[84] Such dust exposures may not be particularly amenable to control by dust cleaning devices in HVAC systems.

Air cleaning can be used to reduce particulate- or gas-phase contaminants (or both) in building air streams and by implication, the air of building spaces as well. Air cleaning techniques for both particulate- and gas-phase materials are described below.

Dust/Particulate-Phase Contaminants

Most HVAC systems are designed to provide some measure of dust control. In many cases, panel or other low efficiency filters are located upstream of fans and other mechanical equipment to protect them from the fouling effects of large fibers, soil particles, etc.[86] Since they are primarily designed to control large particles, their overall collection efficiency is quite low (<20%). Filters and filtration systems of higher efficiency are also available but the value of such systems is usually not appreciated by building designers and managers who are often more concerned with cost than performance.[87]

Higher performance dust cleaning does incur higher (1) initial costs, (2) operating costs due to the increased resistance of higher efficiency filters which require greater fan power, and (3) replacement costs for higher efficiency filters.[86] Higher performance dust cleaning systems do, of course, have significant benefits. These include improved long-term building maintenance, cleanliness of interior building surfaces, and occupant health and comfort.[87]

Inadequacies even in the use of low efficiency dust-stop filters have been reported to have significant impacts on building air quality. Woods et al.[88] and Lane et al.[89] have suggested that the failure to use and adequately maintain HVAC system filters can be a significant contributing factor to problem building complaints. Badly soiled filters, for example, may increase air flow resistance and decrease system flow rates. Missing filters may cause the soiling of mechanical equipment and impair its performance.

A variety of moderate to high efficiency filters and filtration systems are available for dust cleaning in buildings. These include dry-type media panel filters, viscous media panel filters, renewable media filters and extended-surface dry-type filters. In the last case, efficiency is enhanced by increasing the density and thickness of filter media. The surface is extended by pleating the collecting medium or using bag-like construction. By extending the surface, flow losses associated with increased air resistance are reduced. Electronic air cleaners collect particles by imparting positive charges to particles passing through the system. They are then deposited and collected on negatively charged metal plates downstream. Dust cleaners including those utilizing media and electronic filtration have been reviewed in the ASHRAE *Equipment Handbook*[90] and by Godish.[91]

Air Cleaner Performance

The ability of air cleaning systems to effectively control particles in a building depends on the overall cleaner efficiency,[92] particle size,[93] the proper installation of filters/filtration systems,[87] the rate of airflow for electronic air cleaners,[94] and recirculation rates.[92] The effectiveness of air cleaners under real-world conditions will depend on the design efficiency of the filter/filtration system. Filtration efficiencies vary across the particle size spectrum with small particles (circa 0.1–1 μm) being the most difficult to collect.[93] Overall cleaning effectiveness increases each time air passes through the filter. High recirculation rates therefore increase cleaning effectiveness.[92] In electronic air cleaners,[94] cleaning effectiveness decreases with increased volume flow rates.

Other factors affect the performance of filtration systems as well. One of the more neglected of these is the proper installation and sealing of filters/filtration systems so that air does not bypass the filter medium/system.[87] Filtration systems can only remove particles from air that passes through them. "Leakage" around poorly installed filters can seriously compromise cleaning performance.

The effectiveness of air cleaning in occupied spaces can also be affected by spatial factors. If particles are being generated in a space from a localized source (such as an electrostatic photocopier), the effect of the air cleaner on local particle concentrations may be negligible. Particles must remain airborne and be carried to the air cleaner before they can be collected. This may often be a considerable distance. The source may, as a consequence, cause relatively high exposures because of its proximity to occupants. One example of this phenomenon (relative to the distance that particles must travel before they are removed) is when particles are generated in the HVAC system downstream of the filtration unit. This could include biological contaminants produced and amplified in condensate drip pans[95,96] and porous HVAC system insulation[97,99] and shedding of fibrous glass and other particulate phase materials associated with duct materials.[99]

Evaluation of In-Use Performance

A variety of investigators have attempted to evaluate the performance of particle air cleaners under laboratory conditions or in actual building spaces. In the former case, Leaderer et al.[100] evaluated the performance of an electronic air cleaner (EAC) in a large chamber on RSPs produced by tobacco smoking. The EAC was observed to be highly effective, reducing RSPs (by 95%) under relatively high recirculation rates. Jasinghani et al.[101] evaluated the effectiveness of six air cleaners in removing tobacco smoking-generated RSP in a restaurant, a lounge, and in an office using a mass balance model. Their model results predicted that an ionizing electrically stimulated filter (IESF) was the most effective device, followed in descending order by 95% DOP hospital-type filters, a 95% ASHRAE-type filter, EAC, and a high-efficiency particulate absolute (HEPA) filter. Although the HEPA filter had the highest rated collection efficiency, it had the lowest overall predicted particle collection because of reduced flow rates associated with it because of its higher resistance.

Weschler and Shields[102] studied the effectiveness of extended-surface pleated filters (ASHRAE dust spot rating of 80–85%) in a telephone building. Concentrations of fine indoor particles (0.6–3.0 μg/m^3) were observed to be 1/5 to 1/20th of those outdoors (the major source of fine particles in the building). Coarse particles were also very low, with most values below 0.5 μg/m^3.

Roys et al.[103] conducted experimental studies to determine the effectiveness of filtration as well as other dust cleaning activities on SBS symptom prevalence rates. No significant effect on SBS symptoms was observed with either ceiling or desktop cleaning units.

Control of Mold Spores/Particles

Air cleaning is widely used to control mold levels in homes of asthma and allergy patients. Its effectiveness under proper conditions of use has been demonstrated.[104,105] It is also widely used in hospitals to reduce potential mold infections in immune-compromised patients undergoing transplants or treatment of malignancies. Fungal infections cannot be treated by conventional antibiotic therapy. Of particular concern is *Aspergillus fumigatus*. Streifel et al.[106] reported results of a study of a university hospital building which had been designed as a sealed building with two filter systems having greater than 95% collection efficiencies for 1 μm particles. Indoor mold levels in operating rooms, patient wards, etc., were very low (2–8 CFU/m^3) as compared to outdoor levels (745 CFU/m^3), in the main lobby (41 CFU/m^3), and in an older hospital (83 CFU/m^3).

Streifel and Rhame[107] reported results of mold spore testing in a hospital specially designed to minimize mold spore levels throughout the building. It included HEPA filtration of a bone-marrow transplant unit and 90–95% dust spot efficiency for most of the rest of the building. Outside of the lobby area, fungal counts were typically less than 30 CFU/m^3 with *A. fumigatus* less than 0.5 CFU/m^3.

Such hospital environments represent an extreme case (other than clean rooms) for both the control of mold levels and particles. In such instances, a significant effort is made to install, utilize, and maintain high efficiency filtration systems.

It is notable that in many buildings with mechanical air handling units that mold levels are significantly lower indoors than they are outdoors.[108-110] This is evident in the old hospital building described above. The lower indoor levels suggest that mold spores which are entrained in outdoor air brought into the building for ventilation purposes is in good measure cleaned of its mold load. Such reductions may be due to a variety of factors. However, the most likely is the low-efficiency dust stop filters used to protect mechanical equipment. Their apparent effectiveness may be due to relatively high recirculation rates and increasing efficiencies associated with filter soiling.

Gaseous Contaminants

In general, it is more difficult to control gas/vapor-phase substances than particles with air cleaning devices. This is in part due to the fact that many substances have different chemical properties and therefore may not be effectively removed by generic control technologies. Adsorption using activated-carbon sorbents is the most widely used technique for removing gas-phase substances from airstreams. It is particularly useful for higher boiling point/molecular weight non-polar organic substances such as toluene, xylene, chlorinated compounds, etc. It is relatively ineffective in adsorbing and retaining low-molecular weight organic substances such as formaldehyde and ethylene and inorganic gases such as carbon monoxide and oxides of nitrogen.

Because activated-carbon has a long history of use in solvent recovery, industrial gas cleaning, respirators, and odor control in buildings, it has received some attention as a potential control measure in resolving sick building complaints. The use of activated-carbon filters would be expected to be relatively effective in removing many of the VOCs (other than formaldehyde) found in the air of office and institutional buildings.

The use of activated-carbon for controlling VOC levels in buildings has been recently investigated by USEPA and its contractors.[111-114] In studies conducted by Ramanathan et al.,[111,112] activated-carbon used in panel and duct filters appeared to have limited usefulness in indoor VOC control. Under controlled laboratory conditions, a 6-inch carbon bed used for the adsorption of parts per billion concentrations of benzene, acetaldehyde, and methyl chloroform had a reported lifetime of only minutes. These results are at variance with the service life projections of Sleik and Turk[115] for thin-bed (2.5 cm) carbon filters and the use of carbon for VOC removal in respirators in the parts per million range.

Graham and Bayati[116] reported that good quality coconut-shell charcoal is highly effective in adsorbing and retaining VOCs such as benzene up to more than 5% of its weight and that thin-bed (2.5 cm) filter systems are effective in removing trace VOCs from indoor air. Service life was not reported.

Liu[117] reports that thin-bed activated-carbon filters are effective in removing VOCs at the parts per billion concentrations found in buildings. The adsorption capacity of activated-carbon filters was four orders of magnitude higher than reported by Ramanathan et al.[111,112] with a service life of 60–466 days at exposure VOC concentrations of 7.5 and 143 ppb, respectively, at an efficiency of 80%.

Holmberg et al.[118] evaluated four different activated carbon systems for use in AHUs of ventilation systems. Panel filters with bed thickness of 15 and 27 mm arranged in a V configuration were observed to have excessive pressure drops and thus were deemed to be unsuitable for use in AHUs. Acceptable pressure drops were reported for filters consisting of carbon spheres adhered to open-pore polyurethane foam (foam-bed filters) and cyclindral filters in which activated carbon was placed between concentric perforated shells. The foam-bed filter was reported to have an initially high efficiency which dropped to 70% in 18 hours on exposure to 10 ppm xylene. Because VOC concentrations are usually considerably lower than this, the useful life under continuous operation was suggested to range from a minimum of several hundred hours up to 6 to 12 months.

Kinkead[119] reported results of pleated dry-processed carbon composite (DPCC) based adsorption filters on removing gas-phase contaminants from airstreams under laboratory conditions. DPCC adsorbers consist of very fine particles of activated carbon (or chemically treated activated carbon) introduced into a fiber mat structure. Such adsorbers were observed to have 10 times the adsorption capacity and life of large mesh activated-carbon sorbent. Service life was not reported. Liu[120] indicates that the service life of such adsorbers is much shorter than panel types.

Few investigators have attempted to evaluate the effectiveness of activated-carbon filtration systems in mitigating IAQ problems in buildings. Sterling et al.[121] recommended the installation of an activated-carbon filtration system in the AHU of a library in a four-story office complex plagued by odors from a restaurant and lower level parking garage. Odor complaints were reduced but not eliminated. Holmberg et al.[118] used foam-carbon filters in an airport terminal building to control malodors associated with aircraft operation. Filters were reported to reduce peak concentrations effectively and had to be replaced approximately a year after installation in response to occupant complaints of reoccurring odor problems.

The low service life of carbon beds reported in USEPA-sponsored studies are somewhat problematic particularly in light of other studies which indicate that thin-bed and foam-filters have relatively good adsorption efficiencies and service lives. Assuming that gas/vapor sorbing media filters are as effective as reported by some investigators, they nevertheless would have significant limitations relative to their use in mitigating IAQ problems. One of the more significant of these is the differential adsorption of VOC compounds (activated carbon can only effectively adsorb and retain nonpolar compounds with relatively high boiling points). It would be ineffective on substances such as formaldehyde without the use of specially impregnated media. The effectiveness and service life of carbon media can also be compromised by operating in atmospheres of elevated relative humidities.[120] Such air cleaning would also be confronted with problems which reduce the effectiveness of dilution ventilation: increasing emissions with increasing contaminant removal rates and high reduction requirements to significantly reduce irritant responses.

Rodberd et al.[122] describe an air cleaning system consisting of a series of beds which use O_3-laden air to react with contaminants followed by a hydrocarbon-oxidation catalyst which converts hydrocarbons to CO_2 and H_2O. The resultant airstream is then passed over an O_3-destruction bed to reduce O_3 to background levels. The removal of VOCs (as benzene) was observed to be approximately 50% per pass. The system has only been tested under laboratory conditions and its effectiveness and service life under real-world use was not known.

Combined Filtration Systems

Hedge et al.[123] conducted evaluations of a combined filtration system integrated into an office work station to provide breathing zone filtration (BZF). This BZF system consisted of a coarse filter, an activated-carbon bed and a HEPA filter. Before and after measurements of both contaminant levels and occupant questionnaire responses showed that the operation of the BZF system resulted in a significant reduction in particle levels and significant improvements in occupant ratings of lethargy with marginal improvements in breathing difficulty, dizziness, nasal congestion, and dry cough. Overall perceptions of ventilation, IAQ, and thermal comfort appeared to be better in areas with BZF. TVOC levels were quite variable and the effectiveness of the BZF system on TVOCs could not be assessed.

Hedge et al.[124] conducted another *in situ* evaluation of the BZF system in a building in which occupants had expressed no previous dissatisfaction with air quality. The system was observed to increase vertical dispersion and appeared to decrease submicronic (<0.25 μm) particle counts by over 80%. The latter observation was associated with a significant reduction in perceived workplace dustiness. Complaints of poor IAQ decreased by 72%; insufficient ventilation, 74%; too little air movement, 65%; and stale or dusty air, 50%. The work-related symptom prevalence index was observed to decrease from 3.84 to 2.05 symptoms per employee. Significant reductions were reported for complaints of lethargy, dry skin, and headache. Disruptions in productivity and tired, irritated eyes were significantly reduced. A 61% decrease in certified sickness absence was also reported. VOC levels were generally low and no evidence was presented to indicate any beneficial effect of the system in reducing VOC exposures.

Air Cleaners as Contaminant Sources

Though air cleaners are designed to remove contaminants from indoor air, they may also cause indoor contamination. This is true for filters which may be sources of VOCs and microbial contaminants. Molhave and Thorsen[125] reported significant emissions of VOCs from HVAC system filters. Significant emissions of toluene and xylene from oil-treated filters were also reported by Pasanen et al.[126] as well as a variety of odorous aldehydes (e.g., hexanal, heptanal, octanal, and nonanal) and light organic acids from filters infested with microorganisms. Filters can also be a source of fibrous particles shed during HVAC system operation[99] and electronic air cleaners, a source of O_3.[94]

Filters may produce odors which may decrease satisfaction with air quality. Teijonsalo et al.,[127] using a trained panel, observed that odor intensity of panel filters increased rapidly during the first 13 weeks of use with peak intensities at 19 weeks corresponding to 2.5 decipols (about a 30% dissatisfaction rate). Coarse prefilters were observed to lower odor intensities associated with fine filters and extend useful service life of the latter. The use of inexpensive, frequently replaced prefilters was suggested to be a good way of improving the quality of indoor air.

Filters in HVAC systems can capture and retain fungal spores and mycelia fragments[128-130] which subsequently grow on collected dust. Such growth can penetrate filter media contaminating other filters and air in conditioned spaces downstream. Condensation of water on filters and air ducts contributes to fungal growth on filters.[131]

Though a combination of a low-efficiency dust stop and high-efficiency filters (with a dust-spot rating of 80–90%) were reported by Elixmann et al.[132] to be effective in removing mold spores from HVAC system airstreams, the high efficiency filter was observed to be a significant source of fungal allergen activity downstream. Initially, allergen activity was high (Table 10.5) before the dust stop filter (due to mold spore levels in outdoor air), decreased on passing through it, and increased downstream of the high efficiency filter. Though effective in controlling the initial contaminants of concern, filtration was ineffective in mitigating health effects because allergens were produced within filters with subsequent transmission to the supply airstream.

The use of a biostatic, putatively low-toxicity phosphated quaternary amine compound to control fungal growth on a variety of filtration media was evaluated in studies conducted by Price et al.[133] Treated air filter media were reported to significantly inhibit fungal growth in both laboratory challenge studies and in 30-day HVAC system exposures. No data, however, were provided relative to potential emission of the biocide into indoor air. Exposure to biocides from humidifiers has been suggested by Burge and Robertson[134] to be a potential contributing factor to SBS symptoms.

RECAPITULATION

General dilution ventilation is widely used to minimize odor and comfort complaints associated with human bioeffluents and as a generic measure to mitigate SBS symptoms and IAQ complaints. The use of ventilation for contaminant control is based on general dilution theory. General dilution theory

Table 10.5 ELISA inhibition (allergen activity) of air samples associated with a filtration system in an asthma hospital.

Sample	Collection tube[a]	Inhibition (%)
Air before dust stop filter	A	95[b]
	B	43[b]
	C	21
Air after dust stop filter	A	47[b]
	B	25
	C	24
Air before high efficiency filter	A	46[b]
	B	41[b]
	C	34
Air after high efficiency filter	A	94[b]
	B	42[b]
	C	42[b]

[a] A,B,C serially placed.

[b] Significant ELISA inhibition.

From Elixmann, J.H. et al., 1989. *Environ. Int.,* 15:193–196. With permission from Elsevier Science Ltd.

appears to be applicable to contamination problems such as episodic emissions from tobacco smoking or office copiers or relatively constant emission of human bioeffluents. Ventilation based on general dilution theory is less effective in controlling contaminants such as formaldehyde and VOCs which are emitted by diffusion processes.

Mixed results have been reported from attempts to evaluate ventilation rates on building occupant health and comfort. It appears that increasing ventilation rates to 20 CFM (10 L/s) per person results in decreased symptom prevalence with higher ventilation rates being relatively ineffective. The limited effectiveness of dilution ventilation in reducing SBS symptom prevalence rates may in part be due to the fact that irritant responses are a log-linear function of dose. As a consequence, order of magnitude increases in outdoor ventilation rates would be required to achieve significant reductions in irritant symptoms.

Studies designed to evaluate the effectiveness of ventilation on contaminant concentrations in building environments have also reported mixed results. Because of the significant effect of infiltration, investigators could not, in many cases, attain target air exchange rates by varying outdoor volumetric flow rates. Several studies have reported significant associations between outdoor ventilation rates and the natural log of indoor/outdoor contaminant ratios.

Ventilation standards/guidelines have been developed to assist building designers and managers in providing adequate outdoor ventilation rates to achieve and maintain good air quality. Historically, the VR procedure has been the choice of building professionals in establishing minimum ventilation requirements. The VR procedure prescribes outdoor air flow rates to control bioeffluent levels to consensus guideline values. This procedure is the primary determinant of ventilation practice in buildings despite the fact that it has little relationship to contaminants generated in building spaces other than human bioeffluents. The air quality (AQ) procedure has been developed as an alternative ventilation approach. It specifies guideline values for contaminant levels which are to be achieved by the use of both source control and outdoor air ventilation. The AQ procedure has not received any practical use because of the limited information available on health end points for various contaminants and complex demands on building designers who do not have the requisite knowledge to use it. The PAQ procedure attempts to determine the acceptability of indoor air and requisite outdoor air ventilation flow rates by using panels of trained judges. The PAQ approach is intended to provide sufficient outdoor air to maintain human comfort and reduce occupant health complaints associated with irritants. Limited contaminant emission data and associated air quality impacts reduce its practical application.

The effectiveness of general ventilation can be compromised by short-circuiting phenomena in which supply air fails to adequately mix with air of occupied spaces. Ventilation effectiveness depends on the proximity of supply and return inlets and outlets, supply air temperature, ventilation rates, type of supply air diffuser, and type and supply of return air outlets. Ventilation effectiveness has been reported to be poorer in open-concept office spaces compared to closed office spaces.

Flush-out ventilation is a mitigation technique used in new or renovated buildings. By using initially high outdoor ventilation flow rates, the rate of "decay" of source emissions that occur with time is increased, reducing the time period that relatively high VOC levels are present in such buildings. Local exhaust ventilation is used in most buildings. Applications include the control of lavatory and cafeteria odors and exhaust of boiler flue gases and ammonia from blueprint machines.

A variety of innovations are being developed to improve the effectiveness of conventional ventilation systems for contaminant control. Prominent among these are DCV and DV. In the former case, the volumetric outdoor ventilation rate is controlled by sensors located in the return airstream of air handling units. Though sensors are available for different contaminants, most DCV at the present time is based on the use of CO_2 sensors. Displacement ventilation attempts to remove contaminants from a space by convective movement of air in a piston-like fashion from the floor level to exhausts in the ceiling. When properly designed, DV is significantly more effective in removing contaminants from occupied spaces than conventional dilution ventilation.

Ventilation system-related contamination problems such as entrainment, re-entry, and cross-contamination can be avoided or mitigated by proper design and operation of HVAC systems. These include the proper location of intakes and exhausts and maintaining proper pressure balances between different air handling units and between the inside and outside of the building.

Air cleaning is less widely used to control contaminants in buildings than general dilution ventilation. It includes the control of both particulate- and gas-phase substances which require the application of specially designed control technologies. Particle control is more widely applied and a minimum level of particle cleaning is used in most HVAC systems to protect mechanical equipment. A variety of filters and filtration systems are available for particle control in buildings. These vary from low-efficiency dust-stop filters in most HVAC systems to very high-efficiency systems used in hospital environments and clean rooms. Dust or particle filtration systems can be relatively effective in reducing indoor particle and mold spore levels.

Gases and vapors may be removed from indoor air by means of sorption media such as activated-carbon. Mixed results have been reported in studies conducted on the effectiveness and service life of carbon filters. USEPA-sponsored studies indicate that activated-carbon filters have an extremely short service life and as a consequence, would not be effective in controlling contaminant levels in buildings. A number of other studies indicate that carbon filtration can be used effectively with service lives up to 6–12 months. As with general dilution ventilation, a variety of factors are likely to compromise the effectiveness of gas cleaning in controlling contaminants and mitigating occupant complaints.

Filtration systems may themselves be sources of contaminants including VOCs, O_3, fibers, microbial volatiles, fungal spores and allergenic particles. Filters including high-efficiency types may be amplification sites for mold growth producing allergens which pass through even high efficiency filters contaminating the air of occupied spaces downstream. Filters as they become soiled can produce odors which may cause occupant dissatisfaction with air quality.

General dilution ventilation and air cleaning have significant limitations in both controlling contaminants and reducing occupant health and comfort complaints. As a consequence, they should not be viewed and applied generically to solve health and air quality problems.

REFERENCES

1. Seitz, T.A. 1990. "NIOSH Indoor Air Quality Investigations: 1971 through 1988." 163–171. In: Weekes, D.M. and R.B. Gammage (Eds.). *Proceedings of the Indoor Air Quality International Symposium: The Practitioner's Approach to Indoor Air Quality Investigations.* American Industrial Hygiene Association. Akron, OH.
2. Billings, C.E. and S.F. Vanderslice. 1982. "Methods for Control of Indoor Air Quality." *Environ. Int.* 8:497–504.
3. ASHRAE Standard 62–1981. 1981. "Ventilation for Acceptable Indoor Air Quality." American Society of Heating, Refrigerating, Air-Conditioning Engineers. Atlanta.
4. ASHRAE Standard 62–1989. 1989. "Ventilation for Acceptable Indoor Air Quality." American Society of Heating, Refrigerating, and Air-Conditioning Engineers. Atlanta.

5. Matthews, T.G. et al. 1983. "Formaldehyde Release from Pressed Wood Products." In: Maloney, T. (Ed.). *Proceedings of the Seventeenth International WSU Particleboard/Composite Materials Symposium.* Washington State University. Pullman, WA.
6. Tichenor, B.A. and Z.L. Guo. 1991. "The Effect of Ventilation on Emission Rates of Wood Finishing Materials." *Environ. Int.* 17:317–323.
7. Hodgson, A.T. and J.R. Girman. 1989. "Application of a Multisorbent Sampling Technique for Investigations of Volatile Organic Compounds in Buildings." 244–256. In: Nagda, N.L. and J.P. Harper (Eds.). *Design and Protocol for Monitoring Indoor Air Quality.* ASTM STP 1002. American Society for Testing and Materials. Philadelphia.
8. Godish, T. and J. Rouch. 1988. "Residential Formaldehyde Control by Mechanical Ventilation." *Appl. Ind. Hyg.* 3:93–96.
9. Matthews, T.R. et al. 1986. "Preliminary Evaluation of Formaldehyde Mitigation Studies in Unoccupied Research Homes." 389–401. In: Walkinshaw, D.S. (Ed.). *Transactions: Indoor Air Quality in Cold Climates.* Air Pollution Control Association. Pittsburgh, PA.
10. Jewell, R. 1984. "Reducing Formaldehyde Levels in Mobile Homes Using 29% Aqueous Ammonia Treatment or Heat Exchangers." Weyerhauser Corporation, Tacoma, WA.
11. Alarie, Y. 1981. "Dose-Response Analysis in Animal Studies: Prediction of Human Responses." *Environ. Health Perspect.* 42:9–13.
12. Anderson, R.C. 1991. "Measuring Respiratory Irritancy of Emissions." 19–23. In: *Post-Conference Proceedings of IAQ '91: Healthy Buildings.* American Society of Heating, Refrigerating and Air-Conditioning Engineers. Atlanta.
13. Palonen, J. and O. Seppanen. 1990. "Design Criteria for Central Ventilation and Air-Conditioning System of Offices in Cold Climate." 299–304. In: *Proceedings of the Fifth International Conference on Indoor Air Quality and Climate.* Vol. 4. Toronto.
14. Nagda, N.L. et al. 1991. "Effect of Ventilation Rate in a Healthy Building." 101–107. In: *Proceedings of IAQ '91: Healthy Buildings.* American Society of Heating, Refrigerating and Air-Conditioning Engineers. Atlanta.
15. Collett et al. 1991. "The Impact of Increased Ventilation on Indoor Air Quality." 197–200. In: *Proceedings of IAQ '91: Healthy Buildings.* American Society of Heating, Refrigerating and Air-Conditioning Engineers. Atlanta.
16. Farant, J.P. et al. 1990. "Effect of Changes in the Operation of a Building's Ventilation System on Environmental Conditions at Individual Workstations in an Office Complex." 581–585. In: *Proceedings of the Fifth International Conference on Indoor Air Quality and Climate.* Vol. 1. Toronto.
17. Baldwin, M.E. and J.P. Farant. 1990. "Study of Selected Volatile Organic Compounds in Office Buildings at Different Stages of Occupancy." 665–670. In: *Proceedings of the Fifth International Conference on Indoor Air Quality and Climate.* Vol. 2. Toronto.
18. Norback, D. et al. 1990. "Volatile Organic Compounds, Respirable Dust, and Personal Factors Related to Prevalence and Incidence of Sick Building Syndrome in Primary Schools." *Br. J. Med.* 47:733–741.

19. Menzies, R.I. et al. 1990. "Sick Building Syndrome: The Effect of Changes in Ventilation Rates on Symptom Prevalence: the Evaluation of a Double Blind Experimental Approach." 519–524. In: *Proceedings of the Fifth International Conference on Indoor Air Quality and Climate*. Vol. 1. Toronto.
20. Fanger, P.O. 1988. "Introduction of the Olf and Decipol Units to Quantify Air Pollution Perceived by Humans Indoors and Outdoors." *Energy and Buildings*. 12:1–6.
21. Fanger, P.O. 1989. "The Comfort Equation for Indoor Air Quality." *ASHRAE J*. October:33–38.
22. Wang, T.C. 1975. "A Study of Bioeffluents in a College Classroom." *ASHRAE Trans*. 81:32–44.
23. Turiel, I. et al. 1983. "The Effects of Reduced Ventilation on Indoor Air Quality in an Office Building." *Atmos. Environ*. 17:51–64.
24. ASHRAE Standard 62–73. 1977. "Standards for Natural and Mechanical Ventilation." American Society of Heating, Refrigerating and Air-Conditioning Engineers. Atlanta.
25. World Health Organization. 1987. "Air Quality Guidelines for Europe." World Health Organization Regional Publications. European Series No. 23. Copenhagen.
26. Grimsrud, D.T. and K.Y. Teichman. 1989. "The Scientific Basis of Standard 62–1989." *ASHRAE J*. 31:51–54.
27. Grimsrud, D.T. 1990. "Future Directions for Ventilation Standards." 365–369. In: *Proceedings of the Fifth International Conference on Indoor Air Quality and Climate*. Vol. 5. Toronto.
28. Janssen, J. 1984. "Ventilation for Acceptable Indoor Air Quality: ASHRAE Standard 62–1981." In: *Ann. ACGIH: Evaluating Office Environmental Problems*. 10:59–67.
29. Black, M. et al. 1991. "The State of Washington's Experimental Approach to Controlling IAQ in New Construction." 39–42. In: *Proceedings of IAQ '91: Healthy Buildings*. American Society of Heating, Refrigerating and Air-Conditioning Engineers. Atlanta.
30. Fanger, P.O. 1990. "New Principles for a Future Ventilation Standard." 353–363. In: *Proceedings of the Fifth International Conference on Indoor Air Quality and Climate*. Vol. 5. Toronto.
31. Gunnarsen, L. and C. Broals. 1990. "Adaptation and Ventilation Requirements." 599–604. In: *Proceedings of the Fifth International Conference on Indoor Air Quality and Climate*. Vol. 1. Toronto.
32. Sandberg, M. 1981. "What is Ventilation Efficiency?" *Build. Environ*. 16: 123–135.
33. Janssen, J. et al. 1982. "Ventilation for Control of Indoor Air Quality: A Case Study." *Environ. Int*. 8: 487–496.
34. Farant, J.P. et al. 1991. "Impact of Office Design and Layout on the Effectiveness of Ventilation Provided to Individual Workstations in Office Buildings." 8–13. In: *Proceedings of IAQ '91: Healthy Buildings*. American Society of Heating, Refrigerating and Air-Conditioning Engineers. Atlanta.
35. Woods, J.E. 1984. "Measurement of HVAC System Performance." In: *Ann. ACGIH: Evaluating Office Environmental Problems*. 10:77–92.

36. Fisk, W.J. et al. 1991. "Air Exchange Effectiveness of Conventional and Task Ventilation for Offices." 30–34. In: *Post-Conference Proceedings of IAQ '91: Healthy Buildings*. American Society of Heating, Refrigerating, and Air-Conditioning Engineers. Atlanta.
37. Persily, A. and W.S. Dols. 1991. "Field Measurements of Ventilation and Ventilation Effectiveness in an Office/Library Building." *Indoor Air*. 3:229–245.
38. Menzies, R. et al. 1991. "The Effect of Varying Levels of Outdoor Ventilation on Symptoms of Sick Building Syndrome." 90–96. In: *Proceedings of IAQ '91: Healthy Buildings*. American Society of Heating, Refrigerating and Air-Conditioning Engineers. Atlanta.
39. Wallace, L.A. et al. 1991. "Workplace Characteristics Associated with Health and Comfort Concerns in Three Office Buildings in Washington, D.C." 56–60. In: *Proceedings of IAQ '91: Healthy Buildings*. American Society of Heating, Refrigerating and Air-Conditioning Engineers. Atlanta.
40. Nelson, C.J. et al. "EPA's Indoor Air Quality and Work Environment Survey: Relationships of Employees' Self-Reported Health Symptoms with Direct Indoor Air Quality Measurements." 22–32. In: *Proceedings of IAQ '91: Healthy Buildings*. American Society of Heating, Refrigerating and Air-Conditioning Engineers. Atlanta.
41. Levin, H. 1989. "Building Materials and Indoor Air Quality." 667–693. In: Cone, J.E. and M.J. Hodgson (Eds.). *Problem Buildings: Building-Associated Illness and the Sick Building Syndrome. Occupational Medicine: State of the Art Reviews*. Hanley and Belfus, Inc. Philadelphia.
42. Wallace, L. et al. 1987. "Volatile Organic Chemicals in Ten Public Access Buildings." 188–192. In: *Proceedings of the Fourth International Conference on Indoor Air Quality and Climate*. Vol. 1. West Berlin.
43. California Department of Health Services. 1992. "Guidelines (Draft) for Reduction of Exposure to Volatile Organic Compounds (VOC) in Newly Constructed or Remodeled Office Buildings." November.
44. Black, M.S. et al. 1991. "A Methodology for Determining VOC Emissions from New SBR Latex-Backed Carpet, Adhesives, Cushions and Installed Systems and Predicting Their Impact on Air Quality." 267–272. In: *Proceedings of IAQ '91: Healthy Buildings*. American Society of Heating, Refrigerating and Air-Conditioning Engineers. Atlanta.
45. Storbridge, J.R. and M.S. Black. 1991. "Volatile Organic Compounds and Particle Emission Rates and Predicted Air Concentrations Related to Movable Partitions and Office Furniture." 292–298. In: *Proceedings of IAQ '91: Healthy Buildings*. American Society of Heating, Refrigerating and Air-Conditioning Engineers. Atlanta.
46. Berglund, B. et al. 1982. "A Longitudinal Study of Air Contaminants in a Newly Built Preschool." *Environ. Int.* 8:11–15.
47. Hartwell, T.D. et al. 1985. "Levels of Volatile Organics in Indoor Air." In: *Proceedings of the 79th Annual Meeting of the Air Pollution Control Association*. Detroit.
48. Matthews, T.G. 1986. "Modeling and Testing of Formaldehyde Emission Characteristics of Pressed Wood Products. Report XVIII to U.S. Consumer Product Safety Commission." Report no. ORNL/TM-9867. Oak Ridge National Laboratory, Oak Ridge, TN.

49. Matthews, T.G. et al. 1987. "Interlaboratory Comparison of Formaldehyde Emissions from Particleboard Underlayment in Small-scale Environmental Chambers." *JAPCA*. 37: 1320–1326.
50. Nelms, L.H. et al. 1986. "The Effects of Ventilation Rates and Product Loading on Organic Emission Rates from Particleboard." 469–485. In: *Proceedings of IAQ '86: Managing Indoor Air for Health and Energy Conservation*. American Society of Heating, Refrigerating and Air-Conditioning Engineers. Atlanta.
51. Sodegren, D. 1982. "A Carbon Dioxide Controlled Ventilation System." *Environ. Int.* 8:395–399.
52. Trepte, L. 1989. "Demand Controlled Ventilating Systems: New Concepts for Indoor Air Quality and Energy Conservation." 407–411. In: Breva, C.J. et al. (Eds.). *Proceedings Brussels Conference: Present and Future of Indoor Air Quality*. Elsevier Science Publishers. Amsterdam.
53. Strindehog, O. 1991. "Long-term Experience with Demand-Controlled Ventilation Systems." 108–110. In: *Proceedings of IAQ '91: Healthy Buildings*. American Society of Heating, Refrigerating and Air-Conditioning Engineers. Atlanta.
54. Raatschen, W. 1991. "IAQ Management by Demand-Controlled Ventilation." In: Roulet, C.A. (Ed.). *Proceedings of the Workshop on Indoor Air Quality Management Lausanne (Switzerland), May 27–28, 1991*.
55. Wenger, J.D. et al. 1993. "A Gas Sensor Array for Measurement of Indoor Air Pollution — Preliminary Results." 27–32. In: *Proceedings of the Sixth International Conference on Indoor Air Quality and Climate*. Vol. 5. Helsinki.
56. Meckler, M. 1993. "Evaluating Demand Control Strategies for VAV Supplementary Bypass Systems." 79–84. In: *Proceedings of the Sixth International Conference on Indoor Air Quality and Climate*. Vol. 5. Helsinki.
57. Skaret, E. 1985. "Ventilation by Displacement — Characterization and Design Implications." 827–841. In: *Proceedings of the First International Symposium on Ventilation for Contaminant Control*.
58. Sandberg, M. and C. Blomqvist. 1989. "Displacement Ventilation Systems in Office Rooms." *ASHRAE Trans*. 95. Pt. 2:1041–1049.
59. Svensson, A. 1989. "Nordic Experiences of Displacement Ventilation Systems." *ASHRAE Trans*. 95. Pt. 2:1013–1017.
60. Laurikainen, J. 1991. "Calculation Method for Airflow Rate in Displacement Ventilation Systems." 111–115. In: *Proceedings of IAQ '91: Healthy Buildings*. American Society of Heating, Refrigerating and Air-Conditioning Engineers. Atlanta.
61. Koganei, M. et al. 1991. "Applicability of Displacement Ventilation to Offices in Japan." 116–121. In: *Proceedings of IAQ '91: Healthy Buildings*. American Society of Heating, Refrigerating and Air-Conditioning Engineers. Atlanta.
62. Koganei, M. et al. 1993. "Modeling the Thermal and Indoor Air Quality Performance of Vertical Displacement Ventilation Systems." 241–246. In: *Proceedings of the Sixth International Conference on Indoor Air Quality and Climate*. Vol. 5. Helsinki.
63. Kim, I.G. and H. Homma. 1992. "Distribution and Ventilation Efficiency of CO_2 Produced by Occupants in Upward and Downward Ventilated Rooms." *ASHRAE Trans*. 98. Pt. 1.

64. Breum, N.O. 1988. "Air Exchange Efficiency of Displacement Ventilation in a Printing Plant." *Ann. Occup. Hyg.* 32:481–488.
65. Breum, N.O. et al. 1989. "Dilution Versus Displacement Ventilation—An Intervention Study." *Ann. Occup. Hyg.* 33:321–329.
66. Allen, R.J. et al. 1978. "Characterization of Potential Indoor Sources of Ozone." *Am. Ind. Hyg. Assoc. J.* 39:466–459.
67. Selway, M.D. et al. 1980. "Ozone Production from Photocopying Machines." *Am. Ind. Hyg. Assoc. J.* 41:455–459.
68. Braun-Hansen, T. and B. Anderson. 1986. "Ozone and Other Air Pollutants from Photocopying Machines." *Am. Ind. Hyg. Assoc. J.* 47:659–665.
69. Wolkoff, P. et al. 1992. "A Study of Human Reactions to Office Machines in a Climatic Chamber." *J. Exposure Anal. Environ. Epidemiol.* Suppl. 1:71–96.
70. Tsuchiya, Y. 1988. "Volatile Organic Compounds in Indoor Air." *Chemosphere.* 17:79–82.
71. Tsuchiya, Y. et al. 1988. "Wet Process Copying Machines; A Source of Volatile Organic Compound Emissions in Buildings." *Environ. Toxicol. Chem.* 7:15–18.
72. Tsuchiya, Y. and J.B. Stewart. 1990. "Volatile Organic Compounds in the Air of Canadian Buildings with Special Reference to Wet Process Photocopying Machines." 633–638. In: *Proceedings of the Fifth International Conference on Indoor Air Quality and Climate.* Vol. 2. Toronto.
73. Greenwood, M.R. 1990. "The Toxicity of Isoparaffinic Hydrocarbons and Current Exposure Practices in the Non-industrial (Office) Indoor Air Environment." 169–175. In: *Proceedings of the Fifth International Conference on Indoor Air Quality and Climate.* Vol. 5. Toronto.
74. Tencati, J.R. and H.S. Novey. 1983. "Hypersensitivity Angitis Caused by Fumes from Heat-Activated Photocopy Paper." *Ann. Int. Med.* 98:320–322.
75. Skoner, D.P. et al. 1990. "Laser Printer Rhinitis." *New Eng. J. Med.* 332:1323.
76. Brooks, B.O. and W.F. Davis. 1991. *Understanding Indoor Air Quality.* CRC Press, Boca Raton, FL.
77. Gorman, R.W. 1984. "Cross Contamination and Entrainment." In: *Ann. ACGIH: Evaluating Office Environmental Problems.* 10:115–120.
78. Wilson, D.J. and E. Chui. 1985. "Influence of Exhaust Velocity and Wind Angle on Dilution from Roof Vents." *ASHRAE Trans.* 90. Pt. 3.
79. Reible, D.D. et al. 1985. "The Effect of the Return of Exhausted Building Air on Indoor Air Quality." 62–71. In: Walkinshaw, D.S. (Ed.). *Transactions: Indoor Air Quality in Cold Climates.* Air Pollution Control Association. Pittsburgh.
80. Bahnfelt, D.R. and F.A. Govan. 1987. "Effects of Building Airflow on Reentry and IAQ." 185–194. In: *Proceedings of IAQ '87: Practical Control of Indoor Air Problems.* American Society of Heating, Refrigerating and Air-Conditioning Engineers. Atlanta.
81. Ludwig, J.F. et al. 1993. "Assessment of Hood Stack Re-Entrainment as Determined by Real-Time Tracer Gas Measurements." 175–180. In: *Proceedings of the Sixth International Conference on Indoor Air Quality and Climate.* Vol. 5. Helsinki.

82. Skov, P. et al. 1989. "Influence of Personal Characteristics, Job-Related Factors, and Psychosocial Factors on the Sick Building Syndrome." *Scand. J. Work Environ. Health.* 15:286–295.
83. Gravesen, S. et al. 1990. "The Role of Potential Immunogenic Components of Dust (MOD) in the Sick Building Syndrome." 9–13. In: *Proceedings of the Fifth International Conference on Indoor Air Quality and Climate.* Vol. 1. Toronto.
84. Raw, G.J. 1993. "Indoor Surface Pollution: A Cause of Sick Building Syndrome." British Building Research Establishment.
85. Raw, G.J. et al. 1991. "A New Approach to the Investigation of the Sick Building Syndrome." 339–343. In: *Proceedings of the CIBSE National Conference.* London.
86. McDonald, P. 1991. "Impact of Air Filters, Heat Exchangers, Humidifiers and Mixing." In: Roulet, C.A. (Ed.). *Proceedings of the Workshop on Indoor Air Quality Management, Lausanne (Switzerland), May 27–28, 1991.*
87. Beck, E.M. 1990. "Filter Facts." 171–176. In: *Proceedings of the Fifth International Conference on Indoor Air Quality and Climate.* Vol. 3. Toronto.
88. Woods, J.E. et al. 1987. "Indoor Air Quality Diagnostics: Qualitative and Quantitative Procedures to Improve Environmental Conditions." 80–98. In: *Proc. of the Symposium on Design and Protocol for Monitoring Indoor Air Quality, ASTM.* American Society for Testing and Materials. Philadelphia.
89. Lane, C.A. et al. 1989. "Indoor Diagnostic Procedures for Sick and Healthy Buildings." 237–240. In: *Proceedings of IAQ '89: The Human Equation: Health and Comfort.* American Society of Heating, Refrigerating and Air-Conditioning Engineers. Atlanta.
90. American Society of Heating, Refrigerating and Air-Conditioning Engineers. 1983. "Air Cleaners." 10.1–10.12 *Equipment Handbook.* American Society of Heating, Refrigerating and Air-Conditioning Engineers. Atlanta.
91. Godish, T. 1989. "Air Cleaning," *Indoor Air Pollution Control.* Lewis Publishers, Chelsea, MI, pp. 247–282.
92. McNall, P.E. 1975. "Practical Methods of Reducing Airborne Contaminants in Interior Spaces." *Arch. Environ. Health* 30:552–556.
93. Ensor, D.S. et al. 1991. "Particle-Size-Dependant Efficiency of Air Cleaners." 334–336. In: *Proceedings of IAQ '91: Healthy Buildings.* American Society of Heating, Refrigerating and Air-Conditioning Engineers. Atlanta.
94. Hanley, J.T. et al. 1990. "A Fundamental Evaluation of an Electronic Air Cleaner." 145–150. In: *Proceedings of the Fifth International Conference on Indoor Air Quality and Climate.* Vol. 3. Toronto.
95. Morey, P.R. et al. 1984. "Environmental Studies in Moldy Office Buildings: Biological Agents, Sources, and Preventative Measures." *Ann. ACGIH: Evaluating Office Environmental Problems.* 10:21–35.
96. Morey, P.R. et al. 1986. "Environmental Studies in Moldy Office Buildings." *ASHRAE Trans.* 92. Pt. 1.
97. Morey, P.A. and C. Williams. 1990. "Porous Insulation in Buildings: A Potential Source of Microorganisms." 529–534. In: *Proceedings of the Fifth International Conference on Indoor Air Quality and Climate.* Vol. 4. Toronto.

98. Morey, P.R. and C. Williams. 1991. "Is Porous Insulation Inside an HVAC System Compatible with a Healthy Building?" 128–135. In: *Proceedings of IAQ '91: Healthy Buildings*. American Society of Heating, Refrigerating and Air-Conditioning Engineers. Atlanta.
99. Shumate, M.W. and J.E. Wilhelm. 1991. "Air Filtration Media-Evaluations of Fiber Shedding Characteristics Under Laboratory Conditions and in Commercial Installations." 337–341. In: *Proceedings of IAQ '91: Healthy Buildings*. American Society of Heating, Refrigerating and Air-Conditioning Engineers. Atlanta.
100. Leaderer, B.P. et al. 1984. "Ventilation Requirements in Buildings II. Particulate Matter and Carbon Monoxide from Cigarette Smoking." *Atmos. Environ.* 18:99–106.
101. Jarsinghani, R.A. et al. 1989. "The Effectiveness of Air Cleaners Using an Environmental Tobacco Smoke Material Balance Model." 38–45. In: *Proceedings of IAQ '89: The Human Equation: Health and Comfort*. American Society of Heating, Refrigerating and Air-Conditioning Engineers. Atlanta.
102. Weschler, C.J. and H.C. Shields. 1988. "The Influence of HVAC Operation on the Concentrations of Indoor Airborne Particles." 166–181. In: *Proceedings of IAQ '88: Engineering Solutions to Indoor Air Problems*. American Society of Heating, Refrigerating and Air-Conditioning Engineers. Atlanta.
103. Roys, M.S. et al. 1993. "Sick Building Syndrome: Cleanliness is Next to Healthiness." 261–266. In: *Proceedings of the Sixth International Conference on Indoor Air Quality and Climate*. Vol. 6. Helsinki.
104. Kozak, R.P. et al. 1979. "Factors of Importance in Determining the Prevalence of Molds." *Ann. Allergy* 43:88–94.
105. Godish, T. 1993. Unpublished data.
106. Streifel, A.J. et al. 1989. "Control of Airborne Fungal Spores in an University Hospital." *Environ. Int.* 15:221–227.
107. Streifel, A.J. and F.S. Rhame. 1993. "Hospital Air Filamentous Fungal Spore and Particle Counts in a Specially-Designed Hospital." 161–165. In: *Proceedings of the Sixth International Conference on Indoor Air Quality and Climate*. Vol. 4. Helsinki.
108. Holt, G.L. 1990. "Seasonal Indoor/Outdoor Fungi Ratios and Indoor Bacteria Levels in Non-complaint Office Buildings." 33–38. In: *Proceedings of the Fifth International Conference on Indoor Air Quality and Climate*. Vol. 2. Toronto.
109. Harrison, J. et al. 1990. "An Investigation of the Relationship Between Microbial and Particulate Indoor Air Pollution and the Sick Building Syndrome." 149–154. In: *Proceedings of the Fifth International Conference on Indoor Air Quality and Climate*. Vol. 1. Toronto.
110. Morey, P.R. and B.A. Jenkins. 1984. "What are Typical Concentrations of Fungi, Total Volatile Organic Compounds, and Nitrogen Dioxide in an Office Environment?" 67–71. In: *Proceedings of IAQ '89: The Human Equation: Human Health and Comfort*. American Society of Heating, Refrigerating and Air-Conditioning Engineers. Atlanta.
111. Ramanathan, K. et al. 1988. "Evaluation of Control Strategies for Volatile Organic Compounds in Indoor Air." *Envir. Prog.* 7:230–235.

112. Ramanathan, K. et al. 1989. "Air Cleaners for Volatile Organic Compounds in Indoor Air." 33–37. In: *Proceedings of IAQ '89: The Human Equation: Health and Comfort.* American Society of Heating, Refrigerating and Air-Conditioning Engineers. Atlanta.
113. Viner, A.S. et al. 1991. "Air Cleaners for Indoor Pollution Control." 115–134. In: Kay, J.G. et al. (Eds.). *Indoor Air Pollution: Radon, Bioaerosols and VOCs.* Lewis Publishers, Chelsea, MI.
114. Ensor, D.S. et al. 1988. "Air Cleaner Technologies for Indoor Air Pollution." In: *Proceedings of IAQ '88: Engineering Solutions to Indoor Air Problems.* American Society of Heating, Refrigerating and Air-Conditioning Engineers. Atlanta.
115. Sleik, H. and A. Turk. 1953. "Air Conservation Engineering." Connor Engineering Corporation, Danbury, CT.
116. Graham, J.R. and M.A. Bayati. 1990. "The Use of Activated Carbon for the Removal of Trace Organics in the Control of Indoor Air Quality." 133–138. In: *Proceedings of the Fifth International Conference on Indoor Air Quality and Climate.* Vol. 3. Toronto.
117. Liu, R.T. 1990. "Removal of Volatile Organic Compounds in IAQ Concentrations with Short Carbon Bed Depths." 177–181. In: *Proceedings of the Fifth International Conference on Indoor Air Quality and Climate.* Vol. 3. Toronto.
118. Holmberg, R. et al. 1993. "Suitability of Activated Carbon Filters for Air Handling Units." 375–380. In: *Proceedings of the Sixth International Conference on Indoor Air Quality and Climate.* Vol. 6. Helsinki.
119. Kinkead, D.A. 1990. "Pleated Dry Processed Carbon Composite (DPCC) Based Adsorbers, an Inescapable New Technology for HVAC Air Purification." 139–144. In: *Proceedings of the Fifth International Conference on Indoor Air Quality and Climate.* Vol. 3. Toronto.
120. Liu, R.T. 1993. "Model Simulation of the Performance of Activated Carbon Adsorbers for the Control of Indoor VOCs." 423–428. In: *Proceedings of the Sixth International Conference on Indoor Air Quality and Climate.* Vol. 6. Helsinki.
121. Sterling, E.M. et al. 1991. "Case Studies of Ventilation Retrofits Designed to Resolve Air Quality Problems in Public Buildings." 308–318. In: *Proceedings of IAQ '91: Healthy Buildings.* American Society of Heating, Refrigerating and Air-Conditioning Engineers. Atlanta.
122. Rodberd, J.A. 1991. "A Novel Technique to Permanently Remove Indoor Air Pollutants." 311–317. In: *Proceedings of IAQ '91: Healthy Buildings.* American Society of Heating, Refrigerating and Air-Conditioning Engineers. Atlanta.
123. Hedge, A.R. et al. 1991. "Breathing-Zone Filtration Effects on Indoor Air Quality and Sick Building Syndrome Complaints." 351–355. In: *Proceedings of IAQ '91: Healthy Buildings.* American Society of Heating, Refrigerating and Air-Conditioning Engineers. Atlanta.
124. Hedge, A.R. et al. 1993. "Effects of a Furniture-Integrated Breathing-Zone Filtration System on Indoor Air Quality, Sick Building Syndrome, Productivity, and Absenteeism." 383–388. In: *Proceedings of the Sixth International Conference on Indoor Air Quality and Climate.* Vol. 5. Helsinki.
125. Molhave, L. and M. Thorsen. 1991. "A Model for Investigation of Ventilation Systems as Sources for Volatile Organic Compounds in Indoor Climate." *Atmos. Environ.* 25A:241–249.

126. Pasanen, P. et al. 1990. "Emissions of Volatile Organic Compounds from Air-Conditioning Filters of Office Buildings." 183–186. In: *Proceedings of the Fifth International Conference on Indoor Air Quality and Climate.* Vol. 3. Toronto.
127. Teijonsalo, J. et al. 1993. "Filters of Air Supply Units as Sources of Contaminants." 533–538. In: *Proceedings of the Sixth International Conference on Indoor Air Quality and Climate.* Vol. 6. Helsinki.
128. Schata, M. et al. 1989. "Allergies to Molds Caused by Fungal Spores in Air–Conditioning Equipment." *Environ. Int.* 15:177–179.
129. Elixmann, J.H. et al. 1990. "Fungi in Filters of Air-Conditioning Systems Cause the Building-Related Illness." 193–196. In: *Proceedings of the Fifth International Conference on Indoor Air Quality and Climate.* Vol. 1. Toronto.
130. Martikainen, P.J. et al. 1990. "Microbial Growth on Ventilation Filter Materials." 203–206. In: *Proceedings of the Fifth International Conference on Indoor Air Quality and Climate.* Vol. 3. Toronto.
131. Jantunen, M.J. et al. 1990. "Does Moisture Condensation in Air Ducts Promote Fungal Growth?" 73–78. In: *Proceedings of the Fifth International Conference on Indoor Air Quality and Climate.* Vol. 2. Toronto.
132. Elixmann, J.H. et al. 1989. "Can Airborne Fungal Allergens Pass Through an Air-Conditioning System?" *Environ. Int.* 15:193–196.
133. Price, D.L. et al. 1993. "Assessment of Air Filters Treated with a Broad Spectrum Biostatic Agent." 527–532. In: *Proceedings of the Sixth International Conference on Indoor Air Quality and Climate.* Vol. 6. Helsinki.
134. Burge, P.S. and A. Robertson. 1993. "Surveillance of Office Buildings Using the Workforce to Measure the Building Symptom Index (BSI). A Measure of the Sick Building Syndrome." 569–574. In: *Proceedings of the Sixth International Conference on Indoor Air Quality and Climate.* Vol. 6. Helsinki.

INDEX

A

Abortion, spontaneous, video-display terminal hazard, 112
Absenteeism, 17
Acetic acid, sick building, 6
Acetophone, 106
Acrolein, tobacco smoke, 37
Active integrated sampling, measurement, indoor air contaminants, 277
Adhesive, office floor covering, 7, 121–122, 309
Age, occupant-related risk factor, 29
AIHA protocol, 215–217
Air, see also Control of contaminant
  cleaning
    for control of contaminant, 368–375
    dust/particulate phase contaminants, 368–372
    evaluation, 370–371
    filtration system, 374
    gaseous contamination, 372–374
    mold spore control, 371–372
    performance of air cleaner, 370
    as source of contamination, 374–375
  exchange rate, control of contaminant, 350
  flow/movement, 59
  ions, 61–64
  quality
    occupant satisfaction, 68–73
    survey questionnaire, NIOSH, 254–255
Allergic rhinitis, 1, 332–334
  chronic, 177
Allergy, 177–186
  dust mite, 178–180
  macromolecular organic dust, 185–186
  mold, 180–185
All-water system, 246
American Industrial Hygiene Association, see AIHA
American investigative protocols, 207–224
  AIHA protocol, 215–217
  "Building Diagnostics" protocol, 217–221
  California protocol, 211–214
  environmental health/engineering protocol, 221–224
  NIOSH protocol, 207–209
  private consultants, 217
  USEPA/NIOSH protocol, 209–211
American Society of Heating, Refrigeration, and Air-Conditioning Engineers, 352
American Thoracic Society, 2
ASHRAE, see American Society of Heating, Refrigeration, and Air-Conditioning Engineers
Aspergillus, 301
Asthma, 1, 176–177, 332–334
  source control, 310
Atopic history, occupant-related risk factors, 30–31

B

Bacteria, 186–188
"Bad products", control of contaminant, 317
Bake-out, of building, 321–324
Behemic acid, 106
Benzaldehyde, 106
Benzene, tobacco smoke, 37
Bioeffluents, 140–145, 351
Biological aerosol, disinfection, 298
Biological contaminants, 80, 325–334
  allergic rhinitis, 332–334
  allergy, 177–186
    dust mite, 178–180
    marcomolecular organic dust, 185–186
    mold, 180–185
  asthma, 176–177, 332–334
  bacteria, 186–188
  chronic allergic rhinitis, 177
  dust mites, 334
  endotoxin, 173, 188–189
  glucans, 188–189

humidifier fever, 173
hypersensitivity pneumonitis, 171–173, 330–332
Legionella pneumophila, 325–330, see also Legionnaires' disease
Legionnaires' disease, 173–176, 325–330
biocidal treatments, 325–327
cooling water, 174–175
potable water system, 176, 327–328
source modifications, 327
microbial products, 188
mycotoxins, 189–194
Pontiac fever, 172
surface dust, 334–336
Blueprint machines, sick building, 6
Boiler additives, sick building, 6
Bond paper, 97, 100–101
"Building Diagnostics" protocol, 217–221
Building-related illness, usage of term, 1

C

Cadmium, tobacco smoke, 37
California Healthy Building Study, 14
California protocol, 211–214
Canadian investigative protocol, 6–8, 224–227
Ontario Interministerial Committee, 226–227
Public Works Canada protocol, 224–226
Carbon dioxide
measurement, indoor air contaminants, 280–282
tobacco smoke, 37
ventilation rate, 248–250
Carbon monoxide, 1
measurement, indoor air contaminants, 282–283
tobacco smoke, 37
Carbonless copy paper, 16, 93, 308
Carpet, 20, 114–125
adhesives, 121–122
cleaning agent, 124–125
electrostatic shock, 124
source control, 309, 310
Chemical contaminants, 76–77, 309–325
avoidance, 309
"bad products", 310
bioassays, 310
building bake-out, 321–324
design of building, 317–318
criteria, 318–319
emission characterization, 309–310
prevention, 309
private initiatives, 313–315
product labeling, 316–317
source characterization, 309–310
source modification, 324–325
source removal, 319–320
source treatment, 320–321
TVOC, 313–315
USEPA guidelines, 310–311
Washington, state of, initiative, 311–313
Chemical sense, 3
Chronic allergic rhinitis, 177
Cladosporium, 301
"Clean products", control of contaminant, 314
Cleaning agent, carpet, 124–125
Climate survey questionnaire, Danish Building Research Institute, 266–267
Comfort
complaints, 1
sick building, 1
guidelines, use of, in diagnosis of building, 239–243
Commission of European Communities, 2
Computer, 107–113
electromagnetic radiation, 111–112
ergonomic problems, 110–111
job stress, 110–111
reproductive hazards, 112
skin symptoms, 108–110
Computer paper, 100–101
Control of contaminant, 347–387
air cleaning, 368–375
dust/particulate phase contaminants, 368–372
evaluation, 370–371
filtration system, 374
gaseous contamination, 372–374
mold spore control, 371–372
performance of air cleaner, 370
as source of contamination, 374–375
air exchange rate, 350
ammonia, 362
"bad products", 317
"clean products", 314
cross contaminant, 367–368
decipol level, 362
demand-controlled ventilation, 361–363
dilution ventilation, 347–366
contamination concerns, 350–351
health/comfort concerns, 349–350
human bioeffluents, controlling, 351
indoor air quality procedure, 353–354

perceived air quality procedure, 354–358
studies, 349–351
theory of, 348
application, 348–349
exceptions, 349
ventilation
rate procedure, 352–353
standards, 352–358
displacement ventilation, 363–366
duct liner, 318
emission limit, 354
entrainment, 367–368
ergonomic symptoms, 359
flush-out ventilation, 359–361
human odor, 352
humidity, 349
ion exchange rate, 351
local exhaust ventilation, 366–367
nicotine, 350
odor, 366
olf rating, 354–356
open-concept office space, 358
paper, gender factors, 30
piston flow, 364
re-entry, 367–368
sensors, 361
"shift zone", 364
stratification, 359
"stuffiness", 351
temperature, 349, 364
ventilation
effectiveness, 358–359
innovations, 361
Copy paper, carbonless, 93
Copying machines, electrostatic, 103–105
Cotinine, tobacco smoke, 42
Cresol, office materials/equipment, 125
Cross-contamination, 82–83

D

Danish Building Research Institute
indoor climate survey questionnaire, 266–267
protocol, 227–229
Danish Town Hall Study
sick building, 11
skin symptoms, 11
Decipol level, 362
Demand-controlled ventilation, 361–363
Density, of occupants, occupant-related risk factors, 35
Dermatitis, atopic, overview, 15
Diagnosis of building
air quality survey questionnaire, NIOSH, 254–255
Danish Building Research Institute, indoor climate survey questionnaire, 266–267
human resources questionnaire, 259
HVAC system, 243–252
air-water system, 245–246
evaluation, 246–247
types of, 245–246
ventilation rate, 247–252
carbon dioxide technique, 248–250
thermal balance, 250–251
tracer gas, 251–252
industrial hygiene, role of, 206–207
investigative process, 207–233
American investigative protocols, 207–224
AIHA protocol, 215–217
"Building Diagnostics" protocol, 217–221
California protocol, 211–214
environmental health/engineering protocol, 221–224
NIOSH protocol, 207–209
private consultants, 217
USEPA/NIOSH protocol, 209–211
Canadian investigative protocol, 224–227
Ontario Interministerial Committee, 226–227
Public Works Canada protocol, 224–226
European investigative protocol, 227–233
Danish Building Research Institute protocol, 227–229
Nordtest, 229–233
investigative protocol, features of, 233–243
comfort guidelines, 239–243
contaminants, 238–239
heat, ventilation, air conditioning system assessment, 237–238
IAQ guidelines, 239–243
multiple stages of investigation, 233–234
personnel conducting investigation, 234
site visits, 234
source assessment, 238–239
symptoms/complaint assessment, 234–236

occupant health/comfort questionnaire, AIHA, 256–258
Ontario Interministerial Committee questionnaire and survey, 260–265
private group, role of, 205–206
public group, role of, 205–206
Diazo-photocopier, 106
Dilution ventilation, 347–366
contamination concerns, 350–351
health/comfort concerns, 349–350
human bioeffluents, controlling, 351
indoor air quality procedure, 353–354
perceived air quality procedure, 354–358
studies, 349–351
theory of, 348
application, 348–349
exceptions, 349
ventilation
rate procedure, 352–353
standards, 352–358
Direct-read instruments, electronic, measurement, indoor air contaminants, 276–277
Disinfection, biological aerosol, 298
Displacement ventilation, control of contaminant, 363–366
Duct liner, 7
control of contaminant, 318
Duplicating machine, 107
"Dust", 158–163
airborne, 159–160
measurement, indoor air contaminants, 288
concentration, 158–159, 161
organic, marcomolecular, 185–186
settled, measurement, indoor air contaminants, 288
surface, 160–161
Dust mite, 178–180, 334
Dutch Office Building Study, sick building, 11

E

Eczema, occupant-related risk factor, 31
Electric field, 65
Electromagnetic radiation, 16, 22
video-display terminal, 111–112
Electronic direct-read instruments, measurement, indoor air contaminants, 276–277
Electrostatic charge, 64–65
Electrostatic copying machines, 103–105
Electrostatic photocopier, local exhaust ventilation, 367
Electrostatic shock, carpet, 124
Emission characterization, chemical contaminants, 309–310
Endotoxins, 188–189
Energy conservation, 3
Entrainment/re-entry, 80–82
Environmental conditions, 53–66
air flow/movement, 59
air ions, 61–64
electric field, 65
electrostatic charge, 64–65
humidity, 57–59
lighting, 60–61
magnetic field, 65
noise, 61
thermal conditions, 54–57
comfort relationships, 54–55
dissatisfaction with, 55
temperature, 55–57
vibration, 61
Environmental health/engineering protocol, 221–224
Environmental systems
air quality, occupant satisfaction, 68–73
biological contaminants, 80
building studies, 68–75
chemical contaminants, 76–77
cross contamination, 82–83
entrainment/re-entry, 80–82
fiber, man-made, 78–80
field investigations, 67
ventilation system, 68, 73–74
workstation characteristics, 73–75
Epidemiological studies, office materials/equipment as causal agent, 98
Ergonomic symptoms, 359
video-display terminal, 110–111
Erythema, face, 14–15
European investigative protocol, 227–233
Danish Building Research Institute protocol, 227–229
Nordtest, 229–233
Eye strain, computers, 110

F

Fiber, man-made, 78–80
Fiberboard, 7
Fiberglass, 1, 7
Field investigation, 5, 67
Filtration system, for air cleaning, 374
Floor covering, 114–129, see also Carpet

vinyl, 125–129
Flow, of air flow, 59
Flush-out ventilation, 359–361
Formaldehyde, 1, 6, 145–148
measurement, indoor air contaminants, 283–285
source control, 309, 310
tobacco smoke, 37

G

Gas sampling tube, measurement, indoor air contaminants, 275–276
Gas vapor
bioeffluents, 140–145
formaldehyde, 145–148
pheromones, 143–145
volatile organic compounds, 148–158
Gender
as risk factor, 29, 30
tobacco smoke, 30
Glucans, 188–189

H

Hardwood, 7
Hay fever, occupant-related risk factor, 31
Health and Welfare Canada (HWC), 8
Health complaints, sick building, 1
"Healthy Buildings" program, Washington state, 311
Heat, ventilation, air conditioning system, see HVAC system
Human resources questionnaire, 259
Humidifier fever, 1, 171–173, 330–332
Humidity, 4, 59–57
tight building, 4
HVAC system, 243–252
air-water system, 245–246
all-water system, 246
evaluation, 246–247
types of, 245–246
ventilation rate, 247–252
carbon dioxide technique, 248–250
thermal balance, 250–251
tracer gas, 251–252
Hypersensitivity, 1
pneumonitis, 7, 171–173, 330–332
Hysteria, 31–32

I

IAQ guidelines, use of, in diagnosis of building, 239–243
Indoor air quality survey questionnaire, NIOSH, 254–255
Industrial hygiene, role of, 206–207
Infiltration, 4
Instruments, direct-read, electronic, for measurement of indoor air contaminants, 276–277
Integrated sampling, measurement, indoor air contaminants, 277
Investigative protocol
American investigative protocols, 207–224
AIHA protocol, 215–217
"Building Diagnostics" protocol, 217–221
California protocol, 211–214
environmental health/engineering protocol, 221–224
NIOSH protocol, 207–209
private consultants, 217
USEPA/NIOSH protocol, 209–211
Canadian investigative protocol, 224–227
Ontario Interministerial Committee, 226–227
Public Works Canada protocol, 224–226
European investigative protocol, 227–233
Danish Building Research Institute protocol, 227–229
Nordtest, 229–233
features of, 233–243
comfort guidelines, 239–243
contaminants, 238–239
HVAC system assessment, 237–238
IAQ guidelines, 239–243
multiple stages of investigation, 233–234
personnel conducting investigation, 234
site visits, 234
source assessment, 238–239
symptoms/complaint assessment, 234–236
Ion, 61–64
exchange rate, 351

J

Job satisfaction, occupant-related risk factors, 34
Job stress, 33, 110–111
video-display terminal, 110–111

K

Kindling, defined, 21

L

Labeling, of products, chemical contaminants, 316–317
Laser printer, 105–106
  local exhaust ventilation, 367
Legionella pneumophila, 325–330, see also Legionnaires' disease
Legionnaires' disease, 1, 173–176
  biocidal treatments, 325–327
  control measure effectiveness, 328–330
  cooling water, 174–175
  potable water system, 176, 327–328
  source modifications, 327
Library of Congress, 9
Lighting, 60–61
Local exhaust ventilation, 366
  ammonia, 366
  control of contaminant, 366–367
  electrostatic photocopier, 367
  laser printer, 367
  odor, 366
  spirit duplicator, 367
  wet-process photocopier, 367

M

Magnetic field, 65
Marcomolecular organic dust, 185–186
Mass psychogenic illness, occupant-related risk factors, 31–33
MCS, see Multiple chemical sensitivity
Measurement, indoor air contaminants, 273–306
  active integrated sampling, 277
  administrative practices, 280
  airborne dust, 288
  Aspergillus, 301
  biological aerosols, 288–301
  carbon dioxide, 280–282
  carbon monoxide, 282–283
  Cladosporium, 301
  dust, 288
  electronic direct-read instruments, 276–277
  formaldehyde, 283–285
  gas sampling tube, 275–276
  passive integrated sampling, 277
  Penicillium, 301
  quality assurance/calibration, 277–278
  sampling decisions, 278–280
    duration of sampling, 279
    location, 278
    number, 280
    sampling methods, 274–277
    Stachybotrys, 301
    time of sampling, 278–279
  volatile organic compounds, 285–288
Methanol, sick building, 6
Methyl alcohol, 107
Microbial contamination, 7, 188
Microfilm copier, 106
Mite, dust, 178–180, 334
Mold, 180–185
  spore control, air cleaning, 371–372
Movement, of air, 59
Mucous membrane irritation, 2, 6
Multiple chemical sensitivity (MCS), 18–22
Mycotoxins, 189–194

N

National Institute of Occupational Safety and Health, see NIOSH
Neurotoxic health problems, 2, 6
Nicotine, 350
NIOSH, 6
  indoor air quality survey questionnaire, 254–255
  investigation, sick building, 6–8
  protocol, 207–209
Nitrogen oxide, tobacco smoke, 37
Noise, 61
Nordic Ventilation Group, 14
Nordtest, building investigative protocol, 229–233

O

Occupant health/comfort questionnaire, AIHA, 256–258
Occupant-related risk factor
  acrolein, tobacco smoke, 37
  age, 29
  atopic history, 30–31
  benzene, tobacco smoke, 37
  cadmium, tobacco smoke, 37
  carbon dioxide, tobacco smoke, 37
  carbon monoxide, tobacco smoke, 37
  cotinine, tobacco smoke, 42
  eczema, 31
  formaldehyde, tobacco smoke, 37
  gender, 29, 30
  hay fever, 31

hysteria, 31–32
job satisfaction, 34
job stress, 33
mass psychogenic illness, 31–33
nitrogen oxide, tobacco smoke, 37
occupant density, 35
personal characteristics, 29–31
physical work environment, satisfaction with, 35–36
psychosocial phenomenon, 31–37
psychosocial risk factors, 33
significance of, 36–37
seasonal affective disorder, 36
tobacco smoking, 29, 37–46
Odor, 2, 3, 366
human, 352
Office materials/equipment, 93–114
acetone, 121
acetophone, 106
adhesives, carpet, 121–122
behemic acid, 106
benzaldehyde, 106
bond paper, 97, 100–101
carbonless copy paper, 93
carpet, 114–125
cleaning agent, 124–125
electrostatic shock, 124
causal factors, 98–99
complaint investigation, 95
computer, 107–113
electromagnetic radiation, 111–112
ergonomic problems, 110–111
job stress, 110–111
paper, 100–101
reproductive hazards, 112
skin symptoms, 108–110
cresol, 125
diazo-photocopier, 106
"dust", 158–163
airborne, 159–160
concentration, 158–159, 161
surface, 160–161
electrostatic copying machines, 103–105
epidemiological studies, 98
equipment, 101–114
exposure investigation, 95–97
eye strain, 110
floor covering, 114–129
human exposure studies, 113–114
job stress, 110–111
laser printer, 105–106
methyl alcohol, 107
microfilm copier, 106
paratoluene sulfonate, 95
phenol, 121
phenolic resin, 94
product characterization, 93–95
sick building, 6
spirit duplicating machine, 107
styrene-butadiene rubber, 94
terphenyls, 94
thiourea, 106
toluene, 121
toner, 106
TVOC theory, 149–158
video-display terminal, 107–113
electromagnetic radiation, 111–112
ergonomic problems, 110–111
job stress, 110–111
skin symptoms, 108–110
vinyl floor covering, 125–129
wet-process photocopier, 101–103
xylene, 121
Olf rating, contaminant, 354–356
Olfactory-limbic pathway, 21
Ontario Interministerial Committee, 226–227
questionnaire and survey, diagnosis of building, 260–265
Open-concept office space, 358
Organic compounds, volatile, 148–158
Organic dust, marcomolecular, 185–186

P

Paper
bond, 97, 100–101
computer, 100–101
Paratoluene sulfonate, 95
Particleboard, 7
Passive integrated sampling, measurement, indoor air contaminants, 277
Penicillium, 301
Personal characteristics, occupant-related risk factors, 29–31
Personal symptom index, 13
Pesticide, 20
Phenol, 94, 121
Pheromones, 143–145
Photocopier, 16
diazo, 106
wet-process, 101–103
Physical work environment, occupant-related risk factors, satisfaction with, 35–36
Piston flow, 364
Plywood, 7
Pneumonitis, 1
hypersensitivity, 171–173

Pontiac fever, 172
Potable water system, Legionnaires' disease, 327–328
control measure effectiveness, 328–330
Pregnancy, video-display terminal, hazards, 112
Private consultants, protocols, investigation, 217
Private group, role of, in diagnosis of building, 205–206
"Problem building"
  defined, 5
  sick building, distinguished, 5
Product characterization, office materials/equipment, 93–95
Product labeling, chemical contaminants, 316–317
Productivity, 17–18
Psychosocial risk factors, 3
  occupant-related risk factors, 31-37
Public group, role of, in diagnosis of building, 205–206
Public Works Canada protocol, 224–226

Q

Quality assurance/calibration, measurement, indoor air contaminants, 277–278
Questionnaire
  air quality, NIOSH, 254–255
  climate, Danish Building Research Institute, 266–267
  complaint, assessment, diagnosis of building, 236
  human resources, diagnosis of building, 259
  occupant comfort, AIHA, 256–258
  Ontario Interministerial Committee, diagnosis of building, 260–265
  symptom, 10, 236

R

Recirculated air, 4
Reproductive hazards, computers, 112
Rhinitis, allergic, 1, 332–334
  chronic, 177
Rhinolaryngoscopy, 20
Risk factors
  biological origin contaminant, 171–203
  environmental conditions, 53–66
  environmental systems, 66–84
  gas, 139–170
  office materials/equipment, 93–114
  particulate matter, 139–170
  vapor, 139–170

S

SAD, see Seasonal affective disorder
Sampling methods, measurement, indoor air contaminants, 274–277
Satisfaction, with job, occupant-related risk factors, 34
Seasonal affective disorder, occupant-related risk factors, 36
Sensory irritation, 2, 3
"Shift zone," control of contaminant, 364
Sick building syndrome, 1-4
Site visits, diagnosis of building, 234
Skin, symptoms, 2, 3
  computer, 108–110
  Danish Town Hall Study, 11
  sick building, 14–16
  video-display terminal, 108–110
Somatization disorder, 20
Source control, 307–346
  adhesives, 309
  air cleaning, 308
  asthma, 310
  avoidance, 309
  "bad products", 310
  bioassays, 310
  biological contaminants, 325–334
    Legionnaires' disease, 325–330
  building bake-out, 321–324
  carbonless copy paper, 308
  carpet, 309, 310
  chemical contaminants, 309–325
  design of building, 317–318
    criteria, 318–319
  emission characterization, 309–310
  formaldehyde, 309, 310
  "Healthy Buildings" program, Washington state, 311
  mitigation, 308
  prevention, 309
  private initiatives, 313–315
  product labeling, 316–317
  skin symptoms, 308
  source characterization, 309–310
  source modification, 324–325
  source removal, 319–320
  source treatment, 320–321
  threshold limit value, 311
  TVOC, 313–315
  USEPA guidelines, 310–311

Washington, state of, initiative, 311–313
Spirit duplicator, 6, 107
local exhaust ventilation, 367
Spore sampler, biological aerosols, measurement, indoor air contaminants, 293–294
Stachybotrys, 301
Stratification, 6
Stress
job, 110–111
occupant-related risk factors, 33
video-display terminal, 110–111
"Stuffiness", control of, 351
Styrene-butadiene rubber, 94
Survey
Danish Building Research Institute, 266–267
Ontario Interministerial Committee, diagnosis of building, 260–265
questionnaire, air quality, NIOSH, 254–255
Swedish study, sick buildings, 11–12
Symptom prevalence, sick building, 8–9
Symptoms/complaint assessment, diagnosis of building, 234–236
questionnaire, 236

T

Temperature, building, 55–57
tight, 4
Terphenyls, 94
Thermal conditions, 54–57
balance, ventilation rate, 250–251
comfort relationships, 54–55
dissatisfaction with, 55
temperature, 55–57
Thermophilic actinomycetes, 7
Thiourea, 106
Threshold limit value (TLV), 311
Tight building, 4–5
defined, 4
humidity, 4
infiltration, 4
recirculated air, 4
temperature, 4
ventilation, 4
TLV, see Threshold limit value
Tobacco smoking, 4, 17, 37–46
environmental, 2
gender differences, 30
local exhaust ventilation, 366
occupant-related risk factor, 29
sick building, 6
Toluene, 121
Toner, 106
Tracer gas, ventilation rate, determination of, 251–252
Trigeminal nerve, 3, 21
TVOC theory, 149–158

U

U.K. studies, of sick buildings, 10
Ultraviolet light, 22
United States Environmental Protection Agency, see USEPA
Urea-formaldehyde, 6
USEPA, 9
USEPA/NIOSH protocol, 209–211

V

VDT, see Video-display terminal
Ventilation, 3, 4, 73–74
assessment, in diagnosis of building, 237–238
demand-controlled, 361–363
dilution, 347–366
contamination concerns, 350–351
health/comfort concerns, 349–350
human bioeffluents, controlling, 351
indoor air quality procedure, 353–354
perceived air quality procedure, 354–358
studies, 349–351
theory of, 348
application, 348–349
exceptions, 349
ventilation rate procedure, 352–353
ventilation standards, 352–358
displacement, 363–366
effectiveness, for control of contaminant, 358–359
flush-out, 359–361
innovations, 361
local exhaust, 366–367
rate
carbon dioxide technique, 248–250
HVAC system, 247–252
thermal balance, 250–251
tracer gas, 251–252
system, environmental systems, 68
tight building, 4
Vibration, 61

Video-display terminal (VDT), 17, 107–113
  electromagnetic radiation, 111–112
  ergonomic problems, 110–111
  job stress, 110–111
  skin symptoms, 108–110
Vinyl floor covering, 125–129
Volatile organic compounds, 4, 148–158

W

Washington, state of, chemical contaminants, initiative, 311–313
Wet-process photocopier, 101–103
  local exhaust ventilation, 367
Work environment, physical, occupant-related risk factors, satisfaction with, 35–36
Workstation
  characteristics of, 73–75
  control of contaminant, 314
World Health Organization, sick building syndrome, 1

X

Xylene, 121